SPRINGER HANDBOOK OF
AUDITORY RESEARCH

Series Editors: Richard R. Fay and Arthur N. Popper

Springer

New York
Berlin
Heidelberg
Barcelona
Budapest
Hong Kong
London
Milan
Paris
Singapore
Tokyo

SPRINGER HANDBOOK OF AUDITORY RESEARCH

Volume 1: The Mammalian Auditory Pathway: Neuroanatomy
Edited by Douglas B. Webster, Arthur N. Popper, and Richard R. Fay

Volume 2: The Mammalian Auditory Pathway: Neurophysiology
Edited by Arthur N. Popper and Richard R. Fay

Volume 3: Human Psychophysics
Edited by William Yost, Arthur N. Popper, and Richard R. Fay

Volume 4: Comparative Hearing: Mammals
Edited by Richard R. Fay and Arthur N. Popper

Volume 5: Hearing by Bats
Edited by Arthur N. Popper and Richard R. Fay

Volume 6: Auditory Computation
Edited by Harold L. Hawkins, Teresa A. McMullen, Arthur N. Popper, and Richard R. Fay

Volume 7: Clinical Aspects of Hearing
Edited by Thomas R. Van de Water, Arthur N. Popper, and Richard R. Fay

Volume 8: The Cochlea
Edited by Peter Dallos, Arthur N. Popper, and Richard R. Fay

Volume 9: Development of the Auditory System
Edited by Edwin W. Rubel, Arthur N. Popper, and Richard R. Fay

Volume 10: Comparative Hearing: Insects
Edited by Ronald R. Hoy, Arthur N. Popper, and Richard R. Fay

---

Forthcoming volumes (partial list)

Comparative Hearing: Amphibians and Fish
Edited by Arthur N. Popper and Richard R. Fay

Plasticity in the Auditory System
Edited by Edwin W. Rubel, Arthur N. Popper, and Richard R. Fay

Comparative Hearing: Birds and Reptiles
Edited by Robert Dooling, Arthur N. Popper, and Richard R. Fay

Speech Processing in the Auditory System
Edited by Steven Greenberg, William Ainsworth, Arthur N. Popper, and Richard R. Fay

Hearing by Whales and Dolphins
Edited by Whitlow Au, Arthur N. Popper, and Richard R. Fay

Ronald R. Hoy
Arthur N. Popper
Richard R. Fay
Editors

Comparative Hearing: Insects

With 130 Illustrations

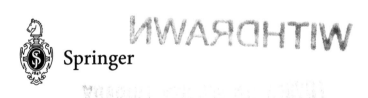
Springer

Ronald R. Hoy
Section of Neurobiology & Behavior
Cornell University
Ithaca, NY 14853

Arthur N. Popper
Department of Zoology
University of Maryland
College Park, MD 20742-9566, USA

Richard R. Fay
Department of Psychology and
Parmly Hearing Institute
Loyola University of Chicago
Chicago, IL 60626, USA

Series Editors: Richard R. Fay and Arthur N. Popper

Cover illustration: Vibrations (acceleration) on banana plant pseudostem due to crawling cockroach (top) and vibrations (displacement) generated by a fly entrapped in the water surface. The complete figure appears on p. 254 of the text.

Library of Congress Cataloging-in-Publication Data
Hoy, Ronald R.
    Comparative hearing. Insects / Ronald R. Hoy, Arthur N. Popper,
Richard R. Fay.
      p. cm. — (Springer handbook of auditory research; v. 10)
    Includes index.
    ISBN 0-387-94682-9 (hardcover: alk. paper)
    1. Hearing. 2. Insects — Physiology. 3. Physiology,
Comparative. I. Popper,
Arthur N. II. Fay, Richard, R. III. Title. IV. Series.
QP461.H69   1998
573.8′9157—dc21   97-45233

Printed on acid-free paper.

Production managed by Terry Kornak; manufacturing supervised by Joe Quatela.
Typeset by Best-set Typesetter Ltd., Hong Kong.
Printed and bound by Maple-Vail Book Manufacturing Group, York, PA.
Printed in the United States of America.

9 8 7 6 5 4 3 2 1 WITHDRAWN

ISBN 0-387-94682-9 Springer-Verlag New York Berlin Heidelberg   SPIN 10527965

Series Preface

The *Springer Handbook of Auditory Research* presents a series of comprehensive and synthetic reviews of the fundamental topics in modern auditory research. The volumes are aimed at all individuals with interests in hearing research, including advanced graduate students, postdoctoral researchers, and clinical investigators. The volumes are intended to introduce new investigators to important aspects of hearing science and to help established investigators to better understand the fundamental theories and data in fields of hearing that they may not normally follow closely.

Each volume is intended to present a particular topic comprehensively, and each chapter serves as a synthetic overview and guide to the literature. As such, the chapters present neither exhaustive data reviews nor original research that has not yet appeared in peer-reviewed journals. The volumes focus on topics that have developed a solid data and conceptual foundation, rather than on those for which a literature is only beginning to develop. New research areas will be covered on a timely basis in the series as they begin to mature.

Each volume in the series consists of five to eight substantial chapters on a particular topic. In some cases, the topics will be ones of traditional interest for which there is a substantial body of data and theory, such as auditory neuroanatomy (Vol. 1) and neurophysiology (Vol. 2). Other volumes in the series will deal with topics that have begun to mature more recently, such as development, plasticity, and computational models of neural processing. In many cases, the series editors will be joined by a co-editor having special expertise in the topic of the volume.

Richard R. Fay, Chicago, IL
Arthur N. Popper, College Park, MD

Preface

A major goal of the study of hearing is to explain how the human auditory system normally functions and to help identify the causes of and treatments for hearing impairment. Experimental approaches to these questions make use of animal models, and the validity of these models will determine the success of this effort. Comparative hearing research establishes the context within which animal models can be developed, evaluated, validated, and successfully applied, and is therefore of fundamental importance to hearing research in general. For example, the observation that hair cells may regenerate in the ears of some anamniotes cannot be evaluated for its potential impact on human hearing without a comparative and evolutionary context within which observations on the animal model and humans can fit and be fully understood, and within which we can generalize observations from one species to another with any confidence. This confidence arises from comparative research that investigates the diverse structures, physiological functions, and hearing abilities of various vertebrate and invertebrate species in order to determine the fundamental principles by which structures are related to functions and to help clarify the phyletic history of the auditory system among animals.

In light of the fundamental importance of comparative hearing research, we have decided to incorporate into the *Springer Handbook of Auditory Research* a series of books focusing specifically on hearing in nonhuman animals. The first in this series was *Comparative Hearing: Mammals* (editors, Fay and Popper) and the companion volume *Hearing by Bats* (editors, Popper and Fay). In addition to this current volume, additional comparative volumes will soon be forthcoming on *Comparative Hearing: Fish and Amphibians* (editors, Fay and Popper), *Comparative Hearing: Reptiles and Birds* (editors, Dooling, Popper, and Fay), and *Comparative Hearing: Whales and Dolphins* (editors, Au, Popper, and Fay). Our feeling is that in providing a comprehensive introduction to comparative hearing, we will help all investigators in our field have a better appreciation for the diversity of animal models that could be incorporated into future research programs.

The purpose of this volume on insects is to introduce the hearing community to the extensive but often unfamiliar literature on the ways that insects detect and process sounds. As described by Hoy in Chapter 1, the mechanisms of hearing in insects are extraordinarily diverse. Hoy provides an invaluable overview of the taxonomy of the insects discussed in this volume and helps the reader understand the complexity of this taxonomy. He also discusses what is understood by an insect "ear," which, unlike in vertebrates, can be found nearly anywhere on the body. In Chapter 2, Michelsen discusses the biophysical principles of sound localization as applied to insects, and all animals, especially small ones. In Chapter 3, Römer considers the ecological constraints on acoustic communications in insects, including some remarkable acoustic adaptations in specific species. The development of the auditory system is treated by Boyan in Chapter 4. The morphological building blocks for the insect hearing organ are shown to be derived from segmental proprioceptors; this is and important link to understanding how the insect ear evolves.

Central processing of signals is dealt with by Pollack in Chapter 5, in which it is seen that the well-known "economy" of invertebrate neural systems extends to the auditory system. Robert and Hoy, in Chapter 6, describe a uniquely specialized auditory system in parasitoid flies, in which a novel mechanism for directional sensitivity appears to have evolved. Vibration detection in spiders is discussed by Barth in Chapter 7. It is important to note that much communication in insects is transmitted by substrate vibrations, not airborne pressure waves. Although spiders are not insects, some of the most interesting work on vibration communication has been conducted by Barth and his colleagues on spiders. Finally, interactions of bats and moths, and the evolution of this system, are discussed by Fullard in Chapter 8. The bat–moth story remains a paradigm within insect bioacoustics, owing to its development over 40 years ago by the late Kenneth Roeder.

A number of the chapters in this volume are closely related to chapters in other volumes of the *Springer Handbook of Auditory Research*. The discussion of development by Boyan parallels a number of chapters in Volume 9 (*Development of the Auditory System*), while the physiology of hearing can be compared with mammals in Volume 2 (*The Mammalian Auditory System: Neurophysiology*). Chapter 8 by Fullard is closely allied to a number of chapters in Volume 5 (*Hearing by Bats*), including those by Grinell, Fenton, Moss, and Schnitzler, and by Simmons and colleagues.

Ronald R. Hoy
Arthur N. Popper
Richard R. Fay
August, 1997

Contents

Series Preface v
Preface vii
Contributors xi

Chapter 1   Acute as a Bug's Ear: An Informal Discussion
             of Hearing in Insects 1
             RONALD R. HOY

Chapter 2   Biophysics of Sound Localization in Insects 18
             AXEL MICHELSEN

Chapter 3   The Sensory Ecology of Acoustic Communication
             in Insects 63
             HEINER RÖMER

Chapter 4   Development of the Insect Auditory System 97
             GEORGE S. BOYAN

Chapter 5   Neural Processing of Acoustic Signals 139
             GERALD S. POLLACK

Chapter 6   The Evolutionary Innovation of Tympanal Hearing
             in Diptera 197
             DANIEL ROBERT AND RONALD R. HOY

Chapter 7   The Vibrational Sense of Spiders 228
             FRIEDRICH G. BARTH

Chapter 8   The Sensory Coevolution of Moths and Bats 279
             JAMES H. FULLARD

Index 327

Contributors

Friedrich G. Barth
Institut für Zoologie, Universität Wien, A-1090 Vienna, Austria

George S. Boyan
Zoologisches Institut, Ludwig-Maximilians-Universität, D-80333 München, Germany

James H. Fullard
Department of Zoology, University of Toronto Erindale College, Mississauga, Ontario L5L-1C6, Canada

Ronald R. Hoy
Section of Neurobiology & Behavior, Cornell University, Ithaca, NY 14853, USA

Axel Michelsen
Institute of Biology, Odense University, DK-5230 Odense M, Denmark

Gerald S. Pollack
Department of Biology, McGill University, Montreal, Quebec H3A-1B1, Canada

Daniel Robert
Section of Neurobiology & Behavior, Cornell University, Ithaca, NY, USA

Heiner Römer
Institute of Zoology, Karl-Franzens-University, A-8010 Graz, Austria

1
Acute as a Bug's Ear: An Informal Discussion of Hearing in Insects

RONALD R. HOY

1. Introduction

It is often said that humans are visual animals. Our ability to get around in this world may depend a lot on our eyes, but our ability to communicate with others and to make sense of the world requires our ears. This surmise applies not just to humans, but equally well to some members of another spectacularly successful class of animals, the Insecta. When suddenly startled, a person is described as having his or her eyes opened as "wide as saucers," or as being "bug-eyed," in reference to the conspicuous eyes of insects such as the familiar houseflies, bees, dragonflies, praying mantises, and even cockroaches. The heads of these insects seem to be nearly "all eyes," and indeed, beneath their compound eyes up to two thirds of an insect's brain volume may be devoted to its visual centers. Yet, as is documented in this book, many kinds of insects hear as well as see to an extent more than the average reader might have suspected, and the purposes to which hearing is put among insects are familiar.

Among terrestrial animals, only vertebrates and insects are widely endowed with a sense of hearing. In this chapter an introduction to the subject of insect hearing is provided to serve as an informal guide for readers who are only casually acquainted with insect biology. Hopefully, these remarks will make the excellent and authoritative chapters that follow in this volume more accessible. The chapter begins, generally, by comparing and contrasting the sense of hearing in insects and vertebrates, particularly mammals. Then an attempt is made to disentangle the "name problem" that comes from the somewhat confusing use of certain common names to describe insects of different taxa. The other volumes of this series of books on hearing attest to the widespread occurrence of the sense of hearing among vertebrate animals, particularly mammals. However, thousands of different species of insects from seven different taxonomic orders can also hear and for them the sense of hearing subserves pretty much the same behavioral functions it does in vertebrates, namely, the detection of predators and prey as well as prospective mates and rivals. This volume recognizes the manifold aspects of insect hearing.

If an animal has one ear, it has another. Ears come in pairs — one on each side of the head — in all vertebrates, anyway. In mammals, one thinks of a pair of external fleshy appendages; in frogs or lizards, a pair of naked "eardrums." In either case, the external ears lead to internalized middle and inner ears. Indeed, the auditory apparatus of all terrestrial vertebrates is bilaterally paired cranial structures, and as far as we know this holds for extinct vertebrates as well, including dinosaurs and therapsids.

In striking contrast, an insect's ears can be located virtually anywhere on its body, albeit also in pairs. Tympanal organs can be found in various locations, in its legs or its mouthparts, in its abdomen or its thorax, and even in its wings (Ewing 1989; Hoy and Robert 1996). Sometimes, as in the praying mantis, the ears may be in such close proximity that the term *auditory cyclops* comes to mind, but even here, the ears are paired (Yager and Hoy 1986). Of course, the location of any animal's ears is a legacy of its evolutionary and developmental origins, as is discussed in subsequent chapters (Fullard, Chapter 8; Boyan, Chapter 4; Robert and Hoy, Chapter 6).

It seems to be the case that almost all vertebrates are endowed with at least a rudimentary sense of hearing, but in insects the answer to "who can hear?" depends on what we define as "hearing." True tympanal hearing, which is the main concern of this book (see later for a definition) is the province of seven different taxonomic orders of insects (out of a possible 27 orders, Evans 1984), and even within an order the occurrence of hearing organs may range from common (in crickets and grasshoppers) to rare (in flies and beetles). It is appropriate here to note a caveat: hearing, as understood in this book, is a subclass of mechanoreception, which is defined as the detection of a mechanical disturbance, propagated through the air, from source to receiver. But many insects produce and detect vibrations that are propagated through the substrate that supports them, such as a branch or leaf of a bush or tree (Markl 1983; Dambach 1989). In fact, many more insects are sensitive to substrate-borne signals than to airborne ones, and it is entirely possible that the latter evolved from the former (Boyan, Chapter 4).

In this volume the decided emphasis is on the hearing of airborne signals, and this coincides with what most people understand as "hearing." However, we should be aware that even among vertebrates, substrate vibrations are by no means rare (Lewis and Narins 1985). In his chapter on spiders, Friedrich Barth provides a fascinating glimpse into their ability to detect vibrations. Although this is not an example of acoustic communication in the sense of detecting minute changes in air pressure that result from the transmission of an acoustic signal from an emitter, the comparison between substrate-borne and airborne signals is both fascinating and important. Strictly speaking, the inclusion of a chapter on spiders in this volume might have prevented its title being *Comparative Hearing: Insects* but instead required the title, *Comparative Hearing: Arthropods*, because spiders are, of course, not insects. However, given that the other eight chapters are on insects, and on their detection of airborne sounds, the title is based on the rule of the majority.

Nonetheless, the subject of substrate signals is important to consider in the context of insect communication signals because many insects also use substrate vibrations for communication or predator detection, probably more species than use the sense of hearing by means of a tympanal organ. As discussed in other chapters of this book, it seems likely that auditory organs, ears, evolved from precursor organs sensitive to substrate vibrations (Boyan, Chapter 4).

Although it may appear that a discussion of the comparative biology of insect hearing is an exercise in qualification and examination of definitions, it is worth pointing out that in the comparative study of vertebrate hearing, which has been vastly more studied, definitional issues arise around what we all understand to be "ears" and "hearing." This can be discerned in several chapters in important volume on the evolution of hearing (e.g., Bullock 1992; Popper, Platt, and Edds 1992). The chapters in Webster, Fay, and Popper (1992) can be read profitably as companion readings to this volume.

2. Just What Is an Insect "Ear?"

For the most part, the tympanal hearing organs of insects can be characterized morphologically by three more or less diagnostic features. First, a localized thinning of the external cuticle at the site of the hearing organ often defines the tympanal membrane, or eardrum. The insect's translucent tympana contrast conspicuously with the surrounding pigmented cuticle in crickets, moths, and grasshoppers, for example. Second, internally and in close apposition to the tympanum is an air-filled sac of tracheal origin, which is sometimes expanded into a chamber. Third, this complex is innervated by a scolopidial (sometimes called a *chordotonal*) organ, which is the sensory organ itself. It is composed of a group of cells, including a sensory neuron, the dendrite of which inserts into a specialized cell called a *scolopale cell*, which is of ciliary origin. The neuron and scolopale cells are, in turn, associated with or enveloped by one or more glial and support cells (McIver 1985).

Sound waves propagating through the air from a vibrating source impinge upon the insect's tympanal eardrums, setting them into vibration. In some insects, sound impinges on the eardrum via an internal as well as external pathway, giving rise to pressure-difference receivers (Bennet-Clark 1983; Michelsen 1992, Chapter 2). The vibrations of the tympana and associated air sac(s) are transferred to the scolopidial sensory organ, where mechano-electrical transduction results in the generation of nerve impulses in the auditory nerve. Thus, unlike vertebrate ears, where airborne vibrations at the eardrum are converted into vibrations within the fluid-filled cochlea by middle ear bones, no such conversion need occur in insects. Without a fluid-filled cochlea, there is no need for impedance matching by an intervening middle ear.

In describing the tympanal hearing organs of insects, the strict adherence to the possession of a frank, clearly differentiated tympanal membrane or eardrum, on the outer (exo)cuticle must be qualified by noting that there are a few exceptions. For example, the scolopophorous sensory hearing organs of the green lacewing, *Chrysopa carnea*, are contained within a (presumably) air-filled wing vein (Miller 1970). Here, two of the three defining criteria are apparent, but the wing does not reveal a localized tympanum; however, it might be argued that the thin membranous wing itself could serve as a tympanum. Recent work on the American cockroach, *Periplaneta americana*, provides evidence that this familiar insect pest possesses a sense of hearing, yet no obvious tympanal eardrum can be observed at the site of its subgenual auditory organs, in its legs (Shaw 1994).

The giant hissing cockroach of Madagascar, *Gromphadorhina portentosa*, is another exception. Its hearing organs appear to have been localized to the scolopophorous subgenual (beneath the "knee") organs of the forelegs and midlegs (Nelson 1980). The relationship of the subgenual organs to tympanal organs in insects that have both is the subject of Chapter 4 by Boyan and will not be further explored here, except to make the point that whereas the subgenual organs of cockroaches are associated with air-filled respiratory tracheae in the leg, there does not appear to be any localized thinning of the cuticle around the organ. Clearly, this is another exception to the requirement that conspicuous and well-differentiated tympana define tympanal hearing organs. However, of the insects known to possess a sense of hearing, it is true that tympana can be found in the vast majority of them, as reported in the literature. It is worth commenting that these exceptions might serve as a cautionary note for physiological entomologists who might presume that the absence of a tympanal eardrum would indicate the absence of hearing. Clearly, this would be a mistake; moreover, it might mean that the sense of hearing is more widespread among insects than we are aware of now.

It should also be noted that in discussing tympanal hearing organs, we refer to mechanosensory receivers that are sensitive to the pressure variations in a propagated sound wave, which is usually referred to as the acoustic *farfield*. Of course, very close to the vibrating sound source (within a distance of one or two wavelengths), the sound wave consists of the bulk movement of air particles, in the acoustic *nearfield*. Near-field detectors are relatively insensitive to pressure variations but are very sensitive to the molecular movements within the enveloping nearby surrounding air; they are particle velocity detectors. In insects such as mosquitoes, the antennae are covered with thousands of filamentous hairs, which serve as near-field detectors. Near-field hearing organs are not a central focus in this volume, but they are described briefly in Michelsen's chapter. This subject is also a relevant issue in vertebrate hearing (Popper, Platt, and Edds 1992).

3. How Widespread Is Hearing Among Insects?

To the nonspecialist, the answer may seem obvious: not very widespread at all. After all, through the auditory "window" of human hearing, the "kinds" of conspicuously noisy animals can be counted on one hand: crickets, katydids, and grasshoppers (order Orthoptera) and cicadas (order Homoptera). Moreover, the totality of these insects is miniscule compared with the legions of the most common insects: the Lepidoptera (butterflies and moths), Coleoptera (beetles), Diptera (flies), and Hymenoptera (bees and wasps), most of which appear not to hear (a caveat: few have been tested). In fact, of the 27 or so recognized orders of insects (Evans 1984), tympanal hearing (defined later) has been reliably identified in 7 orders, including species in the most speciose and widespread orders (Table 1.1). As can be seen from Table 1.1, even within a given order of insects, a sense of hearing may have evolved independently more than once.

Why are most of us unaware that so many insects have a sense of hearing and ears to hear with? For one thing, just because humans don't hear an insect making a conspicuous sound doesn't mean they aren't producing sounds (e.g., ultrasound) or that they can't hear them. Hearing can be used for predator detection as well as for communication, and when predators produce sounds that are beyond the range of human hearing, such as the biosonar signals of insectivorous bats (Roeder 1967), we are oblivious to them without specialized transducers.

4. What's in a Name? When Is a Cricket not a Cricket?

Before going any farther it is appropriate in this book, aimed at the non-entomologist, to provide an informal synonomy of the common names applied to the most conspicuously acoustic insects. There is no confusion when the insects are identified by Linnean nomenclature, but this is not how most people generally identify insects. This section is needed because of the different common names given to insects of the same family or subfamily in different countries. European, British, and U.S. common names are particularly confusing, as is seen in this short section.

The name *cicada*, for example, is relatively unambiguous designation in referring to an acoustically active homopteran insect, except when it is called a "13-" or "17-year locust." Locusts they are definitely not. Even the

TABLE 1.1. Tympanal hearing organs in seven orders of insects.

Order	Superfamily/family	Species	Ear location	Reference
Neuroptera	1. Chrysopidae	*Chrysopa carnea* (Green lacewing)	Wing base	Miller LA (1970)
Lepidoptera	2. Geometroidea Geometridae	*Larentia tristata*	Abdomen	Cook MA, Scoble MJ (1992)
	3. Noctuoidea/Noctuidae		Metathorax	Doe CQ, Goodman CS (1985) development and segmental
	Notodontidae	*Phalera bucephala*	Metathorax	Richards E (1933)
	Arctiidae	*Cycnia tenera*	Metathorax	Scoble MJ (1992)
	4. Papilionoidea Nymphalidae	*Heliconius erato*	Base of forehead or hindwing	Swihart SL (1967)
	5. Hedyloidea/Hedylidae	*Macrosoma hyacinthina*	Forewing base	Scoble MJ (1986)
Coleoptera	6. Cicindelidae	*Cicindela marutha* (Tiger beetle)	Abdomen	Spangler HG (1988)
	7. Scarabaeidae	*Euetheola humilis*	Cervical membranes	Forrest TG (1994)
Dictyoptera	8. Mantodea	*Mantis religiosa* (Praying mantis)	Ventral metathorax	Yager DD, Hoy RR (1989)
	Blattoidea	*Periplaneta americana* (American cockroach)	Metathoracic leg	Shaw SR (1994)
Orthoptera	9. Acrididae	*Locusta migratoria* (Migratory locust)	First abdominal segment	Fullard JH, Yack JE (1993)
	10. Gryllidae	*Gryllus bimaculatus* (Field cricket)	Prothoracic leg	Schwabe J (1906)
Hemiptera	11. Cicadidae	*Cystosoma saundersii*	Abdomen	Young D, Hill KG (1977)
	12. Corixidae	*Corixa punctate*	Mesothorax	Prager (1976)
Diptera	13. Tachinidae	*Ormia ochracea*	Ventral prosternum	Robert D, Read MP, Hoy RR (1994)
	14. Sarcophagidae	*Colcondamyia auditrix*	Ventral prosternum	Soper RS, Shewell GE, Tyrrell D (1976)

relatively clarity of the term *cricket* can be muddled in the mind of the lay reader because "field" cricket, "ground" cricket, "tree" cricket, and "house" cricket all refer to insects of the order Orthoptera and the family Gryllidae. But "bush" crickets (European and British usage) are not crickets. This name is used primarily in Europe and Britain to describe a group of insects that in North America are known as *katydids*, except where they are known as *long-horned grasshoppers* (to distinguish them from *short-horned grasshoppers*, which are "true" grasshoppers of the family Acrididae). Thus, *bushcricket, long-horned grasshopper*, and *katydid* refer to insects of the order Orthoptera in the family Tettigoniidae. But they are neither members of the family Gryllidae (which includes the field crickets) nor members of the family Acrididae (which includes true grasshoppers). Moreover, the implication of the term *bush* might, in this context, be thought of as a habitat signifier, as in "tree" or "field" cricket, but in the case of bushcrickets it is not. Indeed, some bushcrickets can be found in the same habitat as some tree crickets.

Before further confusing the reader, the following informal glossary of acoustically active insects is offered. It is not meant to be authorative, because the author makes no claims to be taxonomist of the Orthoptera. It should also be noted that this listing includes mainly those insects that actually produce sounds. In fact, most insects that "sing" also hear (certainly all those discussed later), but it is worth keeping in mind that many insects that have tympanal hearing organs and are very sensitive to acoustic signals do not themselves produce acoustic signals (that we know of) but do have predators, for example, that detect their prey acoustically, most notably insectivorous bats (Roeder 1967).

4.1. Crickets (exposed eardrums found on forelegs)

Field cricket appears to be the common name in both Europe and North America for orthopteroids of the family Gryllinae, subfamily Gryllinae, genus *Gryllus*. These are the large, black insects with long antennae and shiny, hard wingcovers that, with the first frosts of autumn, retreat into the dark corners of one's garage and can be heard to sing loud "chirps" or "trills" day and night. There is no identity problem here.

Ground cricket and *tree cricket* refer to other members of the subfamilies of the Gryllidae, Nemobiinae and Oecanthinae, respectively. Like field crickets, which are usually the referents for the unqualified use of the term *cricket*, their ears are borne on their forelegs and they stridulate with their forewings. Both are small-bodied insects and can be found just where the names suggest, but elsewhere as well.

House cricket refers to another of the Gryllidae, *Acheta domesticus*. It is of Asian origin and has invaded many other continents, usually in association with human settlement. These crickets are sold comercially as bait for fishermen and as live food for zoo-captive reptiles, amphibians, and insectivorous mammals.

Mole cricket refers to large crickets of the family Gryllotalpinae. These noisy crickets dig deep, often large, horn-shaped burrows. These crickets disperse on the wing at night. They have an affinity for lawn turf, and hence are the bane of the golf course industry in the southeastern United States.

4.2. Katydids (eardrums found on forelegs)

Now comes the problem. As mentioned earlier, these insects of the family Tettigoniidae are referred to in the United States as *katydids*, except in regions where they are called *long-horned grasshoppers*, which is problematic, as is seen later. In Europe, katydids are called *bushcrickets*, and the reader will encounter this designation often in this volume. In the United States, the term *katydid* is used to designate the entire family Tettigoniidae and derives from a single exemplar, the true katydid, *Pterophylla camellifolia*, whose call is euphonized as "katy did, katy didn't; katy did . . ." This is unfortunate because *P. camellifolia* is found in the northeastern and southeastern United States only, and the songs of the many other members of the species of Tettigoniidae are better characterized as a noisy buzz, a raspy trill, a "chip," or a "tick." Another American term for tettigoniids is *long-horned grasshopper*, which is also unfortunate because although they have long antennae ("horns"), they are not grasshoppers (see later). Many of these insects are acoustically active at night.

4.3. Grasshoppers and Locusts (eardrums on first abdominal segment, covered by wings)

These insects are members of the superfamily Acridoidea, and many of the favorites for acoustic study are in the family Acrididae. They are often seen or heard in meadows and grasslands and are acoustically active during the day. Their songs often have a whisper-like or raspy quality. These insects are often referred to as *short-horned grasshoppers*. This is an appropriate designation: they have short antennae ("horns") and they are grasshoppers. Locusts and grasshoppers, unlike crickets and katydids, do have their hearing organs in their metathoracic body segments (the same segment that bears the hindwings and hindlegs), and they stridulate with their hindlegs and wings.

4.4. Cicadas (exposed eardrums on the abdomen)

These are among the most conspicuously noisy insects and are mainly daytime singers. There is not much confusion in what these insects are called, except in a few regions of the United States, where they are referred to as *17-year locusts*! Unlike crickets, grasshoppers, and katydids, cicadas produce sound by means of a specialized tympal mechanism, and not by stridulation (Ewing 1989). Table 1.1 provides summary information about the occurrence of tympanal hearing organs in seven orders of insects, with representative examples from each.

5. Acoustic Behavior: The What and Why of Insect Hearing

5.1. Conspecific Communication

Many crickets, grasshoppers, katydids, and cicadas use acoustic signals to mediate the functions of long-distance (in the range of tens to thousands of body lengths) mate calling, short-distance (in the range of a few body lengths) courtship, and territorial proclamation (which ranges from short to long distance). The signals mediate a coevolved system in which there is a good match between the acoustic properties of the sender's signal and the auditory sensitivity of the receiver. The function of acoustic signals for intraspecific communication is essentially the same as that in many vertebrates, including frogs, birds, and mammals. Because many animals are restricted to particular regions of a biotope, the efficiency of propagation of a given signal through it is subject to biotic and abiotic factors, as is discussed by Römer in Chapter 3.

5.2. Predator Detection

As was demonstrated by Kenneth Roeder 40 years ago, it is clear that insectivorous bats, possessed of formidable ultrasonic biosonar systems, have been a source of selective pressure in the evolution of ultrasound-sensitive tympanal ears. Thus in moths hearing organs appear to have independently evolved in at least four different families (see Table 1.1). Since Roeder's work on moths, many other nocturnal, night-flying insects have been shown to elicit an acoustic startle response when stimulated with batlike pulse trains of loud ultrasound, including green lacewings (Miller 1970), field crickets (Moiseff et al. 1978), praying mantises (Yager and Hoy 1986), locusts (Robert 1989), katydids (Libersat and Hoy 1991), beetles (Spangler 1988; Forrest et al. 1997; Yager and Spangler 1997), and many more moths (see Fullard, Chapter 8). It is clear that predation from bats has been a potent force in the evolution of insect hearing organs.

5.3. Localization of Hosts by Parasites

Over the course of evolution, the reproductive linkage between parasite and host can be as strong and essential for survival as between mates. There are two well-documented examples of acoustic parasitism in insects, in which a parasitic fly detects and locates its host by hearing the calling song of the latter. One involves a tachinid fly (*Ormia ochracea*), which is a cricket parasitoid (Cade 1975), and the other involves a sarcophagid fly (*Colcondamyia auditrix*), which is a parasitoid of cicadas (Soper 1976). Further investigation of some of the parasitic species led to the discovery of tympanal ears in the parasitic flies (Lakes-Harlan and Heller 1992; Robert et al. 1992). This unusual function of hearing is discussed fully by Robert and Hoy (Chapter 6).

5.4. Insect Psychoacoustics

The ability of any sensory modality to discriminate differences in its adequate stimulus along some temporal, spectral, or intensity axis is a traditional line of study in the investigation of human perceptual abilities (Yost, Popper, and Fay 1993). Similar studies have been made in other vertebrates, for example, bats (Popper and Fay 1995) and birds (Dooling 1982), among others. However, similar studies in insects have been rare, in spite of the fact that they are the only invertebrates known to possess an auditory sense. Recently, studies indicate that there are parallels between vertebrate and insect auditory "perception" in such traditionally vertebrate psychoacoustic domains as habituation (May and Hoy 1991), precedence effect (Wyttenbach and Hoy 1993), and categorical perception (Wyttenbach et al. 1996). Although it would be rash to infer higher cognitive function in insects from such studies, it seems reasonable to point out that the auditory behavior of insects shows more plasticity and complexity than might have been expected, when subjected to appropriate behavioral tests. It is also reasonable to suggest that insects might provide interesting models for central auditory processing for phenomena such as categorical perception.

6. Comparing and Contrasting the Ears and Auditory Systems of Vertebrates and Insects

Naturally, whether the listener of an acoustic signal is a moose or a mantis, the physics of sound propagation dominates the hearing process. For example, elephants, crickets, canaries, ormia flies, and humans are all sensitive to a 5-kHz tone, emitted, say, at 75 dB SPL. In fact, as is seen in the following chapters, the ears of insects can rival those of vertebrates in sensitivity, and insects are able to localize sound sources with great accuracy. The ability of each animal to localize the sound source depends on how it processes the binaural cues generated by the interaction of the 5-kHz sound wave with its auditory receivers. There is a critical interaction between the wavelength of sound and the size of the animal (interaural separation), which results in the generation of interaural difference cues in arrival time and sound level, and in our examples the interaural separation varies by over three orders of magnitude. In general, acoustically active insects are considerably smaller than most vertebrates, yet they are sensitive to a common band of sound frequencies, ranging from 3 kHz to about 15 kHz. The problem of sound localization for an animal of small size is discussed in this volume by Michelsen (Chapter 2), Römer (Chapter 3), and Robert and Hoy (Chapter 6).

In the realm of conspecific social signals, one would expect a matching of auditory sensitivity and power in the animal's sound production properties. Similarly, the tuning properties of predators and prey might be overlapping. To produce conspecific social signals, vertebrates use their vocal/respiratory tracts, whereas many insects use frictional stridulatory mechanisms. In spite of this difference, there is a surprising amount of overlap in the frequency bandwidth of sensitivity between many insects and vertebrates. Many insects, including crickets, grasshoppers, katydids, and cicadas, have representative species whose mating calls encompass a frequency range from about 2.5 kHz to 15 kHz. Many of them also can hear well beyond this range into the ultrasound, up to 100 kHz, but the point is that 2 to 15 kHz is well within the hearing range of many vertebrates, including most mammals.

6.1. Signal Diversity

Even a casual listener can observe that the acoustic signals of the most aurally conspicuous insect songsters divide into those that are clearly tonal and those that are "buzzy" or noisy. By and large, the former describe the songs of crickets (family Gryllinae) and the latter the signals of katydids, locusts, and cicadas. The songs of crickets are reminiscent of the mating calls of some frogs and toads. Whereas a few birds may sing monotonously tonal trills, a striking aspect of bird songs is their frequency modulation (FM), and, by and large, frequency modulation (FM) is not a prominent feature in the songs of crickets or other insects of which this author is familiar. This does not mean that the insect ear is insensitive to FM signals, however, because insectivorous bats produce FM-modulated biosonar signals (Faure, Fullard, and Dawson 1993).

6.2. Anatomy

As is discussed in Chapters 3 to 5, there are parallels in the anatomical organization of the auditory system. For example, in field crickets and katydids the auditory receptors of the crista acustica (the sensory organ) in the foreleg form a linear array of cells. This linear anatomical array is the basis of a linear tonotopic coding, such that sensory cells in the proximal portion of the organ code for low frequencies and in the distal portion code for high frequencies, with a systematic gradient of frequency coding from low to high frequencies corresponding to a linear proximal-to-distal anatomical addressing (first reported by Zhantiev and Korsunovskaya 1978, and then by Oldfield 1982, 1985). The terminals of the auditory afferents also make a roughly tonotopic projection in the central auditory neuropil of katydids (Römer 1983), and in crickets there appears to be a tonotopic organization of interneurons (Atkins and Pollack 1987; Chapter 5). The critical link between implementation of frequency-dependent auditory behavior and activation of specific frequency channels in the central nervous system has not yet been made in these insects, but it will be surprising if such a linkage is not found.

7. Why Are the Ears of Insects Found Here, There, and Everywhere?

As illustrated schematically in Figure 1.1 and as documented in Table 1.1, an insect's ears can appear just about anywhere on its body. Unlike vertebrates, the embryological origins of tympanal organs are not tied to an apparently one-time evolutionary event, an apomorphy that can be presumably traced to our piscine origins. Insect tympanal organs are derived from specialized proprioceptive sensory structures, chordotonal organs, which are often repeated segmentally and are often associated with respiratory tubes, the tracheae. The evolutionary development of a tympanal organ

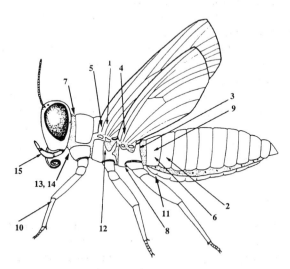

FIGURE 1.1. Cartoon of generic insect displaying the location at which tympanal hearing organs have been identified in the Insecta. The numbers above the arrows refer to the second column of Table 1.1, except for number 15, which points to the pilifer organ of hawkmoths. (From Hoy and Robert, 1996, modified from Yack and Fullard 1993, with permission.)

from a proprioceptor has occurred many times in acoustically active insects and has been plausibly supported in recent developmental studies (see Boyan, Chapter 4).

As described earlier, an insect's tympanal hearing organs can be characterized by a suite of three anatomical features: a localized thinning of the cuticle, an adjacent air-filled sac of tracheal origin, and innervation by a chordotonal sensory organ. An insect's body is covered by cuticle; internally, respiratory tracheae course throughout the body and open out to the surrounding air via spiracles. Respiratory tracheae run along an insect's thoracic and abdominal body walls, and branch extensively; they also extend down into prominent appendages, such as legs. It is not uncommon to find chordotonal organs serving as proprioceptors at the appendicular joints and at intersegmental "joints" between the thorax and abdomen, and in the abdomen itself. Thus, the opportunity for an ear to arise occurs frequently over the body of a typical insect, given its multiple joints and appendages.

For example, James Fullard, in Chapter 8, proposes that the ear of the noctuid moth may have evolved from a proprioceptor that in its nonhearing ancestor may have served as a wing-hinge stretch receptor to monitor flight movements. In crickets (gryllids) and katydids (tettigonids), the foreleg tympanal hearing organ most likely evolved from the nearby subgenual complex. In the parasitoid dipteran fly, *Ormia ochracea*, the tympanal hearing organ appears to have evolved from a proprioceptor of unknown function, but that is found in many higher flies that do not have a sense of hearing (Edgecomb et al. 1995).

8. Why Study Insect Ears?

Comparative studies are always intrinisically interesting for biologists interested in processes and origins, in form and function. The physics underlying hearing in mammals constrains hearing in insects as well. Certainly size, anatomy, and body plan dissimilarities abound. However, unexpected similarities — convergences or parallelisms in form and function — can be found in the insect auditory system. An important take-home message from the study of the insect ear is what it has to tell us about evolution in action. In its origins and structure, the anatomy of the insect ear is very different from its vertebrate counterpart. Yet tonotopic organization of the receptor array is apparent in crickets and katydids, even exhibiting a vertebrate-like linear tonotopy (Oldfield 1982). Tonotopic organization extends into the organization of a central neuropile and tracts (Roemer 1983; Atkins and Pollack 1987). We are a long way from understanding auditory processing in the central nervous system. But we know a little more about the processing of sound in the auditory periphery. Here we hope that the relative numerical simplicity of the insect auditory pathway (in which the number of auditory interneurons are countable in tens to perhaps a few hundred, if those in the brain are eventually described) might reveal principles of auditory processing. Whether we will again find interesting similarities or strongly divergent differences between vertebrates and insects is something that we just don't know yet. I am willing to bet on the former.

Summary

Among terrestrial animals, only insects rival vertebrates in form, function, and diversity of their auditory apparatus. Auditory organs subserve adaptive functions such as predator detection and the mediation of mating signals, in insects just as in vertebrates. Hearing organs may be found virtually anywhere on an insect's body, depending on species. Insect ears can be quite sensitive to the direction of a sound source and can be divided into pressure receivers, pressure-difference receivers, and mechanically coupled receivers. The application of psychoacoustic tests, developed in the context of mammalian auditory investigations, can be applied to insect hearing, with the result that such phenomena as the precedence effect, categorical perception, habituation, and minimal audible angle, can be demonstrated. Although the morphological bases of hearing in vertebrates and insects are widely divergent, there are demonstrable functional and behavioral convergences.

References

Atkins G, Pollack GS (1987) Correlations between structure, topographic arrangement and spectral sensitivity of sound-sensitive interneurons in crickets. J Comp Neurol 266:398–412.

Bennet-Clark HC (1983) Insect Hearing: Acoustics and Transduction. In: Lewis T (ed) Insect Communication. London: Academic Press, pp. 49–82.

Boyan G (1993) Another look at insect audition: tympanic receptors as an evolutionary specialization of the chordotonal system. J Insect Physiol 39:187–200.

Bullock TH (1992) Comparisons of major and minor taxa reveal two kinds of differences: "lateral" adaptations and "vertical" changes in grade. In: Webster DB, Fay RR, Popper AN (eds) The Evolutionary Biology of Hearing. New York: Springer-Verlag, pp. 15–18.

Cade WH (1975) Acoustically orienting parasitoids: fly phonotaxis to cricket song. Science 190:1312–1313.

Cook MA, Scoble JJ (1992) Tympanal organs of geometrid moths: a review of their morphology, function, and systematic importance. Syst Entomol 17:219–232.

Dambach M (1989) Vibrational responses. In: Huber F, Moore TE, Loher W (eds) Cricket Behavior and Neurobiology. Ithaca, NY: Cornell University Press.

Doe CQ, Goodman CS (1985) Early events in insect neurogenesis: development and segmental differences in the pattern of neuronal precursor cells. Dev Biol 111:193–205.

Edgecomb RS, Robert D, Read MP, Hoy RR (1995) The tympanal hearing organ of a fly: phylogenetic analysis of its morphological origins. Cell Tissue Res 282:251–268.

Evans HE (1984) Insect Biology. Reading, MA: Addison-Wesley.

Ewing AW (1989) Arthropod Bioacoustics. Ithaca, NY: Cornell University Press.

Faure PA, Fullard JH, Dawson, JW (1993) The gleaning attacks of the Northern long-eared bat, *Myotis* septentrionalis, are relatively inaudible to moths. J Exp Biol 178:173–189.

Forrest TG (1994) From sender to receiver: propagation and environmental effects on acoustic signals. Am Zool 34:644–654.

Forrest TG, Farris HG, Hoy RR (1995) Ultrasound acoustic startle response in scarab beetles. J Exp Biol 198:2593–2598.

Forrest TG, Read MP, Farris HE, Hoy RR (1997) A tympanal hearing organ in scarab beetles. J Exp Biol 200:601–606.

Fullard JH, Yack JE (1993) Trends Ecol Evol 8:248–252.

Hoy RR (1992) Evolution of ultrasound hearing in insects. In: Webster DB, Fay RR, Popper AN (eds) The Evolutionary Biology of Hearing. New York: Springer-Verlag, pp. 115–129.

Hoy RR, Robert D (1996) Tympanal hearing in insects. Annu Rev Entomol 41:433–450.

Huber F, Moore TE, Loher W (1989) Cricket Behavior and Neurobiology. Ithaca, NY: Cornell University Press.

Kerkut GA, Gilbert LI, eds. (1985) Comprehensive Insect Physiology, Biochemistry, and Pharmacology, Vols. 1–13. New York: Pergamon Press.

Lakes-Harlan R, Heller K-G. (1992) Ultrasound-senitive ears in a parasitoid fly. Naturwissenschaften 79:224–226.

Larsen ON, Kleindienst H-U, Michelsen A (1989) Biophysical aspects of sound reception. In: Huber F, Moore TE, Loher W (eds) Cricket Behavior and Neurobiology. Ithaca, NY: Cornell University Press, pp. 364–390.

Lewis ER, Narins PM (1985) Do frogs communicate with seismic signals? Science 227:187–189.

Libersat F, Hoy RR (1991) Ultrasonic startle behavior in bushcrickets (Orthoptera, Tettigoniidae). J Comp Physiol 169:507–514.

Markl H (1983) Vibrational communication. In: Huber F, Markl H (eds) Neuroethology and Behavioral Physiology. Berlin: Springer-Verlag, pp. 332–353.

May ML, Hoy RR (1991) Habituation of the ultrasound-induced acoustic startle response in flying crickets. J Exp Biol 159:489–499.

McIver SB (1985) Mechanoreception. In: Gibert LI, Kerkut GA (eds) Comprehensive Insect Physiology, Biochemistry, and Pharmacology, Vol. 6, New York: Pergamon Press, pp. 71–132.

Meier T, Reichert H (1991) Embryonic development and evolutionary origin of the orthopteran auditory organs. J Neurobiol 21:592–610.

Michelsen A (1992) Hearing and sound communication in small animals: evolutionary adaptations to the laws of physics. In: Webster DM, Fay RR, Popper AN (eds) The Evolutionary Biology of Hearing. New York: Springer-Verlag, pp. 61–78.

Michelsen A, Larsen ON (1985) Hearing and sound. In: Kerkut GA, Gilbert LI (eds) Comprehensive Insect Physiology, Biochemistry, and Pharmacology, Vols. 1–13. New York: Pergamon Press, pp. 494–556.

Miller LA (1970) Structure of the green lacewing tympanal organ (*Chrysopa carnea*). J Morphol 131:359–382.

Moiseff A, Pollack GS, Hoy RR (1978) steering responses of flying crickets to sound and ultrasound: mate attraction and predator avoidance. Proc Natl Acad Sci USA 75:4052–4056.

Nelson MC (1980) Are subgenual organs "ears" for hissing cockroaches? Soc Neurosci Abstr 6:198.5.

Oldfield BP (1982) Tonotopic organisation of auditory receptors in Tettigonidae (Orthoptera, Ensifera). J Comp Physiol 147:461–469.

Oldfield BP (1985) The tuning of auditory receptors in bushcrickets. Hearing Res 17:27–35.

Popper AN, Fay RR (eds) (1995) Hearing by Bats. New York: Springer-Verlag.

Popper AN, Platt C, Edds PL (1992) Evolution of the vertebrate inner ear: an overview of ideas. In: Webster DB, Fay RR, Popper AN (eds) The Evolutionary Biology of Hearing. New York: Springer-Verlag, pp. 49–56.

Prager J (1976) Das mesothorakale Tympanalorgan von *Corixa punctata* III. (Heteroptera Corixidae). J Comp Physiol 110:33–50.

Richards E (1993) Comparative skeletal morphology of the noctuid tympanum. Entomol Am 13:1–43.

Robert D (1989) The auditory behavior of flying locusts. J Exp Biol 147:279–301.

Robert D, Amoroso J, Hoy RR (1992) The evolutionary convergence of hearing in a parasitoid fly and its cricket host. Science 258:1135–1137.

Robert D, Read MP, Hoy RR (1994) The tympanal hearing organ of the parasitoid fly *Ormia ochracea* (Diptera, Tachinidae, Ormiini). Cell Tissue Res 275:63–78.

Roeder KD (1967) Nerves Cells and Insect Behavior. Cambridge, MA: Harvard University Press.

Römer H (1983) Tonotopic organization of the auditory neuropile in the bushcricket *Tettigoni viridissima*. Nature 306:60–62.

Schwab J (1906) Beitrage zur Morphologie and Histologie der tympanalen Sinnesapparate der Orthopteren. Zoologica 20:1–154.

Scoble MJ (1986) The structure and affinities of the Hedyloidea: a new concept of the butterflies. Bull Br Mus Nat Hist (Entomol) 53:251–286.

Scoble MJ (1992) The Lepidoptera: Form, Function, and Diversity. New York: Oxford University Press.

Shaw SR (1994) Detection of airborne sound by a cockroach "vibration detector": a possible missing link in insect auditory evolution. J Exp Biol 193:13–47.

Soper RS, Shewell GE, Tyrrell D (1976) *Colcondamyia auditrix* nov. sp. (Diptera: Sarcophagidae), a parasite which is attracted by the mating song of its host, *Okanagana rinosa* (Homoptera: Cicadidae). Can Entomol 108:61–68.

Spangler HG (1988) Hearing in tiger beetles (Cicindelidae) Physiol Entomol 13:447–452.

Swihart SL (1967) Hearing in butterflies (Nymphalidae: *Heliconius*, *Ageronia*). J Insect Physiol 13:469–476.

Wyttenbach RA, Hoy RR (1993) Demonstration of the precedence effect in an insect. J Acoust Soc Am 94:777–784.

Wyttenbach RA, May ML, Hoy RR (1996) Categorical perception of sound frequency by crickets. Science 273:1542–1544.

Yack JE, Fullard JH (1993) What is an insect ear? Ann Entomol Soc Am 86:677–682.

Yager DD, Hoy RR (1986) The cyclopean ear: a new sense for the praying mantis. Science 231:727–729.

Yager DD, Hoy RR (1989) Audition in the praying mantis, *Mantis religiosa L.*: identification of an interneuron mediating ultrasonic hearing. J Comp Physiol A 165:471–493.

Yager DD, Spangler HG (1995) Characterization of the auditory afferents in the tiger beetle, *Cicindela marutha* Dow. J Comp Physiol 176:587–599.

Yager DD, Spangler HG (1997) Behavioral response to ultrasound by the tiger beetle *Cicindela marutha* Dow combines aerodynamic changes and sound production. J Exp Biol 200:649–659.

Yost WA, Popper AN, Fay RR, eds. (1993) Human Psychoacoustics. New York: Springer-Verlag.

Zhantiev RD, Korsunovskaya OS (1978) Morphological organization of tympanal organs in *Tettigonia cantans* (Orthoptera, Tettigonidae). Zool J 57:1012–1016.

Young D, Hill KG (1977) Structure and function of the auditory system of the cicada, *Cystosoma Saundersii*. J Comp Physiol 117:23–45.

2
Biophysics of Sound Localization in Insects

Axel Michelsen

1. Introduction

Humans use two mechanisms for detecting the direction of sound waves, based on diffraction and time of arrival, respectively (Shaw 1974; Yost and Gourevitch 1987; Brown 1994). The presence of the body may disturb the sound wave so that the sound pressure at the surface of the body differs from that in the undisturbed sound wave (diffraction). The sound pressure at a particular position on the surface, for example, the location of an ear, varies with the direction of sound incidence. Diffraction occurs when the dimensions of the body (head) are larger than one-tenth the wavelength of the sound. The sound spectra at the two ears differ for most sound directions if the ears are some distance apart. It is thus possible for the brain to estimate the direction of the sound source by comparing the sound spectra at the two ears. This task is easier with broad-band sounds than with pure tones or narrow-band sounds.

Sound travels in air with a velocity of about 344 m/s. In humans (head diameter, 17 cm), the maximum difference in the time of arrival at the two ears is approximately 0.5 ms. Humans can localize long pure tones less than about 1400 Hz by means of this mechanism because there is a cycle-by-cycle following of the sound wave by the nerve impulses (phase detection). At higher frequencies the time of arrival can be detected if the sound is sufficiently modulated in amplitude or frequency content to provide reliable time cues for the brain.

In insects the part of the body carrying the ears is about 10 to 50 times smaller than the human head, and large differences in sound pressure at the ears only exist at high frequencies. Insects and other arthropods are able to process information about the onset of excitation in some of their sense organs. For example, scorpions and spiders receive vibrational signals through their legs, and they turn to the side from which they first receive a stimulus (Brownell and Farley 1979; Hergenröder and Barth 1983). Time differences of 0.2 to 1 ms are necessary, and these values correspond to those expected from the size of the animals and the propagation velocity of the signals. A male grasshopper listening to female song transmitted by two loudspeakers placed to its left and right sides will likewise turn to the side of the earlier signal when the songs are time shifted by 0.5 to 1 ms (Helversen and Helversen 1983). However, the maximum time difference expected when the grasshoppers are locating each other is only 20 to 30 μs, and there is no evidence that the insect central nervous system can use such small time differences as directional cues.

Most insects thus have great difficulty in determining the direction of sound incidence by means of these two mechanisms, and they have to use alternative strategies. One strategy used by many insects is to let the sound waves reach both the outer and inner surfaces of the eardrum. As we are about to see, this may result in excellent directional hearing. Another strategy is to respond not to the pressure component of sound but to the inherently directional movement component. The main purpose of this chapter is to explore the physical mechanisms on which these strategies are based. We concentrate on sound in air and ignore substrate-borne sound.

2. Phonotaxis: Behavioral Evidence for Sound Localization

The ability to localize sound is most easily observed in the directional responses made by flying insects that receive echolocating cries from hunting bats (see Robert and Hoy, Chapter 6). In this section we concentrate on the ability of many insects to use sound signals for attracting and guiding members of the opposite sex during pair formation (Ewing 1989). In some groups (most crickets, cicadas, some bushcrickets, and grasshoppers), the males emit calling songs and the females approach the signaling males. In contrast, in mosquitoes and biting midges, the males are attracted to acoustical females. In other cases (gall midges, some grasshoppers, and bushcrickets), both males and females produce sounds and the walking is done by one or both of the sexes.

The approach of conspecific animals to a singer demonstrates the ability to localize, but it cannot be taken for granted that sound is essential (attractiveness may be caused by looks, as in birds with colorful feathers, rather than by sound). The possible role of sound, thus, has to be demonstrated in experiments in which other factors are excluded or at least controlled. The classic study was by the Slovenian zoologist J. Regen (1913), who showed that female crickets were attracted to a telephone transmitting the calls of male crickets. In designing experiments it is necessary to keep in mind that approaches connected with pair formation are only attempted when the calls are recognized by animals that are ready for copulation (which generally requires a certain physiological state). An absence of phonotaxis may therefore have causes other than a lack of localization ability.

Behavioral experiments on sound localization may be designed in various ways, and the results may reflect not only the properties of the animals but also the experimental conditions. Observations or experiments in the natural habitat tend to be more time consuming and to produce results with more scatter than experiments in the laboratory. To work exclusively in the laboratory is therefore a sensible strategy when studying the role of neural mechanisms for the localization. Laboratory studies are not sufficient, however, if the purpose is to arrive at an understanding of how animals overcome the complexity of the natural habitat (see Section 8 and Römer, Chapter 3).

Arena experiments are the laboratory equivalent of the approach of conspecifics to singing animals in nature. The animals are released into the center of an arena, throughout which the sound from one or more loudspeakers can be heard. The movements of the animal are followed, for example, with a video camera mounted above the arena. The recorded tracks of the movements of an animal show whether it was able to determine the direction of the sound source(s). The main problem for the experimenter is to quantify such behavior. The subjective impression gained through the eye has to be checked against statistical tests (Batschelet 1981). A problem with classical arena experiments is that sound pressure increases in the direction of the loudspeaker. Consequently, more than one explanation is possible for observed approaches (the animal has determined the direction of the loudspeaker or has searched for the loudest sound).

These problems are overcome when the animal is fixed in space when determining the direction of sound. In some animals, this is a part of their natural behavior. For example, in some species of short-horned grasshoppers, the male sings its calling song and waits for a response song from a receptive female. The male then either jumps in the frontal direction or abruptly turns toward the direction of the female song (lateralization). This change in orientation is based on the simple decision of whether the song comes from the left or the right. The male then walks a short distance in the new direction and starts a new sequence by singing again. Using this strategy, the male eventually reaches the female (Helversen and Helversen 1983). In behavioral experiments with males placed between two loudspeakers emitting the female response song, it was found that a difference in sound level of only 1 to 2 dB elicits a turn toward the side stimulated by the louder sound (Helversen and Rheinlaender 1988).

Such lateralization responses in insects are rare. Most insects walk toward the sound source while determining its direction. Partial confinement of the animal in a restricted region along with frame-by-frame television or film analysis is one strategy to follow for the investigator. Alternatively, the animal can be forced to walk and "choose" its direction from one spot without getting anywhere, for example, on a Y-maze globe (Hoy and Paul 1973).

The most elegant method for keeping the animal on one spot while still mimicking normal-approach walking is to use a spherical treadmill in which displacements of the animal on the top of the sphere are compensated by servomotors that turn the sphere (Kramer 1976). The position of the animal is determined in two directions by infrared light reflected from a disc of reflecting foil on the animal's back (Fig. 2.1). The motors then turn the sphere so that the animal is brought back to the center of the scanning field. The signals to the motors are thus a measure of the walking distance and turning angle of the animal.

This method has been perfected by two groups of researchers (Weber, Thorson, and Huber 1981; Schmitz, Scharstein, and Wendler 1982) and used in a large number of studies on sound localization in female field crickets. In contrast to spontaneous walking during periods without presentation of the calling song, acoustically orientated walking is faster, more persistent, and directed toward the sound source. The direction of the animal's body oscillates 30° to 60° around the direction to the sound source (meandering). The amount of meandering and possible excursions away from the direction to the sound source have been used as a measure of the attractiveness of natural and artificial songs. Using this approach, it has

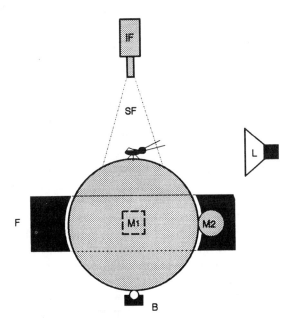

FIGURE 2.1. The locomotor compensator invented by Kramer (1976). IF, infrared light source and scanning camera; SF, scanning field; L, loudspeaker; M1-2, motors with rubber wheels; B, ball bearing; F, frame. (From Wendler and Löhe 1993, with permission.)

been possible to map the female cricket's preferences for the various parameters of the male's song (frequency, duration, and temporal pattern of sound pulses). The results have been used when searching for the neuronal mechanisms in the central nervous system, and excellent agreement between behavior and neuronal properties have been obtained (Schildberger, Huber, and Wohlers 1989; Weber and Thorson 1989).

Although this method has led to a much deeper understanding of both the behavior and neuronal mechanisms involved in song recognition, a word of warning is in order. Several investigators have used the method for monitoring the animal's ability for sound localization after various surgical procedures. Surprisingly, crickets could still localize a sound source, even after losing one ear or the tracheal connection between the ears. Recent studies (see Section 6) have shown that the crickets are left with 1 to 2 dB directionality after these procedures. This may be sufficient in the very homogeneous sound field on top of a Kramer sphere in an echo-reduced laboratory. It is far below the normal directionality, however, and probably quite insufficient in the more complex acoustical conditions in the animals' natural habitats (see Section 8 and Römer, Chapter 3).

Most behavioral experiments on insect sound localization have been performed by observing the reactions of animals to a single loudspeaker emitting recorded conspecific songs or models of such songs. This situation (a male alone with a female) is probably not common in real life. Animals in natural habitats are usually surrounded by many singers. In this situation, it may not be possible for the animals to find a conspecific partner of the opposite sex just by applying the simple rule of "turning to the louder side." Obviously, the chances of a success would increase if conspecific singers were more likely to be located than heterospecific singers. This could be achieved if the recognition of a conspecific song enhanced the efforts of localization in the listening animal. Alternatively, features of conspecific songs might make them easier to hear and thus more easy to locate. Behavioral studies on this complicated problem have revealed that crickets and grasshoppers differ with respect to the processing of sound recognition and sound localization.

The grasshoppers behave as if song recognition and sound localization are parallel processes in the nervous system. Female grasshoppers respond by singing when they recognize the songs of conspecific males. Models of male song may be made in such a manner that two models each contain one half of the components in a song that can release the female response song. Each of these models is thus ineffective when presented alone, but they release a response when presented simultaneously through different loudspeakers. The release of a response does not depend on whether the two loudspeakers are close together or whether the sound arrives at the female from very different directions (Helversen and Helversen 1995).

In contrast, the responses released in crickets in such experiments are critically dependent on the position of the loudspeakers. Crickets seem to possess a neural song recognizer on each side of the central nervous system (Pollack 1988), and they are thus able to choose between two simultaneously active sound sources. When given the choice between a loud and a weak source, a cricket may choose the latter if it emits a more attractive song (Weber and Thorson 1989). Such observations have lead to the conclusion that in crickets the information on recognition and localization are processed sequentially, and not in parallel, as in grasshoppers.

3. Diffraction and Other Properties of Sound

The physics of sound waves is described in many textbooks, but much of the literature is not easily accessible to biologists. Most books on acoustics are filled with equations, and the biologist may get the impression that the solution to all problems in acoustics can be simply obtained by calculation. In theory, this is indeed possible. However, it is seldom possible to find an exact solution to real problems. Most equations in acoustics are approximations and are only valid under the assumptions specified when the equations were derived. Furthermore, animals tend to have a complicated anatomy and mechanical properties, which differ considerably from those assumed under ideal, textbook conditions.

Whereas calculations may lead to misleading conclusions, experiments without a proper theoretical basis are not likely to lead to any real understanding. The ideal approach is a combination of experiments and calculations. Two classical textbooks (Morse 1948; Beranek 1954) contain down-to-earth explanations. More advanced books (e.g., Skudrzyk 1971; Pierce 1981) are useful only for experienced readers. Here, we review only those aspects of acoustics that are relevant for directional hearing.

Sound may be described and measured both as fluctuations in pressure and as oscillations of "air particles" (small volumes of air). The oscillations occur in the direction of sound propagation, and sound waves are therefore said to be longitudinal. Many insects can determine the direction of sound incidence by means of sense organs sensitive to the air oscillations. In contrast, pressure is a scalar (nondirectional) quantity, and the force acting on a small pressure receiver does not depend on its orientation. This may seem to be counterintuitive to a biologist: in the literature one often finds the belief that the force acting on a small eardrum should depend on how the ear is oriented relative to the body.

The dual nature of sound (pressure and movement) allows sound to be measured in two different ways. In a pressure receiver (Fig. 2.2A), sound acts on only one surface of a membrane, which is backed by a closed chamber (see Section 5). A movement receiver, like the thin and lightly articulated hair shown in Figure 2.2C, follows the oscillations of the air particles and is thus inherently directional (see Section 7). A third type of receiver, the pressure-difference receiver (Fig. 2.2B) combines some of the

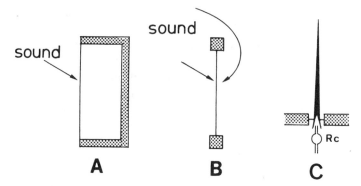

FIGURE 2.2. A: A pressure receiver. B: A pressure-difference receiver. C A movement receiver. Rc, receptor cell. (From Michelsen 1983, with permission.)

properties of the other two (see Section 6). It responds to pressure, but both surfaces of its membrane are exposed to sound. Although the pressure at a particular point in space is a scalar quantity, the change of pressure in the direction toward or away from a sound source is inherently directional. The force acting to move the membrane is proportional to the pressure gradient if the receiver is small relative to the wavelength. Such receivers are therefore called *pressure-gradient receivers*. (The terminology is not fixed. Some authors use the term *pressure-gradient receiver* also for complicated receivers such as those discussed in Section 6).

The directionality of a pressure-difference receiver can be understood by considering a small, plain piece of paper placed at some distance from a sound source emitting a low-frequency tone (with a wavelength larger than the diameter of the paper). The driving force acting to move the paper is at a maximum when the paper is perpendicular to the direction of the sound wave. In this position the amplitudes of the sound pressures are almost the same on the front and back surfaces of the paper, but they are somewhat out of phase because the sound wave reaches the back surface a little later than it reaches the front surface (Fig. 2.2B). The difference between the two pressures will therefore have a certain magnitude and the paper will vibrate. The force is zero, however, when the paper is parallel to the direction of the sound wave (same amplitude and phase on both sides). If the distance from the center of a directional diagram indicates the magnitude of the driving force acting on the receiver, then the directional diagram of a small pressure receiver is circular (Fig. 2.3A), whereas for a small pressure gradient receiver it is a figure eight (Fig. 2.3B). A cardioid directivity pattern (Fig. 2.3C) can be obtained if the sound does not have equal access to the two surfaces. This is the pattern commonly found in insects (see Section 6).

The wavelength is very important when one is dealing with the interaction between sound and solid structures. For example, the efficiency of

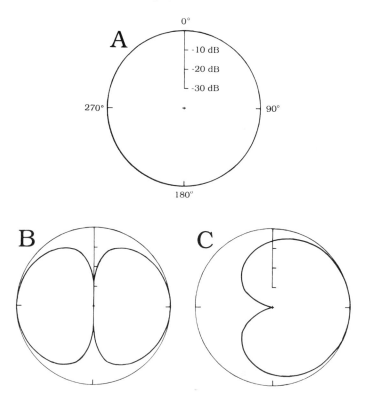

FIGURE 2.3. Three directivity patterns. The distance from the center indicates the force acting on the membrane in the directions indicated. A: Pressure receiver. B: Pressure-difference receiver (the plane of the membrane is in the 0° to 180° direction). C: Cardioid pattern commonly found in insect ears (0° forward direction; 90° ipsilateral to the ear). (Modified from Beranek 1954, with permission.)

sound emission from sound sources increases with the ratio between size and wavelength; the efficiency reaches a constant level when the diameter of the source is equal to or larger than the wavelength. In air, the wavelength is 34 cm at 1 kHz, 34 mm at 10 kHz, and so on. Obviously, insects are generally too small to be efficient sound emitters below 1 kHz.

The ratio between size and wavelength also determines the amount of diffraction of sound around solid objects. In Figure 2.4, this is illustrated for the case of a head, which is assumed to be spherical. The curves show the sound pressure (relative to the pressure that existed before the head was placed in the sound field) at the position of an ear and for various directions of sound incidence. The frequency scale at the bottom is for a sphere of radius $a = 8.75$ cm (the approximate size of a human head). Apparently, a surplus pressure starts to build up above a few hundred hertz at the ear facing the sound source; it increases with frequency and at 5 kHz becomes

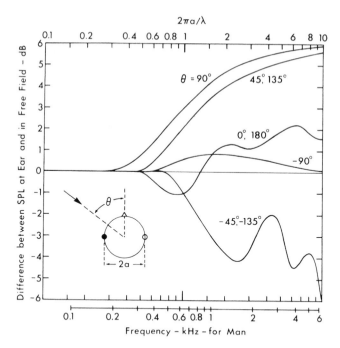

FIGURE 2.4. Calculated transformation of sound pressure level from a free field to a simple ear (point receiver) on the surface of a hard spherical head of radius *a* as a function of $2\pi a/\lambda$ (where λ = wavelength of sound) for various values of azimuth θ of sound. The frequency scale at the bottom is for a = 8.75 cm. (From Shaw 1974, with permission.)

about 6 dB (two times) larger than in the undisturbed sound field. This is not surprising considering the laws of sound reflection from large surfaces. One should be careful, however, not to use commonsense arguments when dealing with the diffraction of sound. The point on the sphere facing directly away from the sound source may also experience a surplus pressure and not the "shadow" that one might expect. At higher frequencies and for heads of more natural shape than that of a sphere, the variation in sound pressure with the angle of incidence is more complicated than shown in Figure 2.4.

The units of the upper scale in Figure 2.4 are π times the ratio between the diameter (2*a*) and the wavelength. This dimensionless scaling allows one to use the figure for other frequencies and for objects of other sizes. For example, one could use the figure for making a rough estimate of the frequencies (wavelengths) that can be used for echolocation of objects (e.g., of insect prey by hunting bats). A word of warning is in order, however. The data in Figure 2.4 were obtained by assuming the object to be inflexible and too heavy to be set into motion by the sound. This is a reasonable assumption in physics but not necessarily in biology (see Section 4).

Before leaving the diffraction phenomenon, we should note that it affects not only hearing but also sound measurements carried out by means of microphones. Ideally, a microphone should be so small relative to the wavelengths of the sounds of interest that it does not disturb the sound. When the microphone is larger than one-tenth of the wavelength, diffraction of sound may affect the sound pressure at the position of the microphone membrane, and thus the result of the measurement.

It is possible to make precise measurements also at higher frequencies (smaller wavelengths), even when the microphone is large. The most common method is to place the microphone membrane at the end of a cylinder, a shape that causes a fairly simple diffraction of sound. Some manufacturers adjust the high-frequency sensitivity of the microphone to compensate for the diffraction. The compensated (free-field type) cylindrical microphone should be pointed toward the sound source, whereas the noncompensated (pressure type) should be pointed perpendicularly to the direction of the source. It is essential to know the type of microphone and to study the diffraction curves in the instruction manual before attempting any measurements.

Microphones used near insect bodies may be too large, not only relative to wavelength but also relative to the insect under study. The problem can be avoided by using a *microphone probe*, a long and thin tube that guides sound to a small chamber in front of the membrane. With an external diameter of 1 mm for the tube, diffraction is not expected to play a role at less than 34 kHz (where the wavelength is 10 mm). However, tubes inherently suffer from resonances such as those exploited in organ pipes. The resonances can make the probes very difficult to use. Fortunately, probes without resonances are now commercially available.

4. Directional Sound Cues

The directional sound cues potentially available to the animals are the variations, with the direction of sound incidence, of the amplitude and phase of sound at the ears. The variations are due to two mechanisms: the diffraction of sound caused by the body of the animal and the time of arrival of the sound at the ears. Not all animals are able to exploit all the cues. Some insects use only amplitude cues for their directional hearing, and other insects can make use of both amplitude and phase cues.

Figures 2.5B and 2.5C show the directional cues available to a small grasshopper at three frequencies (Michelsen and Rohrseitz 1995). The amplitude of the sound pressure at the right ear is shown in Figure 2.5B. The grasshopper was only 14 mm long and 2 to 3 mm wide at the region of the ears. Obviously, it did not cause much diffraction at 5 kHz (wavelength,

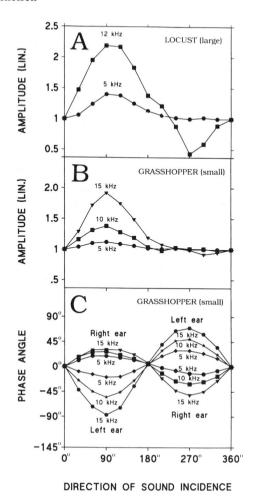

FIGURE 2.5. Sound pressure amplitude in the locust (A) and amplitude (B) and phase angle (C) in the small grasshopper, measured at the eardrum as a function of the angle of sound incidence (0° is frontal and 90° is ipsilateral to the right ear, which was at the center of a carousel carrying the loudspeaker). The amplitude and phase are defined to be 1 and 0°, respectively, when sound arrives at the right ear from the front (0°). The amplitude data have been plotted for the right ear only. The phase angles for the right ear were caused only by diffraction, whereas those for the left ear include the times of arrival of the sound at the two ears. (From Michelsen and Rohrseitz 1995, with permission.)

about 7 cm). At 10 and 15 kHz, however, some surplus pressure was observed when the sound source was ipsilateral to the ear, but very little shadow was seen when sound arrived from the contralateral (210° to 330°) directions. Of course, the sound amplitude at the left ear is a mirror image of that at the right ear. These results are in reasonable agreement with the trends in Figure 2.4 (although the shape of the grasshopper is closer to a cylinder than to a sphere).

So far, we have concentrated on the changes in sound amplitude at the surface of the body, but diffraction of sound involves changes in both amplitude and phase. Phase is a measure of time (here of the arrival of sound at the ear). Small differences in time are conveniently expressed as phase angles, that is, as fractions of the period time, and indicated in degrees (one period time is 360°). Note that the phase angle corresponding to a certain time increases with frequency (because the period time decreases). The changes in phase at each ear caused by diffraction are moderate but not negligible (the curves marked "right ear" in Fig. 2.5C). A larger part of the phase cue is the difference in phase between the ears (or other sound inputs to the ears; see Section 6). This part is due to the different times of arrival of sound at the ears, which in the case of the small grasshopper are 2 to 3 mm apart. In Figure 2.5C, the phase curves marked "left ear" include the effects of both diffraction and time of arrival.

Measurements of these directional cues are in principle quite simple. The tip of a probe microphone is placed at the ear, and sound is made to arrive from various directions. Great care must be taken to create a homogeneous echo-free sound field around the animal because reflections from equipment and other obstacles may cause amplitude changes that are larger than those to be measured (Larsen 1995).

The choice of a reference for the measured data requires careful consideration. Although the changes measured after introducing an animal in a sound field are the true effects of diffraction, such data are not suitable for an analysis of the physics of directional hearing. One needs a reference (e.g., the sound at the right ear when sound arrives from the frontal direction). All other phase angles and sound levels should then be indicated relative to this reference.

Changes in the amplitude of sound pressure cause changes in the vibration of the eardrum and corresponding changes in the neural activity of the sensory cells. The praying mantis has only one ear and does not localize sound sources (Yager and Hoy 1987), although, in theory, it could determine the sound direction by rotating its body while listening. The situation is more favorable when the animal has two ears. The brains of humans and other vertebrates are able to compare both the magnitude and timing of the neural activity arriving from each ear. Most of the difference in the time of arrival of sound at the two ears is caused by the physical distance between the ears, and not by diffraction. In animals with large heads, the time difference may be several hundred microseconds. Some vertebrate brains analyze these time differences with surprising precision. The champions are the barn owls, which are able to detect differences down to 6 μs (Moiseff and Konishi 1981).

In principle, time and amplitude are quite independent properties of sound, but the latency of the nervous response depends upon the magnitude of the force acting on the ear. The minute difference in the time of arrival of the sound at the ears is causing a similar difference in the timing of the nerve impulses from the two ears only if the sound amplitude is the same at both ears. The nervous system has to compare the information about time and amplitude at the two ears if this is not the case. Obviously, a complicated central processing is necessary in order to make use of the time cue for directional hearing. This may be one of the reasons why insects do not seem to exploit the time cues in the same manner as owls or humans. Instead, many insects use the small differences in the time of arrival to change the vibration amplitude of the eardrum. These insects achieve this by guiding sound from the other side of the body, through some internal sound path, to the internal surface of the eardrum (see Section 6).

Recent data from grasshoppers suggest that the diffraction may not be scaled in the simple manner assumed when the data in Figure 2.4 were calculated. Figures 2.5A and 2.5B show the pressure measured at various frequencies at an ear of a large and a small grasshopper, respectively (Michelsen and Rohrseitz 1995). At first sight, the data confirm the expectation of a larger change of pressure, at a certain frequency, in the larger animal. A more detailed comparison of the data allows a guess of how much the two species differ in size: the values for 60°, 90°, and 120° sound direction observed at 10 kHz in the small species are close to those observed at 5 kHz in the large species. This suggests that the size should differ by a factor of 2. In reality, however, the sizes of the animals differ by a factor of 3 to 4. In both species the ear is at the surface of an almost cylindrical body and the anatomy does not vary much with size. The reason for the absence of a simple scaling is not known, but apparently the grasshoppers do not behave as the inflexible and heavy bodies assumed in the calculation of Figure 2.4.

5. Pressure Receivers

In many animals, the difference in the pressure amplitudes at the two ears is sufficient for sound localization and the eardrums need only to receive sound at their external surface. The possible use of the pressure receiver strategy is not determined by the absolute size of the animal but by its size relative to the wavelength of the sounds to be heard. The carrier frequencies of the hunting cries of most echolocating bats are at 30 to 100 kHz (wavelengths 10 to 3 mm). Many insects flying at night are fairly large relative to these wavelengths, and their bodies cause sufficient diffraction of sound to provide pressure differences that can be recorded with pressure-receiving ears.

In a pioneering study by Payne, Roeder, and Wallman (1966), the variation of sound pressure at the ears of moths was estimated from the neural response (most moths have only one or two receptor cells in each ear so the neural response is particularly simple). The position of the loudspeaker could be varied in both the horizontal and vertical directions. As expected, the sound pressure at an ear varied much more at 60 kHz than at 30 kHz. The wings contributed much to the diffraction of sound, and the pressure at an ear depended not only on the direction of the sound source, but also on the position of the wings. With wings above the horizontal plane (seen from the ear), the sound amplitude at an ear was 20 to 40 dB (10 to 100 times) larger for ipsilateral sound sources than for contralateral ones. With wings below the horizontal, a 10 to 25-dB difference in amplitude at the ears was found between sounds coming from above and below, respectively.

In other words, the sound pressure at the ears depends much on the actual position of the wings, and one may speculate that the moth could improve the accuracy of its sound localization if it were able to make a temporal correlation between the wing beats and the neural messages from the ears. In passing, we may note that the position of the wings affects the sound amplitude not only at the ears of the moth but also in the echo from the moth detected by the ears of hunting bats, some of which appear to use the fluttering signal as a sign of live prey (Schnitzler et al. 1983; see also Fullard, Chapter 8). Some nocturnal insects stop flying, fold their wings, and drop to the ground when the bat gets too close, thus reducing the echo reaching the bat and avoiding the fluttering (Roeder 1967).

The differences in sound amplitude at the ears found by Payne and his colleagues were much larger than the 1 to 2 dB necessary for a left-right decision (Section 2). These were due to both the high sound frequencies and the large size of the moths investigated (wing span 60 to 85 mm). In moths of more common size, the interaural difference in pressure amplitude at the frequencies used by bats is around 6 dB (A. Surlykke, personal communication). To my knowledge, the exact scaling has not been studied.

A similar situation exists in many bushcrickets (katydids), in which the wavelengths of the frequencies carrying the songs are of the same order of magnitude as the dimensions of the body. However, the ears are in the thin front legs, far away from the body surface (Fig. 2.6C). Measurements with a thin microphone probe demonstrate that diffraction of sound by the leg does not provide a sufficient directional cue at the frequency range of the song (Michelsen et al. 1994). As expected, however, the diffraction around the thorax is larger and the bushcrickets take advantage of this. A tracheal tube connects the inner surface of the ear drum with the lateral surface of the thorax (Fig. 2.6C), and this trachea acts as a sound guide (Lewis 1983).

In many bushcrickets, the tracheal tube is horn shaped (and known as "the hearing trumpet"). The mouth of the horn is at the eardrum in the leg, and the bell of the horn opens at the body surface. The cross-sectional area thus decreases from the point of sound entry (thoracic surface) to the ear. Although sound suffers some losses when propagating through narrow tubes, the geometry of the trachea acts to increase the sound pressure at the

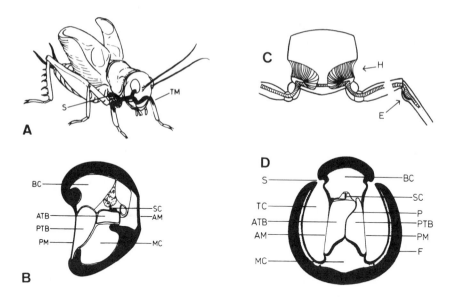

FIGURE 2.6. Schematic diagrams of the hearing organs in crickets (A,B) and bushcrickets (C,D). A: In crickets an H-shaped system of tracheal tubes (shown also in Fig. 2.15C) connects the internal surface of the eardrums (tympanal membrane, TM) of one ear with the other ear and with the ipsilateral and contralateral spiracles (S). B: Ear seen in a cross section of the front leg (symbols as in D). C: In bushcrickets, the horn-shaped trachea (H) from each ear (E) are connected by a thin tube in the thorax. D: Cross section through the ear in the leg. AM, anterior eardrum; ATB, anterior tracheal branch; BC, blood channel; F, flap; MC, muscle channel; P, partition; PM, posterior eardrum; PTB, posterior tracheal branch. SC, sense cells; TC, tympanal cavity. (From Michelsen and Larsen 1978 and Larsen and Michelsen 1978, with permission. A: Redrawn from Michel 1974, with permission.)

narrow end of the trachea (remember that pressure is force per area). Measurements (with methods that are explained in Section 6) show that the pressure gain is well above 1 at frequencies above a few kilohertz and that (organpipe-like) resonances may contribute to the gain. At the frequency of the calling song (typically above 10kHz), the amplitude of the sound carried by the acoustic trachea and acting on the inner surface of the eardrum may be more than 10 times larger than the amplitude of the sound acting at the external surface of the eardrum. The sound at the inner surface thus dominates the ear (the ear is, for all practical purposes, a pressure receiver, although the anatomical arrangement is very far from that shown in Fig. 2.2A). At each frequency in the song, the acoustic trachea faithfully transmits an "amplified" version of the pressure at the body surface to the inner surface of the eardrum. The ears thus obtain the same (diffraction-based) directionality as if they had been situated at the body surface (Lewis 1983).

In the next section we see that crickets also have ears in their front legs. The ears of crickets and bushcrickets obviously have a common evolutionary origin, but crickets have exploited a widely different mechanism for directional hearing.

6. Pressure-Difference Receivers

Many insects have excellent directional hearing at sound frequencies at which the body is much smaller than the wavelength of sound. The vibrations of the eardrum (tympanum) vary much with the direction of sound incidence, but the sound pressure amplitude at the ears is almost constant. The reasons for this apparent paradox are that the sound waves reach both the outer surface of the eardrum (tympanal membrane) and the inner surface. The pathway for sound to the inner surface varies. In many animals, the ears are connected by an air-filled passage. Alternatively, the sound waves may enter the body and reach the inner surface of the eardrum through some other route (e.g., a tracheal tube). Such potentially sound-transmitting pathways are known in several insect groups.

The idea that directional hearing in small animals may be caused by such a mechanism was first proposed by Autrum (1940), who referred to the inherently directional nature of pressure gradient receivers. In recent years, this concept has become the standard explanation for directionality, almost a magic formula, but only little has been done to investigate the problems and limitations of such systems.

The ideal pressure-gradient receiver (Fig. 2.2B) is so small (relative to the wavelength of sound) that is does not cause any diffraction. It is driven by the small difference in phase that is caused by the different times of arrival of sound at its two surfaces, which are equally exposed to sound. The driving force is proportional to the pressure gradient, hence the name. Such devices are not very sensitive, and their directional patterns have the ambiguous shape of a figure eight (Fig. 2.3B). Obviously, these properties are far from those of typical ears.

Animals need sensitive ears with "useful" directional patterns. Directional patterns that emphasize ipsilateral sounds (Fig. 2.3C) can be processed by means of a few neurons in the central nervous system, and such patterns are thus much more useful for small animals than, for example, figure-eight patterns.

The existence of an anatomical air space guiding sound to the inner surface of the tympanum is a necessary prerequisite, but it does not automatically create a very directional pressure-difference receiver. The sound has to arrive at the inner surface of the membrane with a proper amplitude and phase. In addition, the sounds acting on the eardrum should sufficiently reflect the directional cues. One complicating factor in the analysis is that the sound arriving at the inner surface may have entered the body through several auditory inputs.

The clue to a proper understanding of how pressure-difference receivers operate is to measure the change of amplitude and phase that takes place in the sound propagating to the inner surface of the eardrum. Most of the methods used for this purpose in the past suffered from shortcomings, and the results were less reliable. For example, several investigators inserted a thin probe tube connected to a microphone into the air space behind the eardrum in order to measure the sound pressure. The presence of the probe was likely to have a major effect on the sound to be measured (in technical terms, the input impedance of the probe tube was not much larger than the impedance of the air space). Furthermore, the sound measured by such a procedure is likely to include two components: sound transmitted through the drum of the ear considered and sound transmitted from other sound input(s) to the internal surface of the eardrum. Only the latter component is relevant for the directionality of the ear.

Another popular method has been to estimate sound transmission to the inner surface of the eardrum by blocking possible sound inputs and observing the effects on the ear. This is a dubious strategy because blocking can be expected to affect not only the sound transmission but also the mechanics (impedance) of the eardrum. That this is indeed the case has been demonstrated in crickets (Michelsen, Popov, and Lewis 1994). This method may therefore lead to misleading conclusions.

It is possible to circumvent these problems by a combination of two tricks (Michelsen, Heller, Stumpner, and Rohrseitz 1994): Sound is presented locally to one sound input at a time, and the sound pressure at the internal surface of the eardrum is measured by using the eardrum itself as a microphone. It is possible to measure the vibrations of the eardrum without affecting them by focusing a laser beam on the eardrum and analyzing changes in color in the reflected light (laser vibrometry). The "eardrum microphone" is calibrated by driving it with sound only at its external surface and measuring the sound pressure with a probe microphone (Fig. 2.7A). The local sound source and the tip of the probe are then moved to another sound input (Fig. 2.7B). The sound pressure at the internal surface of the tympanum is calculated from the measured vibrations (using the previous calibration) and compared with the sound pressure measured at the sound input. The result (known as the transmission gain) is expressed as a change in amplitude and phase angle.

The transmission gains measured for the narrow tracheal tubes in insects vary much with the needs of the particular insect. Some insects are large enough to cause considerable diffraction in the high-frequency part of their audible range but not at low frequencies. Such animals tend to have a rather large transmission at low frequencies (at which the animal needs its ears to work as pressure-difference receivers) but only little transmission at

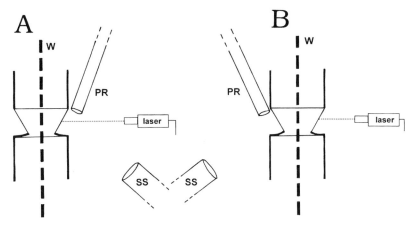

FIGURE 2.7. Measuring the transmission gain (the changes of amplitude and phase) of sounds arriving at the internal surface of the eardrum from the other ear in a grasshopper (see Fig. 2.8). A: Calibration of the "eardrum microphone"; vibrations caused by a local sound source (SS) are measured with laser vibrometry, whereas the sound pressure at the ear is measured with a probe microphone (PR). Sound transmission around the body is attenuated by a wall (W) of beeswax. B: The eardrum is activated by sound from the other ear. The eardrum vibrations reveal the sound pressure at its internal surface and thus the transmission gain. (From Michelsen and Rohrseitz 1995, with permission.)

high frequencies (at which pressure-receiving ears can provide sufficient directionality).

By simple sound transmission through a tube, one expects the sound to suffer some loss in amplitude (more in narrow tubes than in wide ones) and to get a time delay proportional to the distance (length of tube), but the transmission may be more complicated. Tube resonances (like those in organ pipes) may cause the amplitude to vary with frequency. The speed of sound may be smaller inside the tube than in free air, resulting in larger phase angles than expected from the normal speed of about 344 m/s in air. Finally, as we are about to learn, crickets appear to have "invented" a mechanical phase shifter that plays an important role in their directional hearing.

The air space leading to the back surface of the eardrum is often a part of (or connected to) the respiratory pathways. This may have undesirable consequences because the large pressure fluctuations during respiration may affect the ears. In grasshoppers the tympana may be displaced outside their linear range (so that Hooke's law is no longer obeyed). This may affect the threshold for hearing and seriously distort the frequency analysis (Michelsen, Hedwig, and Elsner 1990). Obviously, a reduction of such effects may have been an important factor in the evolution of pressure-difference receivers. In the remaining part of this section we concentrate on the biophysics of sound localization in locusts, grasshoppers, and crickets.

6.1. Locusts and Grasshoppers

Locusts and grasshoppers have an ear on each side of the first abdominal segment. It consists of a sclerotized ring forming a recess in the abdomen and encircling a tympanal membrane, to which 60 to 80 receptor cells are attached. Between the ears are air-filled tracheal sacs, which could permit interaural sound transmission (Fig. 2.8).

Technically primitive measurements performed 25 years ago showed that in locusts the air sacs act as an acoustic low-pass filter, suggesting that the ears were pressure-difference receivers at low frequencies and mainly pressure receivers at high frequencies (Michelsen 1971). This view was supported by later studies that used biophysical (Miller 1977) or electrophysiological methods (Römer 1976). These studies did not reveal, however, how such a mechanism might allow the animal to exploit cues in the incident sounds to determine the direction of sound incidence. In a recent study, more advanced methods have been used for revealing the biophysics of directional hearing, both in the locust *Schistocerca gregaria* and in the

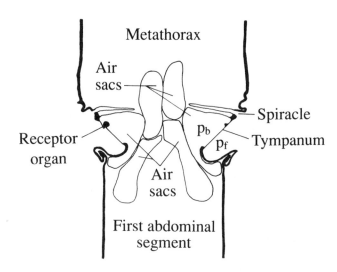

FIGURE 2.8. Horizontal section through the auditory region of a male grasshopper. (Redrawn from Schwabe 1906, with permission.) The ears are connected by tracheal air sacs associated with the spiracles. Each eardrum (tympanum) receives sound both at its external (front) surface (sound component p_f) and at its internal (back) surface (sound component p_b). The anatomy of the tracheal air sacs varies somewhat among species. (From Michelsen and Rohrseitz 1995, with permission.)

three to four times smaller grasshopper *Chorthippus biguttulus* (Michelsen and Rohrseitz 1995).

The directional cues available to these animals were shown in Figure 2.5, except for the phase cues available to the locust, which, as expected, show slightly larger values than those of the grasshopper. The transmission gains for sound propagating through the internal air sacs, from one ear to the internal surface of the other ear, are widely different, however. The measurements in the locust, using the method in Figure 2.7, confirmed the low-pass filter characteristics reported earlier: the amplitude gain was about 0.5 around 4 kHz and 0.1 to 0.2 at 10 to 20 kHz (Fig. 2.9).

A gain of 0.5 means that the amplitude of the sound pressure arriving at the internal surface of an eardrum is 50% of that acting on the external surface of the ear, at which the sound entered the internal air sacs. 0.5 is equal to −6 dB, so one could also say that the sound acting on the external

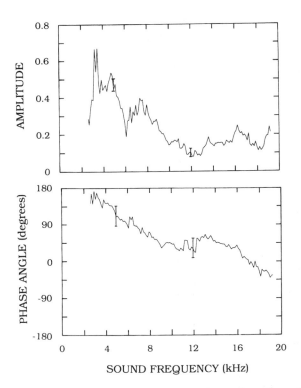

FIGURE 2.9. Transmission gain for sound arriving at the eardrum from the contralateral ear in the locust. An amplitude of 1 and a phase angle of 180° indicate that the amplitude and phase have not changed during the transmission. Averaged data (and standard deviations at 5 and 12 kHz) from three animals. (From Michelsen and Rohrseitz 1995, with permission.)

surface of an eardrum has been attenuated by 6 dB when it arrives at the internal surface of the contralateral ear.

The phase angles shown in Figure 2.9 include 180° that was added by calibrating the "eardrum microphone" with sound at its external surface but using it for measuring sound at its internal surface. At 20 kHz, the phase of sound has changed by about 225° (from 180° to −45°). The corresponding time delay is about 30 μs [equal to (225/360) times the 50-μs period time at 20 kHz]. Sound travels with a speed of about 344 m/s, and a delay of 30 μs thus corresponds to a distance of about 10 mm. This is somewhat above the physical distance between the ears in a locust, and one may therefore guess that the speed of sound in the air sacs connecting the eardrums is less than in free space. A similar trend has also been found in other insects.

It is now possible to calculate the directional sensitivity of an ear, that is, how the amplitude of the eardrum vibrations vary with the direction of sound incidence. It is assumed that the force driving the eardrum is determined only by two components of the sound wave (the two-input model). One component (p_f) propagates directly to the external surface, whereas the other component (p_b), from the other ear, is incident on the internal surface (f and b refer to front and back, respectively). p_f has been measured for 12 directions of sound incidence (Fig. 2.5). p_b can be calculated by multiplying the sound pressure at the other ear by the gain of the interaural sound pathway (Fig. 2.9). Both p_f and p_b can be thought of as acting on the external surface, because the "eardrum microphone" was calibrated with sound at its external surface (adding 180° to the phase of the gain). These two sound pressures (conveniently represented as vectors) can therefore be summed to estimate the total sound pressure (P) driving the eardrum. The calculation is repeated for each of the 12 directions of sound incidence investigated. Figure 2.10 shows an example of such a calculation (at 10 kHz in the grasshopper).

The result for the locust at 5 and 12 kHz is shown in Figure 2.11. The agreement between the calculated directional dependence of the eardrum vibrations and the actual values (measured with laser vibrometry) suggests that the two-input model is a valid description of the acoustics of the ear, both at 5 kHz and at 12 kHz. In other words, there is no need to postulate that sound arriving at the internal surface of the eardrum through other routes (e.g., through spiracles) should play any significant role.

The dotted curves in Figure 2.11 show the amplitude of sound pressure at the external surface of the eardrum (i.e., the amplitude data in Fig. 2.5A). These curves thus show the directionality that would exist in the absence of a pressure-difference mechanism. This mechanism is clearly essential at 5 kHz. At 12 kHz, at which the amplitude of the transmission gain is only about 0.1, it plays a minor but significant role.

Note that, although the pressure-difference mechanism is not essential for the locust at 12 kHz, it still causes a significant decrease in the sensitivity to contralateral sounds. This may seem surprising because the transmission

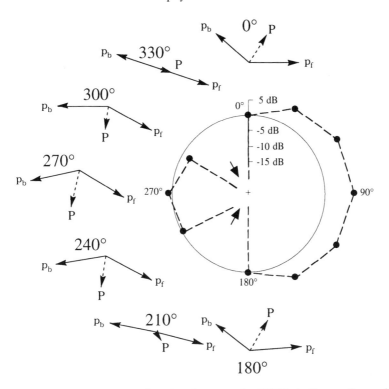

FIGURE 2.10. Calculation of a directional pattern (at 10 kHz in the small grasshopper). The sounds have been drawn as vectors. p_f is the measured sound pressure at the external (front) surface of the eardrum, and p_b is the calculated pressure at the internal (back) surface. The resulting pressure (P) driving the eardrum can thus be calculated simply by summing the two vectors (see the text). At 210° and 330° (arrows within the polar plot) the values of P are −26 dB and −30 dB, respectively. (From Michelsen and Rohrseitz 1995, with permission.)

amplitude is only 0.1. The reason is seen in Figure 2.5A: when 12-kHz sound arrives from the contralateral direction, the contralateral ear experiences a sound amplitude of 2.2 and the sound transmitted to the internal surface of the ipsilateral eardrum is (2.2 × 0.1 = 0.22), which is almost half of the sound at the external surface (about 0.5). Small sound transmissions may thus have significant effects at frequencies at which the diffraction effects are large.

The absence of a "shadow" (at the side contralateral to the sound source) in the small grasshopper means that the sound pressures at the ear at ipsilateral and contralateral directions of sound incidence only differ by a factor of about 2 (6 dB), even at 15 kHz (Fig. 2.5B). In contrast, the ratio is about 5 (14 dB) at 12 kHz in the locust (Fig. 2.5A). The directional hearing

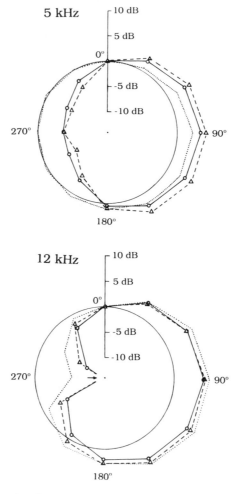

FIGURE 2.11. Directional patterns at 5kHz and 12kHz in the locust. Solid and dashed lines, observed and calculated vibration amplitude of the eardrum, respectively. Dotted lines, amplitude of sound pressure at the external surface of the eardrum (same data as in Fig. 2.5A). (From Michelsen and Rohrseitz 1995, with permission.)

of the small grasshopper is therefore dependent on the ear operating as a pressure-difference receiver at all frequencies.

At 4 to 5kHz, the transmission gain in the small grasshopper is similar to that in the locust, but the amplitude gain increases to about 0.8 at 8 to 18kHz. The development of phase is more moderate than in the locust, reflecting the smaller distance between the ears. A comparison between the calculated and observed directional patterns shows that, like in the locust, a two-input model is a reasonable description of the acoustics of the ear.

The directional patterns obtained in the small grasshopper below about 12 kHz were less useful, however. For example, at 5 kHz the maximum difference in driving force between the ears was above 10 dB in the locust (Fig. 2.10) but less than 5 dB in the small grasshopper. This is sufficient in a free sound field but not always in the natural habitat (see Section 8). At 8 and 10 kHz (Fig. 2.10) the presence of a contralateral lobe in the directional pattern may contribute to the troubles of sound-locating animals. At 12 to 17 kHz, however, the patterns became much like the 12-kHz pattern in the locust (Fig. 2.11).

The intuitive expectation of most investigators has been that the directional hearing of small insects is limited mainly by the modest directional cues available to these animals. This is not necessarily so, however. At 12 kHz, the directional patterns were fairly similar in the locust and the grasshopper, despite the difference in the directional cues. The reason for this paradox became apparent when the amplitude component of the interaural transmission gain was varied in the calculation of the directional pattern for the grasshopper. The calculated directional sensitivity for the forward direction (expressed as the difference between the values for the 30° and 330° directions of sound incidence) was 14 dB. In contrast, this difference became 7 and 4 dB when the transmission was assumed to be 50% and 25% of the measured amplitude (0.34 and 0.17), respectively. Apparently, the large interaural transmission in the small animal compensates for the modest directional cues.

An interesting situation appeared to exist at 5 kHz, at which the amplitude of the interaural transmission was very similar in the two species. The grasshopper is obviously less directionally sensitive than the locust, but the calculations revealed that this was not entirely due to the modest directional cues. By assuming a transmission phase angle similar to that found at 5 kHz in the locust, considerable improvement of directionality was obtained. Apparently, the system of air sacs between the ears does not allow the small grasshopper to obtain a sufficient transmission delay at 5 kHz. At 12 kHz, an increased transmission amplitude caused an improved directionality, but this is not so at 5 kHz. Here, the transmission delay was small, and larger transmission amplitudes would result in a reduced absolute sensitivity of the ear (which would become almost a pressure-gradient receiver).

These observations support the notion that a proper phase relationship between the sounds acting on the eardrum is essential for the directional patterns. The importance of the phase relationships is obvious from Figure 2.10: the amplitudes of the two sound components change very little with the direction of sound incidence, but the variation in relative phase causes the resulting driving force to vary by more than 30 dB. The variation in relative phase is due to the directional cues (mainly to the times of sound arrival at the ears), but the result of the addition is critically dependent on the average phase difference between p_f and p_b.

6.2. Crickets

The calling songs of male crickets (Gryllidae) are almost pure tones of approximately 5 kHz. Like the bushcrickets (Section 5), the crickets have their ears in the front leg tibiae (see Fig. 2.6A and 2.6B), and diffraction around such thin legs does not provide useful directional cues. However, the crickets have solved the problem in a very different way. A pressure-difference receiver mechanism provides the ear with an excellent directionality similar to that of Figure 2.3C, but only at frequencies close to that of the pure tone calling song (Hill 1974). In many species of crickets, the males are calling on the ground in grassland habitats, where sounds below 6 kHz propagate with a minimum of degradation (see Section 8). The calling songs are typically at 4 to 5 kHz. In Section 6.1 it was mentioned that the small grasshopper had not obtained a useful directionality at 5 kHz. Nevertheless, similarly sized crickets appear to have solved the problem.

The internal surface of the cricket eardrum is connected with the ipsilateral acoustic spiracle (IS) on the thorax through a horn-shaped acoustic trachea. The acoustic trachea is fairly independent of the respiratory system and is linked, through a connecting trachea, with the acoustic trachea on the other side of the body. The cricket ear is thus, potentially, a four-input acoustic device. Sound acts on the external surface of the eardrum, and no less than three acoustic inputs may contribute to the sound acting on the internal surface of the eardrum: sound may arrive from the ipsilateral spiracle and, also, across the midline from the contralateral ear and acoustic spiracle.

Very different opinions have been expressed about the importance of the contralateral inputs for creating the directional characteristics of the ear. Some investigators have favored the sound from the contralateral ear (Hill and Boyan 1976; Fletcher and Thwaites 1979), whereas others favored the sound from the contralateral spiracle (Larsen and Michelsen 1978; Schmitz, Scharstein, and Wendler 1983). Finally, the observation that disrupting the connecting trachea does not hinder sound localization (in the homogeneous sound field on a Kramer sphere; see Section 2) has been used as evidence to "toll the death of all such cross-body theories" (Weber and Thorson 1989).

Recently, it has been possible to sort out the relative importance of the acoustic inputs and to explain how the sounds from these inputs interact at the level of the eardrum of the cricket *Gryllus bimaculatus* (Michelsen et al. 1994). The techniques were those mentioned in the previous section. The only assumption was that the sounds acting on the internal surface of the eardrum originate exclusively from the acoustic inputs mentioned. The calculated directional diagram was so close to that obtained by measuring the tympanal vibrations that the assumption must be almost correct.

The amplitude and phase of sound pressure at the four acoustic inputs to the ear, plotted as a function of the direction of sound incidence, are shown in Figure 2.12. It is possible from these data to make some predic-

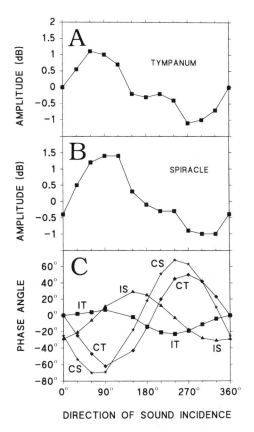

FIGURE 2.12. Directional cues in a cricket: sound pressure at the eardrum (tympanum) and acoustic spiracle at 4.5 kHz (the frequency of the calling song). A: Sound amplitude at eardrum. B: Sound amplitude at spiracle. C: Phase angle at all four acoustic inputs. 0 dB amplitude and 0° phase angle are defined as the sound pressure at the ipsilateral eardrum (IT) when sound arrives from the frontal direction (0°). 90° is the ipsilateral direction. IS, ipsilateral spiracle; CS, contralateral spiracle; CT, contralateral eardrum. (From Michelsen et al. 1994, with permission.)

tions about the mechanism of directional hearing. First, let us assume that the ears of the cricket are simple pressure receivers responding only to the pressure at the outer surface of the eardrum. From Figure 2.12A it is obvious that the amplitude of the sound pressure at the outer surface of the eardrum changes only little with the direction of sound incidence. In the frontal directions (around 0°), which are of prime interest with respect to localization of a sound source, a change of 30° in sound direction would cause the forces driving the two ears to differ by only 1.3 dB (compare the values at 30° and 330°). This difference may be sufficient for a cricket locating a sound source in a very uniform sound field in the laboratory, but not in the field, where the sound field is generally not very uniform (see Section 8).

A pressure-difference receiver is obviously needed for providing more directionality, but from which input should the sound at the inner surface of the eardrum originate? Let us assume that the sound at the inner surface comes exclusively from the ipsilateral spiracle (IS). Furthermore, let us make the favorable assumption that the sound amplitudes at the two surfaces are almost equal in amplitude. From Figures 2.12A and 2.12B, we find that although the amplitudes do change as a function of sound direction, the changes are in the same direction (compare the sound amplitudes at the tympanum and spiracle at 0°, 30°, and 330°). Obviously, the amplitude cue is even smaller than in the case of the two pressure receivers. The phase cue (Fig. 2.12C) follows a similar pattern: the phase angles increase at both inputs when the sound direction changes from 0° to 30° and decrease from 0° to 330°. The difference in phase between the sounds at the inner and outer surfaces of the eardrum changes by only 5° and 2°, respectively. Such small phase changes can cause a directional dependence of the tympanal vibration but only if the driving force is close to zero (approximately the same phase at the inner and outer surface), a situation that would hardly be useful for the animal. In conclusion, *even a well-balanced pressure-difference receiver does not always provide a useful directionality*!

The sounds arriving from the two contralateral inputs are obviously much better potential contributors of directional cues. Both the amplitude and phase of the contralateral sounds change in opposite directions from the values for the ipsilateral sounds when the sound source moves from one frontal direction to another. The change of phase is especially prominent, and thus the most likely contributor to the directionality of the ear. Having made these predictions, let us turn to the actual data on the gain of the tracheal sound guides.

The amplitude part of the transmission gain for sound entering the IS is close to 1 at low frequencies and increases to reach a maximum around 6 to 8 kHz and again at 17 to 19 kHz. With increasing frequency the phase of the sound at the inner surface of the eardrum changes, as one would expect in a transmission line in which the propagation of sound takes a certain time. As in the locust (see Section 6.1), the measured transmission time corresponds to a length of the tracheal tube (about 15.6 mm) that is significantly larger than the actual length (about 12 mm). This means that, as in the locust, the sound propagates with a lower velocity inside the tube than in the air outside. This was first observed by Larsen (1981), who pointed out that the determined value is rather close to that expected for isothermal wave propagation in air (245 m/s). Larsen suggested that an exchange of heat may occur at the tracheal walls, but the lower velocity may also be caused just by the small diameter of the tubes.

The transmission of sound from the *contralateral spiracle* (CS) differs very much from this simple pattern (Fig. 2.13). The amplitude is at a

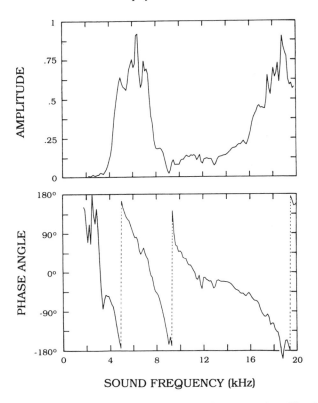

FIGURE 2.13. The cricket ear. Transmission gain (see legend to Fig. 2.9) for the sound arriving at the internal surface of the eardrum from the contralateral spiracle (CS). The phase curve is continuous but has been cut at the position of the broken lines in order to save space. (From Michelsen et al. 1994, with permission.)

maximum at 5 to 7 kHz and again around 18 kHz (much like the sound from IS), but it is virtually zero below 3.5 kHz. In the frequency range of 4 to 5 kHz (around the frequency of the calling song at approximately 4.5 kHz in *Gryllus bimaculatus*), the amplitude of the sound from the CS varies drastically with frequency. Between 4 and 4.5 kHz the amplitude increases by a factor of 4. The strong frequency dependence of the amplitude is accompanied by a large change of phase. From 2 to 10 kHz the phase angle of the sound from the CS decreases by approximately 800° (for comparison, the phase of the sound from the IS shows a decrease of only 180°). Above 10 kHz the rate of decrease approaches that observed in the sound from the IS.

The transmission from the *contralateral tympanum* (CT) follows the pattern observed in the transmission from the CS. However, the amplitude of the sound arriving at the tympanum is considerably smaller at the frequency of the calling song. This component will therefore be ignored in the following analysis.

The measured directional cues at the three auditory inputs and two transmission gains can now be combined in an attempt to account for the dependence of the tympanal vibrations on the direction of sound incidence. The amplitude and phase of the sound at the external surface of the eardrum (IT) for sounds arriving from the frontal direction are, as a matter of definition, 1 and 0°. Typical values for the amplitude part of the transmission gain of the sounds from the IS and CS are 1.5 and 0.44, respectively. The corresponding phase angles are 154° for the IS and 208° for the CS. The calculation proceeds in the same manner as in the locust and grasshopper, but here three sounds, not just two, are acting on the eardrum. The three vectors are added to produce a resulting sound pressure (P), which is proportional to the force that causes the tympanum to vibrate.

The result of this model calculation is shown in Figure 2.14. The directional pattern of the resultant pressure driving the eardrum is given by the solid curve. The polar plot is surrounded by the 12 vector diagrams. The

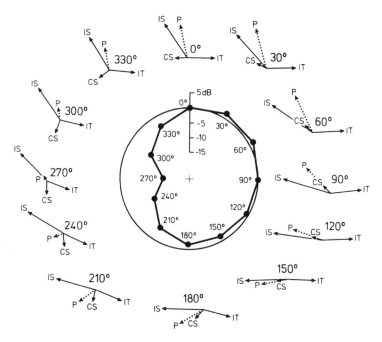

FIGURE 2.14. The cricket ear. The calculated directional pattern of the force acting on the eardrum of the right ear at 4.5 kHz. The force is proportional to a sound pressure, P, which is the sum of three vectors: the sound at the outer surface of the eardrum (IT) and the sounds transmitted to the inner surface from the ipsilateral (IS) and contralateral (CS) spiracles. (From Michelsen et al. 1994, with permission.)

calculated directional pattern has the most important of the features seen in the measured patterns reported in the literature: The driving force is at a maximum at the ipsilateral directions 30° and 60°; the force decreases by approximately 6 dB from 30°, through 0° to 330°; and the force is at a minimum at 270°. The minimum, known as the contralateral null, is 15 dB below the value in the frontal (0°) direction.

When examining Figure 2.14 start by looking at the vector diagram for the 270° direction. Obviously, the null is caused by the fact that the sum of the CS and IT has approximately the same amplitude as IS, but the opposite direction. A deeper minimum (more perfect null, which is often observed) would only require a slight reduction in the amplitude of the sound from the IS. One may now perform a consecutive examination of the vector diagrams for 300°, 330°, 0°, and 30°. The amplitudes of the three vectors change only little, and only little variation is seen in the phase angles for the IS and IT. The only major change is in the phase angles for the CS. At 30° and 60°, the CS has almost the same phase as the IS and the amplitude of P is now at a maximum. Apparently, the slope of the driving force in the forward direction is caused almost entirely by the change in the phase angle of the sound from the CS.

The eardrum and the contralateral spiracle are separated by a large physical distance, and the change of the phase of the sound from the CS is therefore particularly sensitive to the direction of sound incidence. The calculations show, however, that the three vectors must have a proper phase relationship in order to produce the directional diagram of Figure 2.14. This means that the phase part of the transmission gains for the sounds from the IS and CS is critical. The phase of the sound from the IS changes only slowly with frequency. In contrast, the change of phase for the sound from the CS is much larger than expected for sound propagation in a simple delay line (see Fig. 2.13).

Apparently, the right phase relationship between the three vectors exists within a narrow range of frequencies that includes the frequency of the calling song. The tuning of the directionality can be demonstrated in a simple manner by plotting the difference between the eardrum vibrations at 30° and 330° sound direction (Fig. 2.15D). In intact crickets, the difference may be 10 dB at 4.5 kHz (the frequency of the calling song), but only 3 to 4 dB at 4 and 5 kHz (Michelsen and Löhe 1995). The reason for the lower directionality at 4 and 5 kHz is the less favorable phase relationships between the three sound components acting on the eardrum.

The transverse trachea connecting the two acoustic tracheae is not continous but is made up of two tracheal tubes, one from each side. The thin-walled ends of these tubes are very close at the middle of the thorax (Fig. 2.15C). The integrity of this "central membrane" appears to be essential for the directional hearing at sound frequencies around the calling song. A perforation of the central membrane in the transverse trachea can be performed by pushing a human hair through the trachea. This causes a

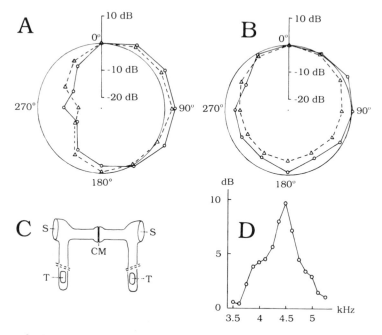

FIGURE 2.15. The cricket ear. Observed (full line) and calculated (broken line) directional patterns in the right ear of an intact cricket (A) and a cricket with a perforated central membrane (B). C: Schematic diagram of the tracheal part of the hearing organ. CM, central membrane; S, spiracle; T, eardrum. D: Frequency tuning of the directional gradient in an intact cricket (i.e., the ratio of the eardrum vibration amplitude at the +30° and −30° [=330°] directions). After the perforation, the ratio becomes 0 to 2 dB at all frequencies. (From Michelsen and Löhe 1995, with permission.)

considerable decrease of directionality. Holes in the central membrane cause the amplitude of the transmitted sound (from the CS) to vary much less with frequency, and the phase around 4.5 kHz to become close to that expected from the physical distance (Michelsen and Löhe 1995). At 4.5 kHz the perforation causes the phase of the transmitted sound to change by about 260°. The phase relationship of the three vectors is now far from the situation in Figure 2.14, and the directional pattern becomes as useless as the 5-kHz pattern in the small grasshopper (compare A and B in Fig. 2.15).

The cricket thus stands out as an animal that has solved a major problem in auditory biophysics — how to delay the sound during the transmission to the other side of the body at frequencies at which the physical distance causes only a moderate phase shift. The physical operation of the phase shifter is not known. Probably, the central membrane interacts with the tracheal tubes to produce the filter effect observed.

7. Other Types of Receivers

A sense of hearing has evolved independently at least a dozen times in different groups of insects, and the ears have further differentiated into a large number of functional types, making use of a wide variety of physical mechanisms. New types of ears are still being discovered. It is possible, therefore, that the three mechanisms for receiving sounds (pressure, pressure difference, and movement receivers) and their directional properties may not be the only methods available to animals for determining the direction of sound incidence. In some cases, the directional hearing has been subjected to a rigorous analysis along the lines outlined in the previous sections. In cicadas, for example, most aspects of the directional dependence of the eardrum vibrations can be explained by such physical mechanisms, but some cannot. This is true, for example, for the large directional sensitivity of the ear around 1.8kHz in the male cicada *Cicada barbara lusitanica* (Fonseca and Popov 1997).

In one case, the ear of ormiine flies, it has been reported that the ear achieves its directional sensitivity in a novel way. The ears of these flies are very close together, and they are joined by a mechanical structure that couples their motion mechanically. It has been postulated that the mechanical coupling also magnifies interaural differences and that the mechanical coupling is the key to this animal's ability to localize sound sources. This remarkable system is dealt with by Robert and Hoy in Chapter 7.

Most hearing insects have to determine the direction of sound incidence with ears sensitive to sound pressure. However, sound may be described and measured both as fluctuations in pressure and as oscillations of "air particles" (see Section 3). An alternative strategy, therefore, is to measure the direction of the oscillatory movements of the air particles. Insects can do this by means of long, lightly articulated sensory hairs protruding from their body surface. Such movement receivers are inherently directional (Tautz 1979). In addition, structural asymmetry of cuticular elements at the base of the hair may favor movements in only one direction. The mobility of the hair may also be affected by the shape of the hair. Finally, the attachment of the sensory cell(s) may also contribute to the directionality.

The airflow-sensitive hair receptors of insects and other arthropods have been described as either sound receptors or wind-sensitive receptors. The terminology is arbitrary. While pressure fluctuations are always associated with airflows, the opposite is not always true. Considered as sound receptors, the hair receptors are only reasonably sensitive close to sound sources. In the farfield (at least half a wavelength away from the sound source), the air velocity is directly proportional to the sound pressure and quite small. Very close to the surface of the sound source the air velocity is much larger and is equal to the velocity of the source. The latter component decreases rapidly with the distance from the source. The near-field is the air space in which both components contribute to the air velocity.

It is therefore not possible to estimate the airflow velocity close to a sound source just by measuring the sound pressure at one spot. In a few cases of nearfields around sound-emitting insects, the sound pressure level and air velocity have both been determined. Close to dancing honeybees, the vibrating wings cause short sound pulses with a sound pressure of about 92 dB sound pressure level (SPL) and an air velocity of up to 50 cm/s (Michelsen et al. 1987). In a farfield, the air velocity corresponding to 92 dB SPL would be about 2 mm/s. The actual air velocity (50 cm/s) in the extreme near field corresponds to that expected at approximately 140 dB SPL in a farfield. A similar difference of about 45 dB has been found in the case of female fruit flies listening at close range to the "love song" of a courting male (Bennet-Clark 1971). The main carrier frequencies in these cases are 280 Hz and 170 Hz, respectively. That is far below the lower limit of a few kilohertz that was indicated in Section 3 for sound radiation into the farfield.

The receptors of the air oscillations in these cases are not mechanorecep-tive hairs but rather the antennae. The flagellum of the antenna is attached to the second segment, which contains a radially oriented chordotonal sense organ. It is named after its discoverer (Johnston 1855), who described it as a hearing organ. In mosquitoes and midges this is indeed the case. Males of the species that mate in swarms are attracted over a short range by the wing-beat flight sound of the female. The Johnston's organ of mosquito males contains no fewer than 30,000 sense cells (Risler and Schmidt 1967). This is twice the number in the human cochlea! Extensive studies per-formed in the 1950s and 1960s (Belton 1974) suggested that males use a process of triangulation based on the planes of vibration of the flagella for localization. Both the resonance frequency of the antenna and the best frequency of Johnston's organ appear to be equal to the wing-beat fre-quency of the female (350 to 400 Hz), which is a few hundred hertz lower than that of the male. The biophysics of these systems should be reinvestigated with modern methods.

Before leaving the strategies that have evolved in insects for determining the direction of sound incidence, it is interesting to compare these solutions with ideal methods. In order to perform an accurate determination of the direction of sound in three-dimensional space, one needs four sound receiv-ers, widely spaced but not placed in the same plane (Møhl and Miller 1976). A computation of the differences in the time of arrival of sound at the receivers would provide the necessary information. Preferably, the receiv-ers should be small and placed far away from large (sound-reflecting) bodies. The receivers would then also be available for making exact records of the temporal and spectral characteristics of the sounds.

Most animals have only two ears, and they are thus inherently not well equipped for solving a three-dimensional problem. Exploiting diffraction around the head/body for directional hearing is, as we have seen, a possibility. Diffractional effects are very frequency dependent, however, and the animal has thus given up some of its ability to determine the exact frequency spectrum of the sounds. One could say that the direction and spectrum are complementary in a system based on diffraction and that one cannot determine both with maximum accuracy when doing a few measurements. In passing, it should be noted that diffraction changes not only the sound pressure but also the phase. Attempts to make the ear more directional are thus blurring the time information.

Pressure-difference receivers are inherently directional. But the price for this useful property is that the force driving the membrane depends on the phase angle between the sounds acting on the two surfaces of the membrane, that is, on the frequency and the angle of incidence of the sound. An optimum performance with respect to directionality requires that the sounds arriving at the two surfaces do not differ much with respect to their amplitude and phase. The driving force will then vary greatly with frequency. So, again, the price for obtaining a reasonable directionality is a lack of ability to perform an accurate frequency analysis of the sounds reaching the animal.

So far, evolution has not produced an ideal solution to the problem of performing a perfect, simultaneous determination of the temporal, spectral, and directional properties of the sounds (e.g., an animal carrying ears at the tip of four tentacles). Arthropods carrying many mechanoreceptive hairs on their body may perhaps be close to the ideal as far as the reception of close range oscillations of the medium are concerned. Most known animals have only two ears, and their performance has to be a compromise. In theory, one should expect the different animals to have arrived at different compromises depending on their needs. Future research may perhaps locate animal specialists in temporal and/or spectral analysis, but in most insects evolution appears to have emphasized a good directional hearing at the expense of temporal and (especially) spectral analysis.

8. Degradation of Directional Cues in the Environment

Since the middle 1970s, several biologists have described the acoustic properties of the habitats of a wide variety of animals that use sounds for communication. Some biologists were also interested in the mechanisms responsible for the different acoustic properties of animal habitats (e.g., Wiley and Richards 1978, 1982; Michelsen 1978; Michelsen and Larsen 1983). The pioneering studies were concerned with the adaptation of sounds to penetrate a habitat over as large distances as possible, but more recent research has focused on how to obtain reliable communication over the required distances, which are not necessarily the longest distances (see Römer, Chapter 3).

Studies of the directional properties of hearing in both vertebrates and insects have long been popular in laboratories, but next to nothing is known about how well the mechanisms of directional hearing work in natural habitats. One exception was the demonstration, by means of a movable neurophysiological preparation of a bushcricket, that sounds received in dense vegetation may have lost their directional cues although other aspects (such as rhythmicity of songs) may still be available for coding by auditory neurons (Rheinlaender and Römer 1986). As mentioned in Section 5, the ears of most bushcrickets are, for all practical purposes, pressure receivers with their input at the body surface.

In the experiment by Rheinlaender and Römer, a bushcricket (*Tettigonia viridissima*) received 20 kHz sound pulses from a loudspeaker that was 1.5 m from ground and 10 m from the animal. Figure 2.16 shows the responses of two interneurons when the preparation was at three different distances from ground. At 1.5 m (the height of the dense bushes) the directional hearing was almost perfect. At 0.75 m two errors occurred. Finally, directional hearing was impossible when the preparation was standing on the ground, although central neurons still coded the rhythmical structure of the song. (Note that the sound level was constant at the loudspeaker and that the sound had been both degraded and attenuated near the ground.)

The degradation of directional cues in a natural grassland habitat of orthopteran insects has recently been studied with a combination of acoustical measurements and behavioral experiments (Michelsen and Rohrseitz 1997). A freshly killed grasshopper was glued to two probe microphones such that the probe tips were at the body surface close to the ears. The preparation and microphones could be rotated, and the amplitude and phase angle of the sounds at the ears could thus be measured at various directions of sound incidence.

Pressure amplitudes measured at 6 kHz at the ears of a grasshopper in a free sound field far from ground are shown in Figure 2.17A. The body length was 30 mm, that is, between the locust (55 mm) and the small grasshopper (14 mm) considered earlier, and the data reflect this (compare with Fig. 2.5). Diffraction of sound caused a surplus pressure at an ear when the sound source was ipsilateral to that ear (90° for the right ear, −90° for the left ear). Only little "shadow" is seen at the other ear.

The central nervous system needs the driving forces at the ears to differ by 1 to 2 dB in order to decide whether the sound source is to the right or the left (Helversen and Rheinlaender 1988). The ratio in sound amplitude at the two ears is shown in Figure 2.17C, which also indicates the expected decisions in a hypothetical animal of the size of the grasshopper, equipped with pressure-receiving ears and supposed to be unable to exploit phase or other time cues. A 1.5-dB difference in level was chosen as the criterion for the left–right decision. A decision to turn to the correct side is indicated with "*." As expected, turns are absent at the directions 0° and 180°, but

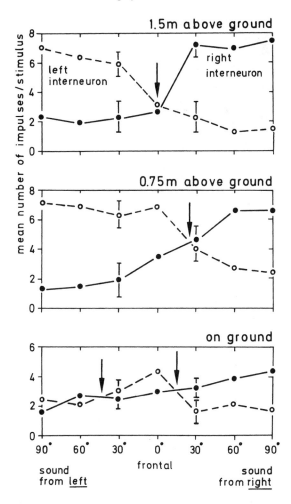

FIGURE 2.16. Responses recorded in a pair of auditory interneurons (T fibers) in a bushcricket at various directions of sound incidence in dense bushland. Further information is supplied in the text. (From Rheinlaender and Römer 1986, with permission.)

they are also absent at −60°, −30°, and 30°. At these three directions, the 1.5-dB criterion is not met, although a turn is expected.

The message of Figure 2.17C is that an animal of the size mentioned, which relies only on measuring the pressure amplitudes at the ears, would make mainly correct decisions when listening to a 6-kHz tone in a free sound field (no turn at 0° and 180°; a turn toward the most stimulated ear at most other directions of sound incidence). At three directions, however, the animal would not be able to decide on the correct turn. These data and other similar observations show that the hypothetical animal would often have troubles at the 30°/−30° directions.

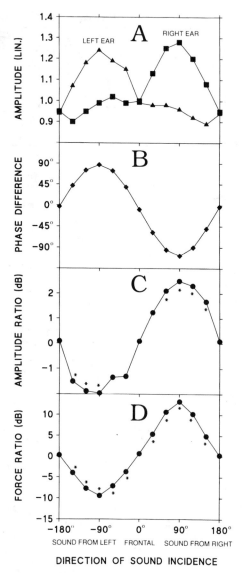

FIGURE 2.17. Directional cues available to a grasshopper in a free sound field at 6 kHz. A: Pressure amplitudes at the two ears at 12 directions of sound incidence. B: The phase cue (the difference in phase between the sound pressures at the ears). C: The ratio of the sound amplitude at the two ears. Predicted correct turns (*) are indicated for an animal that exploits only the amplitude cue (criterion: 1.5 dB difference in level). D: Calculated ratio of the forces at the eardrums when the phase cue is exploited by means of pressure-difference receiving ears. Predicted correct turns as in C. (From Michelsen and Rohrseitz 1997, with permission.)

So far, our hypothetical animal has made use only of amplitude cues. In addition, a phase cue is available (Fig. 2.17B). Grasshoppers can exploit this phase cue in their directional hearing (see Section 6). Calculations (assuming a gain of 0.5 and 60° at 6 kHz; see Section 6) show that a grasshopper of this size would make correct decisions at all 12 directions by exploiting the phase cue with its pressure-difference receiver ears (Fig. 2.17D).

The free sound field used for obtaining the data in Figure 2.17 is the most favorable acoustical situation that one can imagine for directional hearing. Figure 2.18 shows an example of the other extreme situation: a position in the habitat, at which directional hearing is very difficult (2 cm from the ground; 4 m from a 10-kHz sound source; grass of approximately 50-cm height). The amplitude cue (Fig. 2.18A) is much degraded (compare with Figures 2.16 and 2.17A), and the amplitude ratio (Fig. 2.18C) is far from the pattern in Figure 2.17C. Our hypothetical amplitude-perceiving animal makes only five correct turns ("*") plus the correct decision not to turn at the 180° direction. However, it makes four errors (E): a turn at the 0° direction and a turn to the wrong side at the directions −150°, 30°, and 150°. Furthermore, it fails to make a turn at the directions −90° and 60°. This is only a slightly better score than a 50–50 situation, in which sound localization is impossible. The phase cue (Fig. 2.18B) is much less degraded than the amplitude cue, however, and animals that are able to exploit the phase cue may still be able to determine the sound direction with only few errors (Fig. 2.18D, assumed gain 0.5 and 80°).

A close look at the data in Figure 2.18 reveals that the only error in D (at 180°) was caused by a large degradation of the phase cue at that direction (B). Although the degradation of the phase cue may thus be substantial, in general the phase cue is less affected by degradation than the amplitude cue. The robustness of the phase cue probably reflects the fact that both the time of arrival of sound at the ears and the sound diffraction contribute to the phase cue, whereas only the latter process is responsible for the amplitude cue. The finding that ears working as pressure-difference receivers are particularly suited for overcoming the degradation of the directional cues suggests that the possession of such ears may be an adaptation not only to small body size, but also to the kind of habitat.

Further studies in which such predictions were compared with the lateralization responses of live grasshoppers (see Section 2) revealed that both the live grasshoppers (with their pressure-difference ears) and the hypothetical pressure-receiving animal have troubles in the forward direction. This is not surprising, considering the respective mechanisms involved. Both the amplitude cue and the phase cue are zero when sound arrives from the frontal direction, and they are a maximum when the sound source is ipsilateral to one of the ears. Apparently, the scatter in degraded sound fields may be larger than these directional cues at the frontal directions of sound incidence, whereas the ratio between cues and scatter is more favorable in more lateral directions. The dominant role of lateralizations in the

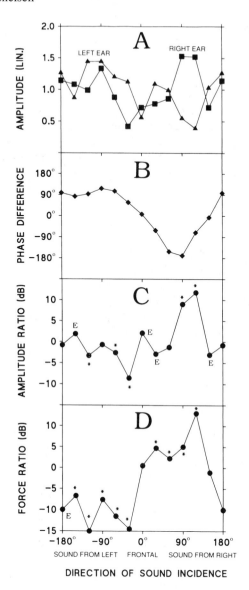

FIGURE 2.18. Degraded directional cues available to a grasshopper receiving 10-kHz sound 2 centimeters from the ground and 4 meters from the source. A: The pressure amplitudes at the ears (cf. Fig. 2.17A). B: The phase cue (cf. Fig. 2.17B). C: The ratio of the sound amplitude at the ears (cf. Fig. 2.17C). Predicted lateralization responses as in Figure 2.17C. D: The ratio of the forces acting on the two eardrums in a pressure-difference receiving grasshopper (cf. Fig. 2.17D). (From Michelsen and Rohrseitz 1997, with permission.)

phonotaxis of some species of grasshoppers may be a strategy for avoiding the critical forward direction.

An analysis of the predictions made on the basis of acoustical measurements showed that the number of errors tended to increase with sound frequency. The trend was weak because the data reflect two opposite trends. The scatter of sound amplitude increases with frequency but so does the diffraction of sound by the body of the grasshopper. The method is thus biased by the size of the animal used during the collection of the primary acoustical data.

Diffraction around a cylindrical body such as that of a grasshopper causes some change in the sound pressure at the surface at all directions of sound incidence, but the change is a minimum at the 0° and 180° directions. A minimum of diffraction means that one may have a better chance to observe variations in the scatter of sound amplitude. This was indeed the case. The degradation of the amplitude cue (indicated as the standard deviation) appeared to be much larger at 5 cm from the ground than at 20 cm from the ground (Fig. 2.19). A clear dependence on sound frequency was also found but only at 5 cm from the ground.

It has long been known that the attenuation of animal calls may be less some distance from ground. For example, male field crickets calling on the ground may be heard by females on the ground at a distance of 6 m; the distance increases to 15 to 22 m if one or both animals move away from

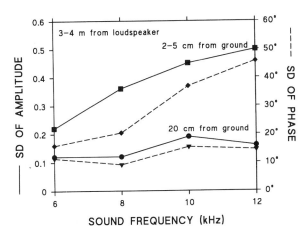

FIGURE 2.19. The effects of sound frequency and distance to ground on the standard deviation of sound amplitude at the ears (full line) and of the interaural phase difference (dashed line) of a grasshopper 3 to 4 m from the loudspeaker. Sound arrived from the 0° and 180° directions. The expected (mean) values of amplitude and phase were 1 and 0°, respectively. (From Michelsen and Rohrseitz 1997, with permission.)

the ground (Paul and Walker 1979). The results of Figure 2.19 show that moving away from ground can also reduce the amount of sound degradation. Furthermore, the results suggest that the species communicating on the ground should prefer rather low frequencies (European field crickets do indeed use 4 to 5 kHz), whereas species communicating at some distance from the ground may have a wider choice. Of course, other aspects than directional hearing may affect the optimum frequency for communication.

The contributions of the physical processes leading to degradation of the directional cues probably differ in different environments. One approach in future work could be to use the methods just described to study a number of different environments. It remains to be learned whether some communication strategies are adapted to minimize the degradation of the directional cues.

9. Summary

Insects are so small that sounds arrive almost simultaneously at their ears, and the time of arrival of sound (one of the cues used by man) is too small to be exploited for directional hearing. Some insects listen to sounds with wavelengths comparable to the size of their bodies (e.g., the ultrasonic cries of hunting bats), and they can obtain useful directional cues from differences in the amplitude of the sounds at the ears caused by diffraction of sound around the body (the other cue exploited by man). Most insects are too small relative to the wavelengths of the sounds of interest, however, and cannot make use of this mechanism either.

One strategy for directional hearing in such insects is to respond, not to the pressure component of sound but to the movement component, which is inherently directional. However, movement receivers are generally not very sensitive and are therefore best suited at close range.

The most commonly used strategy for obtaining a reasonable sensitivity and directionality is to let the sound waves reach both the outer and inner surface of the eardrum. Such pressure difference receivers are common in both insects and some other small animals (e.g., frogs, birds). The directionality is caused by the fact that the phase angles of the sounds arriving at the eardrum vary with the direction of sound incidence. The mechanism at work can be understood by considering an eardrum, which receives sounds with the same amplitude at its two surfaces. At one direction of sound incidence, the sounds at the two surfaces may be totally out of phase and the vibration amplitude of the eardrum a maximum. At another direction, the sounds may have the same phase, which causes the eardrum vibration to stop.

Obviously, both the amplitudes and phase angles of the sounds arriving at the eardrum have to be right for such a mechanism to produce a maximum of directionality. This is almost the case when a female field cricket listens to the pure tone calling song of the male, but not at other frequencies. Some animals live in acoustically favorable environments and do not need a high degree of directionality. In contrast, in the habitats of animals such as field crickets the directional components of sound tend to degrade, and crickets thus need ears with a large directionality.

Acknowledgments. I am grateful to the editors of the book and to my colleague Ole Næsbye Larsen for comments on an earlier version of this review. The Centre for Sound Communication is financed by the Danish National Research Foundation.

References

Autrum H (1940) Über Lautäusserungen und Schallwahrnehmung bei Arthropoden. II. Das Richtungshören von Locusta und Versuch einer Hörtheorie für Tympanalorgane vom Locustidentyp. Z Vergl Physiol 28:326–352.

Batschelet E (1981) Circular Statistics in Biology. London: Academic Press.

Belton P (1974) An analysis of direction finding in male mosquitoes. In: Barton Browne L (ed) Experimental Analysis of Insect Behaviour. Berlin: Springer-Verlag, pp. 139–148.

Bennet-Clark HC (1971) Acoustics of insect song. Nature 234:255–259.

Beranek LL (1954) Acoustics. New York: McGraw-Hill (new edition published by the American Institute of Physics 1986).

Brown CH (1994) Sound localization. In: Fay RR, Popper AN (eds) Comparative Hearing: Mammals. New York: Springer-Verlag, pp. 57–96.

Brownell P, Farley RD (1979) Orientations to vibrations in sand by the nocturnal scorpion *Paruroctonus mesaensis*: mechanisms of target localization. J Comp Physiol 131:31–38.

Ewing AM (1989) Arthropod Bioacoustics: Neurobiology and Behavior. Ithaca, NY: Cornell University Press.

Fletcher NH, Thwaites S (1979) Acoustical analysis of the auditory system of the cricket *Teleogryllus commodus* (Walker). J Acoust Soc Am 66:350–357.

Fonseca PJ, Popov AV (1997) Physical analysis of directional hearing in the cicada *Cicada barbara lusitanica*. J Comp Physiol A, 180:417–427.

Helversen D von, Helversen O von (1983) Species Recognition and Acoustic Localization in Acridid Grasshoppers: A Behavioral Approach. In: Huber F, Markl H (eds) Neuroethology and Behavioral Physiology. Berlin: Springer-Verlag, pp. 95–107.

Helversen D von, Helversen O von (1995) Acoustic pattern recognition in orthopteran insects: Parallel or serial processing? J Comp Physiol A 177:767–774.

Helversen D von, Rheinlaender J (1988) Interaural intensity and time discrimination in an unrestrained grasshopper: a tentative behavioral approach. J Comp Physiol A 162:333–340.

Hergenröder R, Barth FG (1983) Vibratory signals and spider behaviour: how do the sensory inputs from the eight legs interact in orientation? J Comp Physiol A 152:361–371.

Hill KG (1974) Carrier frequency as a factor in phonotactic behaviour of female crickets *Teleogryllus commodus*. J Comp Physiol 93:7–18.

Hill KG, Boyan GS (1976) Directional hearing in crickets. Nature 262:390–391.

Hoy RR, Paul RC (1973) Genetic control of song specificity in crickets. Science 180:82–83.

Johnston C (1855) Auditory apparatus of the *Culex* mosquito. Q J Microsc Sci 3:97–102.

Kramer E (1976) The orientation of walking honeybees in odour fields with small concentration gradients. Physiol Entomol 1:27–37.

Larsen ON (1981) Mechanical time resolution in some insect ears. II. Impulse sound transmission in acoustic tracheal tubes. J Comp Physiol 143:297–304.

Larsen ON (1995) Acoustic equipment and sound field calibration. In: Klump GM, Dooling RJ, Fay RR, Stebbins WC (eds) Methods in Comparative Psychoacoustics. Basel: Birkhäuser Verlag, pp. 31–45.

Larsen ON, Michelsen A (1978) Biophysics of the Ensiferan ear. III. The cricket ear as a four-input system. J Comp Physiol 123:219–227.

Lewis B (1983) Directional cues for auditory localization. In: Lewis B (ed). Bioacoustics, a Comparative Approach. London: Academic Press, pp. 233–257.

Michel K (1974) Das Tympanalorgan von *Gryllus bimaculatus* Degeer (Saltatoria, Gryllidae). Z Morph Tiere 77:285–315.

Michelsen A (1971) The physiology of the locust ear. Z Vergl Physiol 71:49–128.

Michelsen A (1978) Sound reception in different environments. In: Ali MA (ed) Sensory Ecology. New York: Plenum Press, pp. 345–373.

Michelsen A (1983) Biophysical basis of sound communication. In: Lewis B (ed) Bioacoustics. London: Academic Press, pp. 3–38.

Michelsen A, Larsen ON (1978) Biophysics of the Ensiferan ear. I. Tympanal vibrations in bushcrickets (Tettigoniidae) studied with laser vibrometry. J Comp Physiol 123:193–203.

Michelsen A, Larsen ON (1983) Strategies for acoustic communication in complex environments. In: Huber F, Mark H (eds) Neuroethology and Behavioral Physiology. Berlin: Springer-Verlag, pp. 321–331.

Michelsen A, Löhe G (1995) Tuned directionality in cricket ears. Nature 375:639.

Michelsen A, Rohrseitz K (1995) Directional sound processing and interaural sound transmission in a small and a large grasshopper. J Exp Biol 198:1817–1827.

Michelsen A, Rohrseitz K (1997) Sound localization in a habitat: an analytical approach to quantifying the degradation of directional cues. Bioacoustics, 7:291–313.

Michelsen A, Towne WF, Kirchner WH, Kryger P (1987) The acoustic near field of a dancing honeybee. J Comp Physiol A 161:633–643.

Michelsen A, Hedwig B, Elsner N (1990) Biophysical and neurophysiological effects of respiration on sound reception in the migratory locust *Locusta migratoria*. In: Gribakin FG, Wiese K, Popov AV (eds). Sensory Systems and Communication in Arthropods. Basel: Birkhäuser Verlag, pp. 199–203.

Michelsen A, Heller K-G, Stumpner A, Rohrseitz K (1994) A new biophysical method to determine the gain of the acoustic trachea in bushcrickets. J Comp Physiol A 175:145–151.

Michelsen A, Popov AV, Lewis B (1994) Physics of directional hearing in the cricket *Gryllus bimaculatus*. J Comp Physiol A 175:153–164.

Miller LA (1977) Directional hearing in the locust *Schistocerca gregaria*. Forskål (Acrididae, Orthoptera). J Comp Physiol 119:85–98.

Moiseff A, Konishi M (1981) Neuronal and behavioral sensitivity to binaural time differences in the owl. J Neurosci 1:40–48.

Morse PM (1948) Vibration and Sound. New York: McGraw-Hill (new edition published by the American Institute of Physics 1981).

Møhl B, Miller L (1976) Ultrasonic clicks produced by the peacock butterfly: a possible bat-repellent mechanism. J Exp Biol 64:639–644.

Paul RC, Walker TJ (1979) Arboreal singing in a burrowing cricket, *Anurogryllus arboreus*. J Comp Physiol 132:217–224.

Payne R, Roeder KD, Wallman J (1966) Directional sensitivity of the ears of noctuid moths. J Exp Biol 44:17–31.

Pierce AD (1981) Acoustics: An Introduction to Its Physical Principles and Applications. New York: McGraw-Hill (new edition published by the American Institute of Physics 1989).

Pollack GS (1988) Selective attention in an insect auditory neuron. J Neurosci 8:2635–2639.

Regen J (1913) Über die Anlockung des Weibchens von *Gryllus campestris* L. durch telephonisch übertragene Stridulationslaute des Männchens. Pflügers Arch 155:193–200.

Rheinlaender J, Römer H (1986) Insect hearing in the field. I. The use of identified nerve cells as "biological microphones." J Comp Physiol A 158:647–651.

Risler H, Schmidt K (1967) Der Feinbau der Scolopidien im Johnstonschen Organ von *Aedes aegypti* L. Z Naturforsch 22B:759–762.

Roeder KD (1967) Nerve Cells and Insect Behavior, rev. ed. Cambridge, MA: Harvard University Press.

Römer H (1976) Die Informationsverarbeitung tympanaler Rezeptor-elemente von *Locusta migratoria* (Acrididae, Orthoptera). J Comp Physiol 109:101–122.

Schildberger K, Huber F, Wohlers DW (1989) Central auditory pathway: neural correlates of phonotactic behavior. In: Huber F, Moore TE, Loher W (eds). Cricket Behavior and Neurobiology. Ithaca, NY: Cornell University Press, pp. 423–458.

Schmitz B, Scharstein H, Wendler G (1982) Phonotaxis in *Gryllus campestris* L (Orthoptera, Gryllidae). I. Mechanisms of acoustic orientation in intact female crickets. J Comp Physiol A 148:431–444.

Schmitz B, Scharstein H, Wendler G (1983). Phonotaxis in *Gryllus campestris* L. (Orthoptera, Gryllidae). II. Acoustic orientation of female crickets after occlusion of single sound entrances. J Comp Physiol 152:257–264.

Schnitzler H-U, Menne D, Kober R, Heblich K (1983). The acoustical image of fluttering insects in echo-locating bats. In: Huber F, Markl H (eds). Neuroethology and Behavioral Physiology. Berlin: Springer-Verlag, pp. 235–250.

Schwabe J (1906) Beiträge zur Morphologie und Histologie der tympanalen Sinnesapparate der Orthopteren. Zoologica 20:1–154.

Shaw EAG (1974) The external ear. In: Keidel WD, Neff WD (eds) Handbook of Sensory Physiology, Vol. V/1. Auditory System, Anatomy, Physiology (Ear). Berlin: Springer Verlag, pp. 454–490.

Skudrzyk E (1971) The Foundations of Acoustics. Berlin: Springer-Verlag.

Tautz J (1979) Reception of particle oscillation in a medium — an unorthodox sensory capacity. Naturwissenschaften 66:452–461.

Weber T, Thorson J (1989) Phonotactic behavior of walking crickets. In: Huber F, Moore TE, Loher W (eds) Cricket Behavior and Neurobiology. Ithaca, NY: Cornell University Press, pp. 310–339.

Weber T, Thorson J, Huber F (1981) Auditory behavior of the cricket. I. Dynamics of compensated walking and discrimination paradigms on the Kramer treadmill. J Comp Physiol A 141:215–232.

Wendler G, Löhe G (1993) The role of the medial septum in the acoustic trachea of the cricket *Gryllus bimaculatus*. I. Importance for efficient phonotaxis. J Comp Physiol A 173:557–564.

Wiley RH, Richards DG (1978) Physical constraints on acoustic communication in the atmosphere: Implications for the evolution of animal vocalizations. Behav Ecol Sociobiol 3:69–94.

Wiley RH, Richards DG (1982) Adaptations for acoustic communication in birds: Transmission and signal detection. In: Kroodsma DE, Miller EH (eds). Acoustic Communication in Birds, Vol. I. New York: Academic Press, pp. 131–181.

Yager DD, Hoy RR (1987) The midline metathoracic ear of the praying mantis, *Mantis religiosa*. Cell Tissue Res 250:531–541.

Yost WA, Gourevitch G, eds. (1987) Directional Hearing. New York: Springer-Verlag.

3
The Sensory Ecology of Acoustic Communication in Insects

Heiner Römer

1. Introduction

Why should we consider ecological aspects in the context of acoustic communication at all? Ecology may be defined as the study of the interaction of an organism with its environment, including other organisms. Behavioral ecologists, for example, focus their interest on the interactions between an animal and its conspecifics, its predators and prey, food resources, territories, etc. Like ethologists, they are primarily concerned with the adaptive value and/or the evolution of behavior rather than the mechanisms being used. By contrast, sensory ecologists are more concerned with the mechanisms that enable an animal to produce or utilize signals and how the information about identity or location of the sender is transmitted to the receiver(s).

Here we have to rely rather heavily on data collected by physiologists, but those who measure auditory physiology in the lab are usually more interested in the mechanism of sensation rather than in the signals that are actually present in the animal's natural environment. Thus, in order to estimate the auditory performance of an insect in the real world, physiologists face a dilemma; they have to leave their properly controlled anechoic chambers and deal with the tremendous variability of the signal in the transmission channel resulting from many different factors acting on it and that are almost impossible to quantify simultaneously. For this reason, the sensory ecology of hearing in insects is still in its infancy, but there is increasing acknowledgement of the importance of ecological factors for the evolution of acoustic communication systems (Michelsen 1978; Michelsen and Larsen 1983; Römer 1993; see also Michelsen, Chapter 2).

Furthermore, acoustic insects share problems such as frequency-dependent sound transmission with any other terrestrial sound communication system and may overlap their song carrier frequencies with those of frogs, birds, and higher vertebrates. We can therefore base our ideas about insect sound communication on a framework of empirical and theoretical studies performed either on the birdsong system pioneered by Morton (1975) and Wiley and Richards (1978, 1982) or on many studies on sound propagation outdoors related to the human hearing system (reviewed by Embleton 1996).

Figure 3.1 summarizes the potential selective forces that may shape the form of the signal or the signaling behavior. Mating signals are both attenuated and degraded by physical properties of the environment. A potentially informative character, for example, the interval between sound pulses, is degraded by reflection or echoes off structures. This is more of a problem in spatially complex habitats, such as a bushy thicket. These conditions suggest that a signal that functions efficiently in one environment may not be ideal in another. Therefore populations (of a species) inhabiting different habitats may diverge in important properties of their mating signals (see Endler 1992 for an excellent review on visual signals).

A conspicuous signal may be advantageous in mate attraction but may also have costs. For example, predators and parasitoids are known to home in on acoustic signals. Parasitoid tachinid flies use the acoustic signals of crickets or bushcrickets to find their hosts (Cade 1975, 1981; Lakes-Harlan and Heller 1992; see Robert and Hoy, Chapter 6). This and other examples involve direct selection on signal characters, and if the suite of predators and parasites varies among localities, there is the potential for divergence of the signal and/or the sensory organ to result (Hoy 1992; see Fullard, Chapter 8). Sexual selection through male–male competition or through female choice is another powerful driving force for the evolution of signals (West-Eberhard 1984). Furthermore, because female receptors or the central nervous system may be biased in a particular direction (e.g., to detect a potential predator or prey), this also can exert

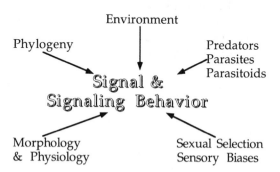

FIGURE 3.1. Evolution of the physical properties of a sound signal or the signaling behavior of an insect may result from various selective forces and constraints, one of which is the environment with its biotic and abiotic factors influencing sound transmission and perception. (From Forrest 1994, with permission.)

significant selective forces on signals or signaling behavior (see Pollack, Chapter 5).

2. Near-Field and Far-Field Sound

This chapter deals with the sensory ecology of far-field sound. Sound is a mechanical vibration that travels as alternating waves of high and low pressure. Pressure variation is accompanied by movements of the particles of the medium. Close to a sound source that is small relative to the produced wavelength, the energy due to the particle displacements may be much greater than the sound pressure. The amplitude of the particle displacement decreases with $1/r^2$ or $1/r^3$ (r = distance from the source) for a monopole or dipole sound source, respectively, in contrast to the propagated pressure variation, which decreases with only $1/r$. The region where the particle displacement is greater than the pressure component is called *near-field sound*, and when the pressure becomes the dominating component it is called *far-field sound*. It is obvious, then, that near-field sound can act only at short distances, even if a receiver is equipped with very sensitive receptors.

We should consider, however, that the same motor act (e.g., a stridulating cricket) may cause three different mechanical disturbances that propagate to different distances: (1) a pressure variation of air that propagates as far-field sound, detectable by tympanal hearing organs over tens or even hundreds of meters; (2) a near-field medium motion caused by the stridulatory action of the forewings, which travels distances of some centimeters and may be detected by the wind-sensitive filiform hairs on the cerci of a receiver; and (3) a substrate vibration, which propagates as torsional and/or bending waves perpendicular to the surface of the substrate. This vibration can be detected by vibration receptors (e.g., subgenual organs in the legs) in a receiver standing on the same substratum, at distances from a few centimeters up to about 2 meters (see also Barth, Chapter 7). However, there is no clear-cut border in terms of the communication distances achieved by far-field sound or vibration.

When a male grasshopper, *Chorthippus biguttulus*, produces a calling song with a sound pressure level of 65 dB at a distance of 10 centimeters, strong excess attenuation (EA) from environmental factors ensures that the signal will not activate tympanal receptors of a female more than 1 m away in its usual microhabitat (Werner and Elsner 1995; Lang 1996). On the other hand, small-sized insects may cause their host plants to vibrate at high acceleration amplitudes, and these vibrations propagate as bending waves with little loss in energy, so that they can be detected by other insects over a few meters (Michelsen et al. 1982). The restriction to far-field, airborne sound is therefore somewhat deliberate, but it may best illustrate the impact of ecological constraints on long-range acoustic communication.

3. Sound Transmission and Range of Communication

3.1. Effect on Signal Amplitude: Geometric Spreading

When a sound wave propagates away from a source, its pressure inevitably decreases with increasing distance. This is a result of the geometry of the air space occupied by the sound energy. Intensity is power/unit area, and if this energy is spread over a greater area when the sound propagates, the acoustic power/unit area decreases. This geometric spreading causes a 6-dB decrease for each doubling of distance (6 dB/dd) and is the minimum attenuation an animal could possibly obtain broadcasting a signal under ideal conditions (for interesting exceptions, see later). Therefore, if selection has been acting on a signal for maximum range of detection, any factors that produce attenuation in excess of geometrical (or spherical) spreading should be minimized or eliminated. Such EA may result from three processes: atmospheric absorption, scattering, and interference of the sound wave with boundaries.

3.1.1. Excess Attenuation from Absorption

Excess attenuation as a result of absorption is probably the least important of the three processes, except for insects producing very high frequencies. As a result of dissipation of sound energy into heat, sound pressure levels decrease at a specific number of decibels per meter, this value depending strongly on sound frequency, temperature, and relative humidity. From laboratory and field measurements conducted by Harris (1966) and Piercy, Embleton, and Sutherland (1977), one can calculate absorption coefficients of about 0.5 dB/100 m for a frequency of 1 kHz and 10 dB/100 m for 10 kHz. As we will see later, attenuation due to scattering and boundary interference may limit communication distances more severely, but for frequencies as high as 100 kHz attenuation due to absorption may amount to 10 dB/m. Although such high frequencies may not be used by many insects, it has nevertheless some indirect effect, at least for insects in flight. High absorption coefficients have been shown to influence the evolution of echolocation signals of insectivorous bats that hunt in open space and use much lower frequencies than those searching for prey close to, or within, vegetation (Neuweiler 1989). This is clearly an adaptation to the greatly reduced range of high-frequency signals caused by absorption in open space.

3.1.2. Excess Attenuation from Scattering by Vegetation and Turbulence

Many insects call within, or close to, meadows, bushland, or forests. Such vegetation includes surfaces in which the dimensions are the same order of magnitude as the wavelength of sound and may redirect sound in all directions in a process termed *scattering* (or *diffraction*). Again, there is a strong correlation between sound frequency and the degree of excess attenuation due to scattering. Leaves that are 5 cm scatter sound with a wavelength of 33 cm (1 kHz), much less than those of 3.3 cm (10 kHz). As a result, coniferous forests exhibit less excess attenuation than deciduous forests, and these effects are more pronounced at higher frequencies (Morton 1975; Marten and Marler 1977).

One should also consider that insects are usually not ideal sound sources that radiate sound energy equally well in all directions. Their sound output, particularly for high frequencies, is often highly directional (Keuper and Kühne 1983; Bailey 1985). For example, in the bushcricket, *Metrioptera sphagnorum*, a high-frequency component of the song at 33 kHz is up to 15 dB more intense from the dorsal than that from the ventral surface (Römer and Morris, unpublished). Scattering reduces the intensity along this beam by redirecting the sound energy outward. Because this effect is not linear with distance, it requires measurements at various distances from the source.

Another possible source of scattering is from atmospheric turbulences as a result of wind shear over irregular surfaces and temperature gradients. Although no experimental data are available, calculations predict strong excess attenuation for higher than for lower frequencies (2.6 dB/100 m for 1 kHz compared with 3.8 dB/m for 10 kHz; Wiley and Richards 1982).

Obtaining empirical data on biologically relevant high-frequency sound propagation is difficult because high-frequency sensitive microphones and insect ears differ substantially with respect to their absolute sensitivity, directionality, and temporal characteristics. As a result, the microphone-transduced signal recorded at a given position in the habitat may be different from what the insect actually hears. One solution is to record the activity of single, identified auditory neurons in the field, thereby using the central nervous system of the insect itself as a receiver to study the transmission and reception of biologically relevant sound (Rheinlaender and Römer 1986). This technique has been successfully used to study frequency-dependent sound attenuation and filtering (Römer and Lewald 1992) or directional hearing of grasshoppers in natural habitats (Werner and Elsner 1995).

An attempt to measure the composite effect of the various sources of EA in a bushcricket habitat is shown in Figure 3.2, with a plot of EA over distance (i.e., the attenuation caused by geometrical spreading subtracted from the total attenuation) when the receiver was placed within the vegetation. If homogeneous, isotropic scattering of sound were the only source of EA, then one would expect to see 12 dB per doubling of distance relationship, both from theoretical and experimental evidence (Meister and Ruhrberg 1959). The data presented in Figure 3.2 favor an EA of mixed origin. They indicate a tendency for a 12 dB/dd relationship at higher fre-

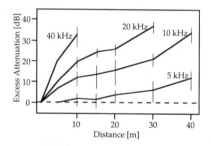

FIGURE 3.2. Excessive attenuation of sound amplitude for pure-tone sound pulses of 5, 10, 20, and 40 kHz in the habitat of a bushcricket. Field recordings of an interneuron of the bushcricket *T. viridissima* responding to sound were used to evaluate attenuation values over distance, with the receiver placed within vegetation (Römer and Lewald 1992). Data averaged from 10 experiments, with the threshold of the neuron at a distance of 1 meter as a reference.

quencies, in particular for 20 kHz, whereas a dB/m relationship appears to exist for 5 kHz. This is consistent with the assumption that multiple scattering within the vegetation represents the major source of EA at higher frequencies. In addition, for frequencies greater than 20 kHz, other sources of attenuation add to the attenuation caused by geometrical spreading and multiple scattering, so that EA can amount to more than 30 dB at a distance of 10 meters with a carrier frequency of 40 kHz. Scattering of sound by atmospheric turbulences or dissipation of sound energy into heat might constitute additional sources of sound attenuation (Lighthill 1953; Griffin 1971; Piercy, Embleton, and Sutherland 1977). In practice, it will be difficult to dissociate total EA into absorption and scattering components.

3.1.3. Excess Attenuation from Sound Interference in Stratified Environments

Sound waves are reflected when they meet a change in the acoustic impedance of the medium, such as air–water or air–soil interfaces, but also with volumes of air that differ in temperature, humidity, or wind velocity. Vertical gradients of temperature occur regularly in natural habitats; lapse conditions during the day, in which the atmospheric temperature decreases with height from the ground, reflect sound in an upward direction and decrease sound levels near the surface. This situation creates shadow zones at some distance from the source where no direct sound can penetrate (Fig. 3.3A). By contrast, when temperature increases with height, sound waves are reflected downward, which increases the sound pressure level (SPL) at some distance from the source (Fig. 3.3B).

Such enhanced SPLs around sunset and sunrise, corresponding to the times of formation and decay of temperature inversions, have been observed by Canard-Coruna et al. (1990), whereas excess attenuation as a

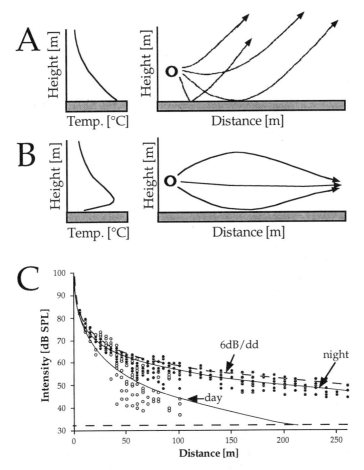

FIGURE 3.3. A,B: Effect of an upward-refracting temperature gradient (A) and downward-refracting gradient (B) on sound propagation above the ground. Left: Temperature gradient (h, height above ground). Right: Rays perpendicular to the wavefront from the source. (Modified from Wiley and Richards 1978, with permission.). C: Changes in sound pressure level of the male call of a primitive African grasshopper, *Bullacris membracioides*, over distance in open grassland indicate that nighttime transmission (solid circles) is consistently close to ideal (6 dB/dd) and suffers little excess attenuation, but during the day (open cirles) measures are highly variable and decrease significantly at about 50 meters, indicating a sound shadow zone as a result of temperature gradients of 4.5°C/10 m above ground level. Extrapolating from the night trajectory and assuming a mean hearing threshold for a conspecific female of 32 dB SPL to the male call (horizontal dashed line), females can hear conspecific male calls at distances between 1.4 and 1.9 kilometers. (From van Staaden and Rómer 1997, with permission.)

result of the shadow zone can increase by 20 to 30 dB, largely independent of sound frequency (Piercy, Embleton, and Sutherland 1977). By modeling these atmospheric situations for infrasound communication in elephants, Garstang et al. (1995) suggested that elephants may achieve acoustic communication distances of up to 10 km. These estimates are larger than those of 2 to 4 km reported by Langbauer et al. (1991) as a result of playback experiments, but the latter study was performed with elephants when they came to drink, without a control of the actual atmospheric conditions.

It is surprising how little attention has been paid in the past to these atmospheric conditions with respect to long-range sound communication of insects. There is only one report on a primitive African grasshopper of the family Pneumoridae, which appears to adopt a calling behavior that controls for such atmospheric conditions (Römer and van Staaden, 1996; Fig. 3.3C). Males of *Bullacris membracioides* produce mating calls with extremely high SPLs using an abdominal resonator (98 dB SPL at a distance of 1 m; re 20 μPa), and females possess a series of serially homologous ears in the abdominal segments A1 to A6, the first of which exhibits a high sensitivity of about 35 dB SPL to the male call.

The rate of signal transmission through the natural environment of the insect was determined for two atmospheric conditions, during the day at about 5 p.m. in the afternoon and at midnight (Fig. 3.3C). During the afternoon, temperature decreased with height above ground by approximately 4.5°C at a height of 10 meters. This produced an upward refracting sound condition and a sound shadow zone, with a consequent marked drop in SPL starting at a distance of about 50 meters. By contrast, temperature increased with height above ground at night; inversions of only 1.2°C were downward refracting, resulting in a tunnel effect with the sound caught between these zones of different temperature and the ground. Therefore the male call approaches almost ideal transmission conditions (compare with the 6 dB/dd relationship in Fig. 3.3C) at a nocturnal time when male and females actually communicate by sound. Due to these rather different atmospheric conditions, hearing distances for the male signal are 120 to 200 meters in the afternoon, but between 1.4 and 1.9 kilometers at night, arguably the largest hearing distance yet reported for insects.

Another problem of boundary interference may occur when the sender and receiver are elevated above ground level and the sound reaches the receiver through two different paths: a direct path and one reflected from the ground (solid and dashed lines in Fig. 3.4A). The spatial arrangement of sender and receiver, sound frequency, and the impedance of the ground all determine the SPL at the position of the receiver because the direct and reflected sound waves interact with each other depending on their phase relationship. The plots in Figures 3.4B and 3.4C model the effect of two ground impedances (a "soft" forest ground and a "hard" boundary, such as a water surface) on sound attenuation, as well as on different heights of

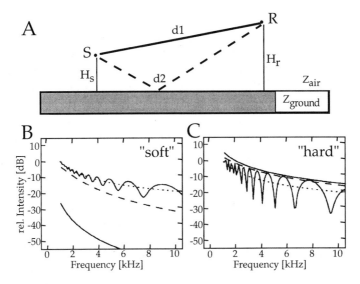

FIGURE 3.4. Propagation of sound over a boundary. A: Sound from the source (S) reaches the receiver (R) through a direct (d1) and a reflected (d2) path. The sender and receiver are elevated above ground level by heights H_s and H_r. Z, impedance of air and the ground. B,C: SPL of a 4-kHz tone as a function of distance over a "soft" boundary (left; e.g., forest floor) and a "hard" boundary (right; e.g., water) with different spatial arrangements of the sender and receiver. The dotted line is the attenuation expected by geometric spreading alone. EA is then the difference between this dotted line and the other curves. Dashed lines, sender elevated by 1 meter and receiver on the boundary surface; solid, wavy lines, sender and receiver elevated by 1 meter; solid, smooth line, sender and receiver on boundary surface. (After Forrest 1994, with permission.)

sender and receiver (Forrest 1994). Over a hard surface, and with the sender and receiver elevated, the interference of the direct and reflected sound waves cause areas of high and low SPL with changing distance. This effect is less pronounced over a soft surface (wavy lines in Fig. 3.4B,C).

Such calculations, and measurements of sound propagation and filtering over grass (e.g., Embleton, Piercy, and Olson 1976), are often taken as evidence that animals elevated above ground should avoid communicating with frequencies below 10 kHz because of such interference problems. A receiver approaching an elevated sender would get misleading information about distance by moving through areas of high and low SPL (compare wavy lines in Fig. 3.4B,C). Oecanthine tree crickets, for example, use pure-tone, low-frequency sound between 2 kHz and 3 kHz and call from elevated positions but, nevertheless, seem to be little affected by destructive interference. Two reasons may account for this: (1) the actual ground impedance of their habitat attenuates the reflecting sound wave to such a degree that destructive interference is irrelevant (Marten and Marler 1977), and/or (2) the coherence of direct and reflected wave is lost due to scattering by air turbulence. The latter would represent an interesting situation in which a disturbance on the transmission channel would have a positive effect on sound communication because it allows insects to avoid ground-wave attenuation or EA in scattering vegetation, without a necessary trade-off with destructive interference at some height above the ground.

3.1.4. Excess Attenuation from Sound Propagation on the Ground

It is an astounding phenomenon for acousticians to see ground-dwelling insect species communicating by sound signals over some distance. These insects face a complicated situation, where the sound waves interact with the ground in a grazing angle that is almost zero; sound waves travel almost parallel to the ground. The wave reflected from the ground will be out of phase with the incident wave at some distance, and when the two interact they cancel each other. Therefore sound propagation parallel to the ground is often referred to as "the forbidden mode of propagation" (Piercy, Embleton, and Sutherland 1977) because it creates another type of sound shadow zone. The broadcast range of a male *Gryllus campestris* calling song (approximately 102 dB SPL at 5 cm) would thus be limited to just 1 or 2 meters, at least from theoretical calculations. This is far from the 42 meters that one would expect if the sound is attenuated by 6 dB/dd (geometric spreading alone). However, although a field study by Popov et al. (1972) confirms attenuation in excess of geometric spreading, the distance over which a female would probably hear the calling song (based on neurophysiological thresholds) was about 10 to 15 meters, and thus considerably larger than the forbidden mode of propagation would suggest. The discrepancy cannot be explained by the so-called ground wave, which describes sound transmission due to the interaction of a spherical wave front with the ground. The ground wave results in transmission of sound with a low-pass filter characteristic below 0.5 to 1 kHz (Embleton, Piercy, and Olson 1976).

3.2. Frequency Filtering

The processes responsible for EA outlined earlier are strongly frequency dependent, with the result that the transmission channel acts essentially as a low-pass filter, with important consequences for insect acoustic communication. Information contained in high-frequency components of a broadband song will not be available for a receiver except at very short distances. As shown in Figure 3.2, at a distance of 10 meters and with the receiver on the ground, the SPL at 40 kHz has dropped by about 50 dB relative to a distance of 1 meter and is thus well below the threshold level of auditory receptors in the hearing organ.

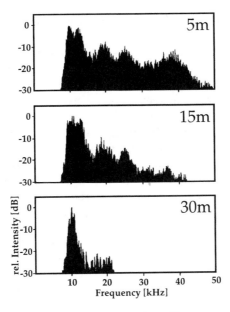

FIGURE 3.5. Frequency spectrum of the male calling song of the bushcricket *Tettigonia viridissima*, recorded at distances of 5, 15, and 30 meters from the male. Note the attenuation of high-frequency components in the song as a result of the low-pass filter characteristic of the communication channel. The heights of the male and microphone are both 50 centimeters.

Similarly, high-frequency components in the calling song at 20 kHz and 30 kHz will be lost at somewhat greater distances (Keuper and Kühne 1983; Fig. 3.5). Therefore, reported preferences for acoustic signals that contain high-frequency components will be strongly affected by the low-pass filtering property of the vegetation (Latimer and Lewis 1986; Latimer and Sippel 1987; Bailey and Yeoh 1988; Bailey et al. 1990). However, such a preference for high-frequency song components may have nothing to do with an active choice of a certain signal character. A signal containing less intense high-frequency components may just indicate a more distant calling male as a result of the environmental low-pass filter. Or alternatively, because high-frequency components often provide better directional cues, females may prefer the high-frequency signal because it enables better localization in a complex environment and thus a reduced risk of predation. As with the preference for more intense signals (see later), there may be no active choice at all but just a movement to the closest signal. This is a situation referred to by Parker (1983) as *passive attraction*. (For the consequence of frequency filtering for directional hearing, see later and Michelsen, Chapter 2).

3.3. Signaling and Information Transfer

Attenuation of a signal in the course of propagation through a natural environment is only one form of limitation for long-range communication, because communication usually requires more than simply detecting the presence (or absence) of a signal. The receiver must distinguish between signals that differ in frequency content and temporal pattern. Consequently, the degradation of these patterns on the transmission channel reduces the probability of signal discrimination for a receiver. Information about species identity is primarily encoded in the temporal structure of the song in insects (Elsner and Popov 1974; von Helversen and von Helversen 1983; Doherty and Hoy 1985), and any degradation of temporal parameters would therefore impose limitations for long-range communication.

Reverberations and amplitude fluctuations induced by wind or atmospheric turbulence produce distortions, particularly in the time domain. Amplitude fluctuations vary in magnitude with weather conditions, time of day, and carrier frequency (Richards and Wiley 1980). Their effect on the perceived signal may be twofold: (1) they may affect the distance over which the receiver can detect the sound, or (2) when superimposed on amplitude-modulated signals, they may increase the difficulty of identifying the signal. As with the development of temperature gradients, fluctuations and wind are greater during midday and less pronounced in the evening. The frequency at which the fluctuations modulate a tonal signal were limited to below 50 Hz, which would hamper the identification of an insect sound with amplitude modulation rates at these frequencies. With even slight wind of less than 10 km/h, amplitude fluctuations in the perceived signal increase dramatically with peak-to-peak values of 40 dB (Richards and Wiley 1980).

Reverberations take the form of miniature echoes following a sound pulse. In sound transmission experiments in a deciduous forest, reverberations depended on the carrier frequency of sound pulses, the presence or absence of foliage, and the directionality of the sender (Richards and Wiley 1980). At higher frequencies (>8 kHz) reverberations took the form of a lengthy period of decay after the sound pulse, because at these frequencies the size of possible reflectors for sound are in the same order of magnitude as the wavelength of sound. Most of the energy of the echoes was concentrated in the first 20 to 50 ms after the sound pulse; therefore, reverberations would interfere most strongly with acoustic signals that use intervals of song elements between 20 and 50 ms, resulting in a smearing of the fine temporal pattern at some distance from the source. Although no systematic study has been performed at higher or even ultrasonic frequencies, one might expect a more severe effect with increasing frequency.

Although there is no doubt about the importance of amplitude modulation for acoustic communication in insects, there are reported cases where females are attracted to temporally unpatterned song or to song in which

1 male

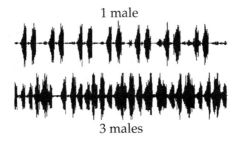

3 males

FIGURE 3.6. Microphone recording at the position of a silenced focal male *Tettigonia viridissima* when only one neighbor male was calling at a distance of 4.5 meters or when the three closest neighbors were calling at the same time. Note that the species-specific temporal pattern of the series of double syllables is largely obscured in the chorus situation.

the normal temporal sequence of song elements had been shuffled (Morris, Kerr, and Fullard 1978; Pollak and Hoy 1979). A female bushcricket *Conocephalus nigropleurum*, for example, oriented to random noise broadcast from a loudspeaker. The rationale behind such an experiment was that for a female from a position outside a chorus of many singing males, she will receive an unpatterned sound due to temporally random overlap of many amplitude-modulated songs (compare with Fig. 3.6). The only cue for the species specifity of the song would thus be the carrier frequency. In the case of *Conocephalus nigropleurum*, females prefer the amplitude-modulated song of a single male given a choice between it and an unpatterned song. Female bladder cicadas *Cystosoma saundersii* are also attracted to a chorus of many males calling from a single tree or bush (Doolan and MacNally 1981), although this chorus sounds from a distance like an unpatterned signal centered at the carrier frequency of about 900 Hz. When females approach a tree by flight and land somewhere within the chorus, they still have the problem of discerning a species-specific, amplitude-modulated pattern (see later). The fact that males differ in their mating success as a result of calling position and the time they occupy this position indicates that females within such a chorus do indeed discriminate among males (Doolan 1981; Doolan and MacNally 1981).

3.4. Sound Localization

In most biological situations, a receiver must be able to both identify and localize the sound source. However, the physical properties of the transmission channel complicate sound localization for a receiver in two different ways. Within vegetation, multiple reflections and scattering will turn the sound field around a listening insect into a diffuse sound field in which sound waves arrive with almost equal intensity from all directions. Moreover, when insects rely on the diffraction mechanism of sound around their body to establish interaural intensity differences, particularly at high frequencies, and these frequencies suffer from strong attenuation (see earlier), directional hearing at some distance from the source will be based on very poor directional cues. Physiological measurements, using directionally sensitive auditory interneurons in the field, indicate that in unfavorable receiver positions the directional cues of a sound signal may be completely lost (Rheinlaender and Römer 1986). Hence, information about the identity of the sender (the temporal structure) may be available but not its location. The receiver may compensate for this loss of information by seeking better positions in the habitat from which to hear. (For a more detailed account on directional hearing in insects, see Michelsen, Chapter 2).

We may expect that directional hearing will be under strong selection in nature, because failure or delays in locating a sound source will have energetic consequences for the searching insect as well as risks of exposure to predators (Heller 1992). It has been argued that unfavorable conditions for directional hearing in the habitat may have consequences for the evolution of neuronal networks capable of handling these difficulties in an appropriate way (Römer 1992).

4. Selection on Signaling and Hearing Caused by Masking Sounds: "Sound Windows" in Time and Space?

4.1. Interspecific Interference

Given the temporal variation in sound transmission properties during day and night outlined earlier, it is no surprise that acoustically communicating animals adapt the broadcast times of their signals to these conditions. Diurnal species, in particular, are more dependent on atmospheric conditions for sound propagation. They should avoid large temperature and wind gradients above the ground that give rise to shadow zones. Indeed, these animals tend to concentrate their singing activity either in the first hours of the day or in the evening. Well-known examples are the dawn choruses of birds and insects (Henwood and Fabrik 1979). Nocturnal animals, on the other hand, find more regular conditions for sound transmission and often call throughout the night. However, this constraint for long-range sound transmission is similar for many species of mammals, amphibians, and insects, and the result is a complex sonic and ultrasonic background over which these animals must communicate.

It is evident, then, that there is a trade-off between optimum broadcast conditions and the problem of detecting and discriminating its own signals in the noise produced by the other species. Measures of the range of communication, or of sound discrimination abilities, are therefore of little relevance without considering the level of masking noise in the environment. For example, in a Bornean mixed dipterocarp forest, the SPL increases considerably at dusk (Young 1981; Gogala and Riede 1995). This "dusk community" consists of a well-defined ensemble of cicada, cricket, and frog species, in which the first half-hour from 1800 to 1830 is dominated by cicadas, and the second half-hour by crickets (Grilloidea) and frogs (*Anura*). Furthermore, the signaling activity of a given cicada species exhibits a narrow temporal segregation in the range of minutes (Fig. 3.7).

Temporal segregation of calling activity may represent one solution to reduce song interference, but in situations with a high species diversity such as tropical rain forests, time sharing alone in mixed choruses of many individuals would probably not permit calling for long enough periods to establish sites and/or to attract mates. An additional solution is to partition calling frequency, thus allowing for "private channels" of communication, analogous to the separation of frequencies in the electromagnetic spectrum

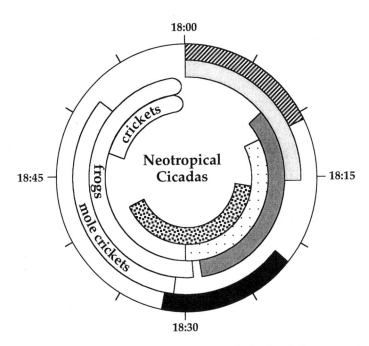

FIGURE 3.7. Temporal separation of calling time within the dusk community of a tropical rain forest (Kinabalu Park, Sabah, Malyasia) between 1800 and 1900 local time. Different cicada species are indicated by differently dotted and shaded areas. The singing activity of many cicada species is limited to a narrow daily time window. Most cicada species at this location sing only at dusk, but there are also species acoustically active at dawn, or within a species-specific time window during the day. (Modified from Gogala and Riede 1995, with permission.)

by radio or televesion stations. This is one mechanism suggested by Narins (1995) for a frog community in Puerto Rico.

Riede's recordings and analysis of chorus sounds of acoustic insects (mainly crickets) in a canopy area of a Bornean rain forest is consistent with this idea (Riede 1995). He analyzed sonograms of sound recordings in different frequency bands, and what appeared to be a mixture of sounds in the oscillogram before filtering turned out to be a well-separated temporal song pattern of an individual cricket male after processing the chorus sound. As in electronic devices, the degree (sharpness) of tuning in the ears of respective receivers determines how much of the noise of heterospecifics is filtered out. This mechanism is likely to function in those insects that concentrate all their sound energy in narrow frequency bands, as do many crickets, although even their ears are sensitive not only to the carrier frequency of the conspecific signal, but also to other, usually higher, frequencies as well (see Pollack, Chapter 5). It remains an interesting but untested hypothesis that under strong acoustic competition for broadcast frequencies in areas of high species diversity, the evolution of auditory systems resulted in more narrowly tuned receivers.

In those insects that use broad-band carrier frequencies in their calling song, substantial sensory interference takes place when several species communicate acoustically at the same time (Greenfield 1988, 1990). This interference is shown in the omega-neuron activity of a bushcricket (*Mygalopsis marki,*) recorded in its natural habitat at a distance of 6.5 meters to another singing conspecific male (Figure 3.8; Römer, Bailey, and Dadour 1989). A male of another bushcricket (*Hemisaga denticulata*) sang sporadically at a distance of 7.0 meters, and a male cricket (*Eurogryllodes* spp.) sang almost continuously only 2 meters from the preparation (Fig. 3.8A). The bursting spike activity of the omega neuron copies the repeated syllable pattern of the conspecific song, but there is also spike activity in response to the discrete chirps of the male *H. denticulata* (see arrows in Fig. 3.8) because the song spectra and hearing sensitivities of both species overlap. In the presence of additional high noise levels produced by the cricket (Fig. 3.8B), the conspecific temporal song pattern can still be detected in the nervous response because the ear of *M. marki* is quite insensitive to the 3.5-kHz carrier frequency of the cricket (compare with Fig. 3.8A), but again there is strong interference with the song of *H. denticulata*, causing a disruption of the clear bursting pattern of the neuron.

We should also expect behavioral consequences as a result of such sensory interference, when two or more species compete for acoustic communication channels. Figure 3.8C shows measurements of the singing activity of both bushcricket species in a bushland site between 1600 and 2200 hours. The number of calling males of *H. denticulata*, sampled at 30-minute intervals, remained almost constant between 1600 and 1900 hours. Thirty minutes later (at approximately 1930 hours), the number of

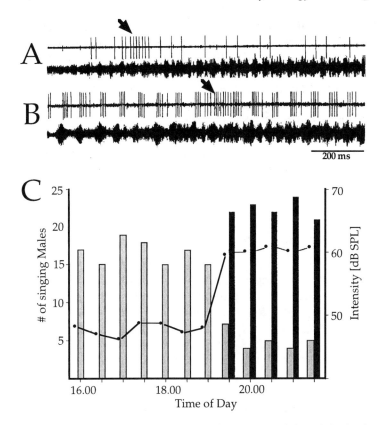

FIGURE 3.8. A,B: Field recording of the omega-neuron activity of the bushcricket *Mygalopsis marki* in a bushland habitat. A singing male of *M. marki* was replaced by the omega-preparation (upper trace, omega activity; lower trace, microphone recording of sound close to the preparation). Arrowheads point to bursts of neuronal activity evoked by the chirps of a male *Hemisaga denticulata*. The high-pass filter characteristic of the hearing organ of *M. marki* avoids interference with low-frequency sound produced by the cricket, but not with the broad-band carrier of another bushcricket. C: Number of calling males of *H. denticulata* (grey bars) and *M. marki* (black bars) and mean background noise intensity (dots and lines) between 1600 and 2200 hours in the same habitat. Note the inhibition of the singing activity of *H. denticulata* with the onset of singing activity of *M. marki*. When all singing males of *M. marki* were removed at about 2100 hours, *H. denticulata* males commenced singing within 15 minutes. (From Römer, Bailey, and Dadour 1989, with permission.)

singing males had declined to four or five insects. This decline coincided with the onset of singing activity of male *M. marki*, which commenced 15 minutes before sunset. The overall noise level in the habitat increased from 48 dB SPL to 60 dB SPL during this time (Römer, Bailey, and Dadour 1989).

Two lines of evidence suggest that the observed change of singing activity of *H. denticulata* is not due to some kind of diurnal periodicity, triggered by sunset, but rather is induced by the acoustic interaction of the other singing species. First, the number of singing *H. denticulata* at another site remained constant throughout the study period. No significant change in the singing activity occurred after sunset, and the overall noise level varied only between 40 and 45 dB SPL. Second, direct experimental evidence was obtained by manipulating song interference in such a way that all singing males of *M. marki* were removed from the population at 2100 hours. Only 15 minutes after the removal did the number of singing males of *H. denticulata* increase to the presunset level as a result of this manipulation (Römer, Bailey, and Dadour 1989).

4.2. Intraspecific Interference

Insects may also sing in dense populations of conspecifics, with many signalers within the hearing range of a receiver. The temporal overlap of several songs, arriving from different directions in space, may result in a contamination of individual temporal song patterns at the position of the receiver (see Fig. 3.6). In such a situation any estimation of the species-specific temporal properties, for example, syllable or chirp duration, or intersyllable intervals, may be very hard to accomplish. The question, therefore, is how many of the signals within hearing range of a receiver, if any, can be evaluated in such choruses.

Pollack (1988) and Römer (1992) described a gain-control mechanism in the afferent auditory pathway of crickets and bushcrickets that, in a competitive situation of several calling males, would always favor the representation of the loudest signal in the brain of a receiver (see also Pollack, Chapter 5). This mechanism enhances intensity differences of 2 to 5 dB so that the acoustic background is reduced. Thus there is a tendency in females for louder songs, which increases selection on the male signal or signaling behavior (Ryan and Keddy-Hector 1992; see Fig. 3.1). The evolution of signals results from selection to increase their effectiveness in changing the response of receivers. The gain-control mechanism suggests a "competition for representation in the CNS of a receiver," and if a male is unable to produce signals of reasonable loudness, or to establish and defend a preferred singing position, the representation of this signal in the auditory pathway of a receiver would be inhibited by those of other nearby males. The male would waste time and energy in calling, and probably do better if it adopted an alternative strategy, perhaps acting as a silent satellite male (Cade 1975).

There is also an implication for female behavior in the field because the gain-control mechanism restricts her hearing range to the one or two closest males, and in order to assess the quality of more males she would have to move throughout the population, thus increasing the risk of predation (Heller 1992). A better strategy may therefore be to move to the closest male. In fact, the majority of female bushcricket *T. viridissima* moved from their release sites to one of the closest singing males (Arak, Eiriksson, and Radesäter 1990). More distant males attracted females only when one of the closest males did not sing, or sang very little, during the period in which females were moving. This behavior is consistent with the properties of the gain-control mechanism found in the same species (Römer 1992).

5. Consequences for Acoustic Communication in Insects

5.1. Selection for Increased Loudness?

I have argued so far that acoustic signals, and/or the acoustic behavior of insects, are adaptive for long-range sound transmission. But is it really of any consequence for a male to call a little louder, or to broadcast his signal more efficiently than other males? We would expect that such males will attract a greater number of females simply because the active range of their signal is greater. Note that the decrease in SPL of a calling song at higher distances is rather flat (compare with Fig. 3.3C), and that a small increase in loudness (or efficiency in transmission) of a few decibels may therefore result in a relatively large increase in broadcast range.

Furthermore, we would expect selection to act on males to produce or broadcast more and more intense signals until the benefit of increased loudness is balanced by the costs of producing the signals, whatever these costs are (see Fig. 3.1). Differential attraction of females to louder calling songs is known for insects, and also for frogs and toads (review by Ryan and Keddy-Hector 1992). Figure 3.9 summarizes results from one of the few field studies, measuring the attraction of flying male and female mole crickets (*Scapteriscus acletus*), to calling males in outdoor arenas (Forrest and Green 1991).

In this study, male mole crickets were kept in outdoor cages and the number of females and males each male was able to attract was sampled. Bars in Figure 3.9 represent the number of individuals attracted per male, as a function of the male's SPL relative to the loudest male, which was set to 0 dB. Clearly the number attracted per male was influenced by the SPL of the song relative to others calling in the arena. Males calling more than 2 dB below the loudest male attracted fewer individuals per male than the overall average (dashed horizontal line). Comparable results have been obtained in many laboratory-based choice experiments.

It has been argued from studies of sound transmission in terrestrial habitats that the maximum range of detection probably is not the primary selection pressure on animal vocalizations (Wiley and Richards 1978). However, virtually all studies performed on acoustic insects support the

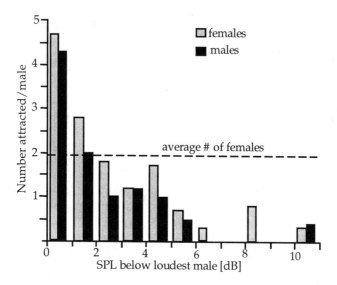

FIGURE 3.9. Attraction of male and female mole crickets *Scapteriscus acletus* (Gryllotalpidae) to calling males in outdoor arenas. Bars represent the number of individuals attracted per male as a function of the males' SPL below the loudest male (0 dB) calling on the same night. The horizontal dashed line is the overall average number of females attracted per male. Note that males calling more than 2 dB below the loudest male attracted fewer individuals than the overall average. (After Forrest and Green 1991, with permission.)

idea that there is strong selection on the signals or signaling behavior to maximize broadcast range. Fruitflies (Partridge, Hoffmann, and Jones 1987), mole crickets (Forrest 1983), crickets (Shuvalov and Popov 1973; Thorsen, Weber, and Huber 1982), and bushcrickets (Latimer and Sippel 1987; Bailey and Yeoh 1988) preferentially approach the louder of two conspecific signals of different intensity. To achieve an increased sound output, selection has favored the use of resonators, amplifying burrows, and baffles.

The mole cricket *Scapteriscus acletus*, for example, uses his underground burrow to increase sound output. By matching the dimensions of the burrow exactly to the carrier frequency of his call, a male can increase the call intensity by 18 dB and significantly increase its broadcast range (Bennet-Clark 1987). The independent development of an abdominal resonator in males of the bladder cicada *Cystosoma saundersii* (Young and Hill 1977) and males of the primitive Acridid family Pneumoridae (also called bladder grasshoppers; van Staaden and Römer 1997) are two other examples for selection on increased loudness. The South African tree cricket *Oecanthus burmeisteri* chews a hole in a leaf and stridulates with its forewings and body filling this space. By using such a self-made acoustic baffle, the insect increases the efficiency of sound production by 10 dB (Prozesky-Schulze et al. 1975).

5.2. Redundancy: A Mechanism for Reliable Information Transfer or a Result of Intraspecific Competition?

An improvement in the receiver's performance to detect and/or to discriminate a signal can be achieved by repeating an identical signal many times. This is particularly helpful with a noisy communication channel, or with communication between widely separated individuals. Rheinlaender and Römer (1986) and Römer and Lewald (1992) used the response of an auditory interneuron in the field to quantify the net effect of amplitude fluctuations and reverberations on the perception of conspecific song patterns in a bushcricket. Their data show that the temporal song pattern is represented in the central nervous system of a receiver with remarkable accuracy at distances well beyond the nearest neighbor distance. This is despite the large-amplitude fluctuations superimposed on the broadcast signal that cause the variability of the neuronal response at some distance from the source (Fig. 3.10B; see also Richards and Wiley 1980).

There are two features of the calling song of the investigated insect that counteract these fluctuations and improve the perception of the temporal pattern. First, like many other tettigoniids, *T. viridissima* does not use pure tones as a carrier, but rather a broad-frequency spectrum. This reduces the variability in the received signal to a large extent (Fig. 3.10A,C). Morton (1975) investigated sound propagation in different habitats and correlated his findings with the frequencies of bird song in these habitats. The data on the perception of bushcricket song confirm Morton's suggestion that amplitude-modulated song patterns with broad-band carriers are more effective for long-range communication than pure tones, at least in open habitats. Second, this particular bushcricket produces a highly redundant signal, in which simple phonatomes are repeated at a high rate, often many hours per day or night. These highly stereotyped repetitions allow the receiving insect (and the investigating zoologist in the field!) to predict the entire signal when part of it is lost on the transmission channel due to signal degradation.

However, to explain signal redundancy purely on the basis of reliable information transfer would exclude other possibilities as to why highly repetitive signals might have evolved. For example, almost continuous male calling patterns could also result from intraspecific competition for representation in the central nervous system of females. These females possess filter mechanisms for the most intensely singing male, but due to the extremely long time constant of the inhibition responsible for the filter, males producing only short bouts of singing would activate the auditory pathway of a female significantly less than a continuous signaler (Pollack 1988;

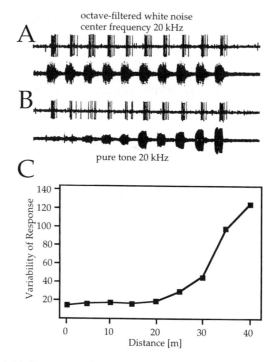

FIGURE 3.10. A,B: Two examples of simultaneous recordings of the microphone signal (lower traces) and the response of an auditory interneuron in the field (upper traces) when the broadcast signal was either a train of pure tones (B; carrier frequency 20 kHz) or a train of octave-band filtered white noise (A; center frequency 20 kHz). Recordings were made with the same preparation and separated less than 3 minutes in time. Distance to the speaker, 20 meters; elevation from the ground, 2 meters. The microphone was placed 10 centimeters away from the preparation. The broadcast intensity was adjusted to 20 dB above the threshold of the neuron for both types of signals. Note that the variability in both the recorded microphone signal and the neuronal responses between different pulse trains is much higher with a pure tone than with the noisy sound pulses. C: Variability of the neuronal response at increasing distance from a speaker broadcasting a conspecific calling song at a constant intensity of 94 dB SPL (height above ground for sender and receiver, 2 meters). The coefficient of variation was calculated from a total of 300 consecutive responses to individual syllables in the song. Note the reduced variability in the representation of the song up to a distance of about 25 meters. (From Römer and Lewald 1992, with permission.)

Römer 1992). These and other features of the circuits underlying the auditory pathway must not necessarily represent features adaptive for acoustic communication. In order to understand why the auditory pathway of insects is organized as it is, we also have to consider its history, because the nervous system has evolutionarily conservative, nonadaptive features that may con-strain an underlying behavior much in the same way as other constraints (see Fig. 3.1; Dumont and Robertson 1986).

We should keep in mind, however, that there are also insect species producing signals at extremely low rates, for example, as a result of selection pressure by acoustically orienting predators (Belwood and Morris 1987). We can only speculate as to how these insects cope with degradation without the advantage of processing redundant signals. Given that other parameters such as SPL and hearing sensitivity are similar, we might predict that species lacking redundant signals would maintain smaller interindividual distances. An informal survey of the genus *Neocono-cephalus* and other species of Tettigoniidae indicates that indeed discontinuously (less redundant) singing species experience high-density populations more frequently than do continuously singing species (Greenfield 1990).

Furthermore, different types of communication systems between male and female, such as in phaneropterine bushcrickets, may cope with signal degradation in quite a different way. Here, pair formation is achieved by duetting; the male elicits an acoustic reply in the female to which the male then responds by phonotaxis (Zhantiev and Dubrovin 1977; Heller and von Helversen 1986; Robinson, Rheinlaender, and Hartley 1986). The female reply is extremely short, sometimes less than 1 ms, leaving the male with the problem of identifying a click of 0.5-ms duration as species specific, although it cannot contain species-specific amplitude modulations. However, the female responds to the call of the male after a very short delay time, which is species specific and remarkably constant for each individual female (Heller and von Helversen 1986; Robinson, Rheinlaender, and Hartley 1986). This time delay of the female could be used by the male as a temporal feature for recognition.

Indeed, experimental variation of the time delay revealed that the female response must occur within a certain time window in order to elicit phonotaxis by the male (Fig. 3.11A). The time window is also species specific and matches the species-specific female delay time (Heller and von Helversen 1986; Robinson, Rheinlaender, and Hartley 1986). It has been argued that using extremely brief, low information signals, combined with the narrow time window of the male and the corresponding delay time of the female, may be adaptive in an environment in which the temporal degradation of a signal is high. By listening for only a short time period for a rather unspecific female signal, the chance of confusion caused by random events acting on the transmission channel is greatly reduced. Species identification can be achieved with a signal that otherwise offers little chance of identification.

We know little either about the ability of receivers to resolve finer temporal details in acoustic signals (such as the tooth impact rate or small gaps on the order of a few milliseconds; but see Ronacher and Römer 1985) or about the behavioral relevance of this information. For example, in the

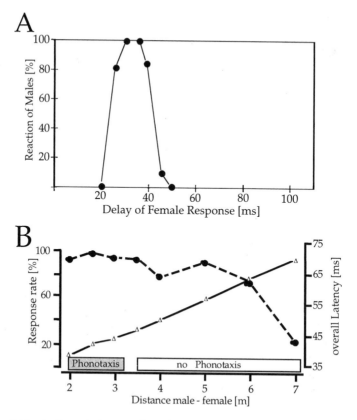

FIGURE 3.11. A: Reaction curve (based on successful approaches to a speaker) of male phaneropterine bushcrickets, *Leptophyes punctatissima*, to a female signal presented at different time delays. Note that males perform phonotaxis only when the delay of the female falls within a critical, narrow time window between 25 and 50 ms. B: Sound parameters of a duetting pair of *Leptophyes punctatissima* at different distances. The horizontal bars indicate when the male performed phonotaxis. Note the linear increase in the overall delay time of the female reply (time for the male sound signal to travel from male to female, and for the female signal back to the male, plus the net response delay of the female, which is about 27 ms). Neither the singing rate of the male (not shown in the figure) nor the response rate of the female constitute limiting factors for the distance over which the male performed phonotaxis. (Modified from Robinson, Rheinlaender, and Hartley 1986; after Zimmermann, Rheinlaender, and Robinson 1989, with permission.)

grasshopper *Chorthippus biguttulus* the gross temporal pattern of male and female song is the same, but the pulse shape differs in the rise time between the sexes. Sawtooth-shaped pulses are the critical feature for the recognition of female song by the male (von Helversen 1993). Interestingly, although pulse duration, pulse interval, and the steepness of pulses do change for a receiver at greater distances, this does not affect the male's response, because steepness below a critical value appears to be important for recognition. It is expected, however, that fine temporal elements such as gaps between syllables will be the first to disappear as the signal passes through the habitat, primarily as a result of reverberations and scattering. Hence, this information may only have relevance for communication over short distances, or alternatively, may represent a cue for estimating the distance to the sender (Simmons 1988).

5.3. Male Signaling Sites and Spacing

Given that size restricts the generation of the signal, forcing most insects to use higher frequencies, and the effect of EA, one consequence for the evolution of acoustic signaling should be that the animal makes use of optimal spatial positions in its environment for signaling. In fact, behavioral adaptations for efficient sound transmission may represent the only mechanism permitting long-range sound communication (Paul and Walker 1979; Doolan and MacNally 1981; Gwynne and Edwards 1986; Römer and Bailey 1986). Males often sing from higher elevations or occupy perches on plants that are higher than the surrounding vegetation. Excess attenuation as a result of scattering within the vegetation decreases markedly with increasing height above the ground. Thus, the range of detection for the fundamental frequency of the male *T. viridissima* song was estimated at about 60 meters for males singing >1 meter above the surrounding vegetation, but only 8 meters for males singing in the middle of dense reed beds (Arak and Eiriksson 1992).

By recording the neurophysiological response of an auditory interneuron in the field, Römer and Bailey (1986) directly determined the broadcast areas of male bushcrickets *Mygalopsis marki*. Broadcast area was defined as the area within which the SPL of a male call exceeded the auditory threshold of a receiver. Males singing on top of homogeneous vegetation doubled their broadcast areas in relation to those singing close to the ground, or within dense vegetation. Similarly, male cricket *Anurogryllus arboreus* that call from an elevation of about 1 meter on tree trunks may increase their broadcast areas 14 times compared with males singing on the ground (Paul and Walker 1979). However, in contrast to the two bushcricket examples, these crickets are more likely to avoid strong ground attenuation rather than excess attenuation due to scattering. The best-documented example of active choice in selecting calling sites that correlate with mating success is in the bladder cicada, *Cystosoma saundersii*, in which males call for a short period of 45 minutes just after dusk (Doolan 1981; Doolan and MacNally 1981). They usually choose positions close to the top of medium-sized bushes, and field observations established that perch height are important for mating success: males that called from between 1.3 and 1.7 meters from the ground obtained more matings than those from higher or lower positions, and those that stayed for a longer period of time at these preferred positions gained more matings than those that stayed for a short period (Doolan 1981).

Careful examination of broadcast heights of singing males may reveal, however, that males establish positions that are less than ideal for maximal sound transmission. Male *T. viridissima* could have increased the detection range of their signal almost threefold by singing from even higher positions than those observed in the field (Dadour and Bailey 1985; Arak and Eiriksson 1992). This probably reflects the common situation of a trade-off between the costs of singing at higher, exposed elevations (risk of predation by visually hunting predators during the day, or by gleaning bats at night, and/or aggression from conspecific neighbors) and the increased detection range of the signal. Therefore, less than ideal acoustic behavior of insects with respect to ecological constraints usually reflects the complex situation outlined in Figure 3.1, in which the observed behavior or the structure of the signal may be a compromise between various selection pressures.

It is often found that signaling males maintain a certain distance from one another, either as a result of physical interactions or by using acoustic cues (Campbell and Clark 1971; Thiele and Bailey 1981; Schatral, Latimer, and Broughton 1984; Römer and Bailey 1986; Arak, Eiriksson, and Radesäter 1990). A regular distribution of males within such aggregations may result from attraction to distant males and repulsion from males calling near by. One simple interpretation of such spacing behavior is that it allows each male in a chorus to broadcast its signal within a zone free from competing conspecifics. By manipulating the degree of aggregations within clumps of calling bushcrickets, Arak, Eiriksson, and Radesäter (1990) demonstrated that regular spacing increases the ability of males to attract females.

6. Hearing Distances, Communication Distances, Response Distances: What Are the Limitations?

The distance over which a receiver can detect the presence of a signal depends on the broadcast SPL of the signal, the rate of signal attenuation and distortion on the transmission channel, the level of ambient noise, and the sensitivity of the receiver for the signal. However, if we define communication as a change in behavior of a receiver as a result of perceiving the signal from a sender, then maximum hearing distances may not correlate at all with communication distances. This is because behavioral thresholds to sound signals are generally higher than neurophysiologically determined hearing thresholds.

One extreme example is the communication system of the bladder grasshopper *Bullacris membracioides* (compare with Fig. 3.3C), in which neurophysiological hearing thresholds indicate that the conspicuously loud male call can be detected by a female at distances between 1.4 and 1.9 kilometers! The female, however, neither responds to the male call with an acoustic reply nor shows any other kind of behavioral response, until the male approaches to within 60 to 100 meters, by randomly searching for females. This distance corresponds to a SPL of the male call of about 60 dB, which is about 25 to 30 dB above the hearing threshold of the female, and decreases the effective communication distance 20 times relative to the hearing distance (Römer and van Staaden 1996).

Interestingly, this is not true for the male listening to the female reply, which is about 40 dB softer than the male call. The hearing threshold of the male for the female reply in the field is between 60 and 80 meters, and this is also the behavioral threshold for the male to perform flight phonotaxis to the female. Thus hearing and communication distances are the same for the female signal, but not so for the male signal.

A similar sexual dimorphism in loudness of male and female signals is found in many phaneropterine bushcrickets, but another interesting feature may represent a further limitation for communication distance. A male *Leptophyes punctatissima* only performs phonotaxis to the responding female if her reply falls within a critical "time window" of 25 to 50 ms after the onset of its own song (Fig. 3.11A; Robinson, Rheinlaender, and Hartley 1986). Therefore the traveling time of sound through air (about 3 ms/m) is a significant fraction of the delay time of the female's reply as perceived by the male. Figure 3.11B shows the linear increase of overall delay of the female reply with distance for one duetting pair. The male performed phonotaxis up to a distance of 3 meters; a further increase by 0.5 m resulted in complete cessation of phonotaxis (Zimmermann, Rheinlaender, and Robinson 1989). Because the male calling rate remained at a high level at all distances and the female response rate was also fairly constant up to 6 meters (Fig. 3.11B), there are only two possible limiting factors for communication distance: the increase in overall delay from 44 to 47 ms, with an increase from 3 meters to 3.5 meters (arrow in Fig. 3.11B), or the decrease in SPL of the female reply from 52 to 50 dB SPL at the males' position.

In a series of experiments, Zimmermann, Rheinlaender, and Robinson (1989) established that limitations for maximum phonotaxis distance were due to either both delay time *and* intensity or intensity alone but never to the delay time alone. The authors pointed out that, unlike the situation in the soundproof room where these experiments were performed, in the real habitat of the insects the high-frequency song would suffer strong EA, and therefore the perceived intensity of the female is more likely to form the only limiting factor for male phonotaxis. Note, however, that communication distance for the female, as indicated by her acoustic reply, is between 6 and 7 meters, whereas for the male it is only 3 meters.

7. Conclusions

The sensory ecology of hearing in insects is still in its infancy, and many of the predictions on sound propagation outdoors are based on measurements in the laboratory with a steady, homogeneous, and isotropic atmosphere. Insects (and small vertebrates as well) are particularly vulnerable to environmental variables, such as temperature, humidity, radiation, and wind, because of their small size and relatively large surface area. They can compensate by exploiting microniches with favorable conditions (Willmer 1982), which in turn largely determine the conditions under which signaling and hearing must take place. However, these macroclimatic and microclimatic conditions and their effect on sound propagation are little understood, and there is not a single case study that demonstrates their influence on sound communication in a quantitative manner. The rich diversity of insect acoustic communication systems may indeed, to some extent, be a secondary result of exploiting finely graded microclimatic habitats.

Compared with vertebrates, insects explore a wide range of signal carrier frequencies, from below 1 kHz to more than 100 kHz. Consequently, the amount of excess attenuation, and thus the range of communication, differs largely between species and habitat. However, because the same physical rules hold for long-range sound transmission in both insects and vertebrates, future field studies in any given habitat should reveal some similarity in signal structure or acoustic behavior across taxa. Differences between insects and vertebrates, such as the use of amplitude versus frequency modulation for reliable long-range sound transmission, are likely to be constrained by the different mechanisms of vocalization and stridulation, rather than different properties of their environments. This review, therefore, emphasizes comparative field measurements and an organismic approach to animal communication systems, in which behavioral, biophysical, physiological, phylogenetic, and environmental approaches are combined to explain proximate and ultimate levels of communication.

8. Summary

The environment plays a crucial role for the evolution of sensory systems. In this chapter the role of both physical and biotic components for insect signaling and hearing is examined under field conditions and placed in the context of the possible selection pressures that may have shaped the character of the signals and the different types of acoustic communication systems. The high sonic or ultrasonic frequencies of the signals result in strong frequency filtering and attenuation of the signal in excess of the spherical spread of sound, especially when insects call within stands of vegetation.

Behavioral mechanisms exist to compensate for the signal loss by choosing appropriate broadcast positions and atmospheric conditions and the best time for broadcasting the signal. Distortions of acoustic signals in the time domain may be severe, but using broad-band, highly stereotyped and redundant signals, the species-specific temporal pattern can be received over remarkable distances. Frequency-dependent scattering and redirection of sound waves also limits the locatability of insect songs in the field. Masking of the conspecific temporal pattern as a result of communicating in aggregations of several conspecific and heterospecific individuals is described, as well as behavioral and neurophysiological mechanisms to avoid this masking.

Acknowledgments. Research on environmental effects on communication systems has been initiated in collaboration with J. Rheinlaender, and later W.J. Bailey, who focused my interest on evolutionary aspects of acoustic signaling and hearing. I also thank M. van Staaden, G.K. Morris, C. Gerhardt, I. Dadour, and D. Gwynne for numerous discussions during our field work. Research for this article was supported by grants from the Deutsche Forschungsgemeinschaft (Ro 728/2-1 and Heisenberg-Programm) and the Austrian Science Foundation (P09523-BIO).

References

Arak A, Eiriksson T (1992) Choice of singing sites by male bushcrickets (*Tettigonia viridissima*) in relation to signal propagation. Behav Ecol Sociobiol 30:365–372.

Arak A, Eiriksson T, Radesäter T (1990) The adaptive significance of acoustic spacing in male bushcrickets *Tettigonia viridissima*: a perturbation experiment. Behav Ecol Sociobiol 26:1–7.

Bailey WJ (1985) Acoustic cues for female choice in bushcrickets (Tettigoniidae). In: Kalmring K, Elsner N (eds) Acoustic and Vibrational Communication in Insects. Paul Parey, pp. 107–111.

Bailey WJ, Yeoh PB (1988) Female phonotaxis and frequency discrimination in the bushcricket *Requina verticalis*. Physiol Entomol 13:363–372.

Bailey WJ, Cunningham RC, Lebel L, Weatherilt C (1990) Song power, spectral distribution and female phonotaxis in the bushcricket *Requena verticalis* (Tettigoniidae, Orthoptera): active female choice or passive attraction. Anim Behav 40:33–42.

Belwood J, Morris GK (1987) Bat predation and its influence on calling behavior in neotropical katydids. Science 238:64–67.

Bennet-Clark HC (1987) The tuned singing burrow of mole crickets. J Exp Biol 128:383–409.

Cade WH (1975) Acoustically orienting parasitoids: fly phonotaxis to cricket song. Science 190:1312–1313.

Cade WH (1981) Field cricket spacing, and the phonotaxis of crickets and parasitoid flies to clumped and isolated cricket songs. Z Tierpsychol 55:365–375.

Campbell DJ, Clark DJ (1971) Nearest neighbour tests of significance for non-randomness in the distribution of singing crickets [*Teleogryllus commodus* (Walker)]. Anim Behav 19:750–756.

Canard-Coruna S, Lewy S, Vermorel J, Parmentier G (1990) Long range sound propagation near the ground. Noise Control Eng 34:111–119.

Dadour IR, Bailey WJ (1985) Male agonistic behavior of the bushcricket *Mygalopsis marki* Bailey in response to conspecific song (Orthoptera: Tettigoniidae). Z Tierpsychol 70:320–330.

Doherty J, Hoy RR (1985) Communication in insects: III. The auditory behavior of crickets: some views of genetic coupling, song recognition, and predator detection. Q Rev Biol 60:457–472.

Doolan JM (1981) Male spacing and the influence of female courtship behavior in the bladder cicada *Cystosoma saundersii* Westwood. Behav Ecol Sociobiol 9:269–276.

Doolan JM, MacNally RC (1981) Spatial dynamics and breeding ecology in the cicada *Cystosoma saundersii*: the interaction between distributions of resources and intraspecific behavior. J Anim Ecol 50:925–940.

Dumont JPC, Robertson RM (1986) Neuronal circuits: an evolutionary perspective. Science 233:849–853.

Elsner N, Popov AV (1978) Neuroethology of acoustic communication. Adv Insect Physiol 13:229–355.

Embleton TFW (1996) Tutorial on sound propagation outdoors. J Acoust Soc Am 100:31–48.

Embleton TFW, Piercy JE, Olson N (1976) Outdoor sound propagation over ground of finite impedance. J Acoust Soc Am 59:267–277.

Endler JA (1992) Signals, signal conditions, and the direction of evolution. Am Nat 139:125–153.

Forrest TG (1983) Calling songs and mate choice in mole crickets. In: Gwynne DT, Morris GK (eds) Orthopteran Mating Systems: Sexual Competition in a Diverse Group of Insects. Boulder, CO: Westview Press, pp. 185–304.

Forrest TG (1994) From sender to receiver: propagation and environmental effects on acoustic signals. Am Zool 34:644–654.

Forrest TG, Green DM (1991) Sexual selection and female choice in mole crickets (Scapteriscus: Gryllotalpidae): modelling the effects of intensity and male spacing. Bioacoustics 3:93–109.

Garstang M, Larom D, Raspet R, Lindeque M (1995) Atmospheric controls on elephant communication. J Exp Biol 198:939–951.

Gogala M, Riede K (1995) Time sharing of song activity by cicadas in Temengor Forest Reserve, Hulu Perak, and Sabah, Malaysia. Malay Nat J 48:297–305.

Greenfield MD (1988) Interspecific acoustic interactions among katydids (*Neoconocephalus*): inhibition-induced shifts in diel periodicity. Anim Behav 36:684–695.

Greenfield MD (1990) Evolution of acoustic communication in the genus *Neoconocephalus*: discontinuous songs, synchrony, and interspecific interactions. In: Bailey WJ, Rentz DCF (eds) The Tettigoniidae: Biology, Systematics and Evolution. Bathurst: Crawford House Press, pp. 71–98.

Griffin DR (1971) The importance of atmospheric attenuation for the echolocation of bats (Chiroptera). Anim Behav 19:55–61.

Gwynne DT, Edwards ED (1986) Ultrasound production by genital stridulation in *Syntonarcha iriastis* (Lepidoptera, Pyralidae): long-distance signalling by male moths? Zool J Linn Soc 88:363–376.

Harris CM (1966) Absorption of sound in air versus humidity and temperature. J Acoust Soc Am 40:148–159.

Heller KG (1992) Risk shift between males and females in the pair-forming behavior of bushcrickets. Naturwissenschaften 79:89–91.

Heller KG, Helversen D von (1986) Acoustic communication in phaneropterid bushcrickets: species-specific delay of female stridulatory response and matching male sensory time window. Behav Ecol Sociobiol 18:189–198.

Helversen D von, Helversen O von (1983) Species recognition and acoustic localization in acridid grasshoppers. A behavioral approach. In: Huber F, Markl H (eds) Neuroethology and Behavioral Physiology. Berlin: Springer-Verlag, pp. 95–102.

Helversen D von (1993) "Absolute steepness" of ramps as an essential cue for auditory pattern recognition by a grasshopper (Orthoptera; Acrididae; *Chorthippus biguttulus* L.). J Comp Physiol A 172:633–639.

Henwood K, Fabrik A (1979) A quantitative analysis of the dawn chorus: temporal selection for communicatory optimization. Am Nat 114:260–274.

Hoy RR (1992) The evolution of hearing in insects as an adaptation to predation from bats. In: Webster DB, Fay RR, Popper AN (eds) The Evolutionary Biology of Hearing. New York: Springer-Verlag, pp. 115–129.

Keuper A, Kühne R (1983) The acoustic behavior of the bushcricket *Tettigonia cantans*. II. Transmission of airborne sound and vibration signals in the biotope. Behav Processes 8:125–145.

Lakes-Harlan R, Heller KG (1992) Ultrasound sensitive ears in a parasitoid fly. Naturwissenschaften 79:224–226.

Lang F (1996) Grasshopper habitats: sound attenuation and acoustic communication distance. In: Elsner N, Schnitzler HU (eds) Proceedings of the 24th Göttingen Neurobiology Conference, Stuttgart. New York: Thieme-Verlag, p. 160.

Langbauer WR, Payne KB, Charif RA, Rapaport L, Osborn F (1991) African elephants respond to distant playbacks of low-frequency conspecific calls. J Exp Biol 157:35–46.

Latimer W, Lewis DB (1986) Song harmonic content as a parameter determining acoustic orientation behavior in the cricket *Teleogryllus oceanicus* (Le Guillou). J Comp Physiol A 158:583–591.

Latimer W, Sippel M (1987) Acoustic cues for female choice and male competition in *Tettigonia cantans*. Anim Behav 35:887–910.

Lighthill MJ (1953) On the energy scattered from the interaction of turbulence with sound or shock wave. Proc Cambridge Soc 49:531–551.

Marten K, Marler P (1977) Sound transmission and its significance for animal vocalizations. I. Temperate habitats. Behav Ecol Sociobiol 2:271–290.

Meister F-J, Ruhrberg W (1959) Der Einfluss von Grünanlagen auf die Ausbreitung von Geräuschen. Lärmbekämpfung 1:5–11.

Michelsen A (1978) Sound reception in different environments. In: Ali MA (ed) Sensory Ecology. New York: Plenum Press, pp. 345–373.

Michelsen A, Larsen ON (1983) Strategies for acoustic communication in complex environments. In: Huber F, Markl H (eds) Neuroethology and Behavioral Physiology. Berlin: Springer-Verlag, pp. 322–332.

Michelsen A, Fink F, Gogala M, Traue D (1982) Plants as transmission channels for insect vibrational songs. Behav Ecol Sociobiol 11:269–281.

Morris GK, Kerr GE, Fullard JH (1978) Phonotactic preferences of female meadow katydid (Orthoptera: Tettigoniidae: *Conocephalus nigropleurum*). Can J Zool 56:1479–1487.

Morton ES (1975) Ecological sources of selection on avian sounds. Am Nat 108:17–34.

Narins P (1995) Frog communication. Sci Amer 273:62–67.

Neuweiler G (1989) Foraging ecology and audition in echolocating bats. Trends Ecol Evol 4:160–166.

Parker GA (1983) Mate quality and mating decisions. In: Bateson P (ed) Mate Choice. Cambridge, UK: Cambridge University Press, pp. 141–164.

Partridge L, Hoffmann A, Jones JS (1987) Male size and mating success in *Drosophila melanogaster* and *D. pseudoobscura* under field conditions. Anim Behav 35:468–476.

Paul RC, Walker TJ (1979) Arboreal singing in a burrowing cricket, *Anurogryllus arboreus*. J Comp Physiol 132:217–233.

Piercy JE, Embleton TFW, Sutherland LC (1977) Review of noise propagation in the atmosphere. J Acoust Soc Am 61:1402–1418.

Pollack GS (1988) Selective attention in an insect auditory neuron. J Neurosci 8:2635–2639.

Pollack GS, Hoy RR (1979) Temporal pattern as a cue for species-specific calling song recognition in crickets. Science 204:429–432.

Popov AV, Shuvalov VF, Svetlogorskaya ID, Markovich AM (1972) Acoustic behavior and auditory system of insects. Rhein-Westf Akad Wiss 53:281–306.

Prozesky-Schulze L, Prozesky OPM, Anderson F, van der Merve GJJ (1975) Use of a self-made sound baffle by a tree cricket. Nature 255:142–143.

Rheinlaender J, Römer H (1986) Insect hearing in the field. I. The use of identified nerve cells as "biological microphones." J Comp Physiol A 158:647–651.

Riede K (1995) Diversity of sound producing insects of a Bornean lowland rain forest. Proceedings of the International Conference on Tropical Rainforest Research, Brunei, Darussalam.

Richards DG, Wiley RH (1980) Reverberations and amplitude fluctuations in the propagation of sound in a forest: implications for animal communication. Am Nat 115:381–399.

Robinson D, Rheinlaender J, Hartley JC (1986) Temporal parameters of male-female sound communication in *Leptophyes punctatissima*. Physiol Entomol 11:317–323.

Römer H (1992) Ecological constraints for the evolution of hearing and sound communication in insects. In: Webster DB, Fay RR, Popper AN (eds) The Evolutionary Biology of Hearing. New York: Springer-Verlag, pp. 79–93.

Römer H (1993) Environmental and biological constraints for the evolution of long-range signalling and hearing in acoustic insects. Trans R Soc Lond [B] 226:179–185.

Römer H, Bailey WJ (1986) Insect hearing in the field. II. Male spacing behavior and correlated acoustic cues in the bushcricket *Mygalopsis marki*. J Comp Physiol 159:627–638.

Römer H, Lewald J (1992) High-frequency sound transmission in natural habitats: implications for the evolution of insect acoustic communication. Behav Ecol Sociobiol 29:437–444.

Römer H, Bailey WJ, Dadour I (1989) Insect hearing in the field. III. Masking by noise. J Comp Physiol 164:609–620.

Ronacher B, Römer H (1985) Spike synchronization of tympanic receptor fibres in a grasshopper (*Chorthippus biguttulus* L., Acrididae). A possible mechanism for detection of short gaps in model songs. J Comp Physiol 157:631–642.

Ryan MJ, Keddy-Hector A (1992) Directional pattern of female mate choice and the role of sensory biases. Am Nat 139:S4–S35.

Schatral A, Latimer W, Broughton B (1984) Spatial dispersion and agonistic contacts of male bushcrickets in the biotope. Z Tierpsych 65:204–214.

Shuvalov VF, Popov AV (1973) Significance of some of the parameters of the calling songs of male crickets *Gryllus bimaculatus* for phonotaxis of females [in Russian]. J Evol Biochem Physiol 9:177–182.

Simmons LW (1988) The calling song of the field cricket, *Gryllus bimaculatus* (De Geer): constraints on transmission and its role in intermale competition and female choice. Anim Behav 36:380–394.

Staaden van MJ, Römer H (1997) Sexual signalling in bladder grashoppers: tactical design for maximizing calling range. J Exp Biol 200:2597–2608.

Thiele DR, Bailey WJ (1980) The function of sound in male spacing behavior of buschcrickets (Tettigoniidae: Orthoptera). Aust J Ecol 5:275–286.

Thorson J, Weber T, Huber F (1982) Auditory behavior of the cricket. II. Simplicity of calling song recognition in *Cryllus*, and anomalous phonotaxis at abnormal carrier frequencies. J Comp Physiol 146:361–376.

Werner A, Elsner N (1995) Directional hearing of the grasshopper *Chorthippus biguttulus* (L.). II. A biological microphone. In: Elsner N, Menzel R (eds). Proceedings of the 23rd Göttingen Neurobiology Conference, Stuttgart. New York: Thieme-Verlag, p. 277.

West-Eberhard MJ (1984) Sexual selection, competitive communication and species-specific signals in insects. In: Lewis T (ed) Insect Communication. London: Academic Press, pp. 283–324.

Wiley RH, Richards DG (1978) Physical constraints on acoustic communication in the atmosphere: implications for the evolution of animal vocalizations. Behav Ecol Sociobiol 3:69–94.

Wiley RH, Richards DG (1982) Adaptations for acoustic communication in birds: sound transmission and signal detection. In: Kroodsma DE, Miller EH, Quellet H (eds) Acoustic Communication in Birds. New York: Academic Press, pp. 131–181.

Willmer PG (1982) Microclimate and the environmental physiology of insects. In: Berridge MJ, Treherne JE, Wigglesworth VB (eds) Advances in Insect Physiology. London: Academic Press, 16: pp 1–57.

Young AM (1981) Temporal selection for communicatory optimisation: the dawn-dusk chorus as an adaptation in tropical cicadas. Am Nat 117:826–829.

Young D, Hill KG (1977) Structure and function of the auditory system of the cicada, *Cystosoma saundersii*. J Comp Physiol 117:23–45.

Zhantiev RD, Dubrovin NN (1977) Sound communication in the genus *Isophya* (Orthopters, Tettigoniidae) [in Russian]. Zool Zurnal 56:40–51.

Zimmermann U, Rheinlaender J, Robinson D (1989) Cues for male phonotaxis in the duetting bushcricket *Leptophyes punctatissima*. J Comp Physiol A 164:621–628.

4
Development of the Insect Auditory System

George S. Boyan

1. Introduction

The insect auditory system is the product of two historical processes. The first is the evolutionary history of the species, evident in the pattern of gene expression that directs neurogenesis, growth, and connectivity. The second is the developmental history of the individual animal in which differential gene expression, environmental interactions, and processes such as activity-dependent competition between neurons for synaptic sites all combine to produce interindividual variation in neuronal structure and function. Over the past decade there have been impressive advances in our understanding of the cellular and molecular mechanisms of some of these developmental processes (for reviews, Campos-Ortega and Knust 1990; Grenningloh et al. 1990; Goodman and Doe 1994). Furthermore, our ability to determine a neuron's lineage from an identified precursor cell in the early embryo means that the basic organization of the insect auditory system can now be understood ontogenetically and is one of the reasons why this system is being studied by developmental neurobiologists.

In this chapter the structure and development of the tympanic (auditory) system of the insect are examined in light of these recent developments. Following a brief overview of insect development and nervous system organization (Section 2.1), the different strategies employed in the development of the auditory system in three representative insect groups (the grasshopper, the bushcricket, and the cricket) are described. Development of the peripheral auditory system begins in the early embryo with the differentiation of the receptor cells and the cuticular structures to which they are attached (Section 2.2.1); then follows the phase of axogenesis and pathfinding by the receptor cells as their axonal processes navigate their way to the central nervous system (Table 4.1; Section 2.2.2).

Although the full complement of receptors is present at hatching in most insects, as in vertebrates there is an extensive phase of postembryonic development during which, for example, the cuticular structures of the ear and the attachment sites of the receptors differentiate (Section 2.2.3). Only

TABLE 4.1. List of abbreviations used in the text.

A1,2,3	abdominal neuromeres 1,2,3
aRT	anterior ring tract
CNS	central nervous system
DC I–VI	dorsal commissures I–VI
DIT	dorsal intermediate tract
DMT	dorsal median tract
DUM	dorsal unpaired median neuron
VCLII	dorsal part of ventral commissure loop II
FETi	fast extensor tibia motoneuron
GMC	ganglion mother cell
IN	interneuron
LDT	lateral dorsal tract
LVT	lateral ventral tract
MDT	median dorsal tract
MNB	median neuroblast
MVT	median ventral tract
NB	neuroblast
PVC	posterior ventral commissure
SMC	supramedian commissure
T1	prothoracic neuromere
T2	mesothoracic neuromere
T3	metathoracic neuromere
TR	trachea
VAC	ventral association center
VC I,II	ventral commissure I,II
VIT	ventral intermediate tract
VLT	ventral lateral tract
VMT	ventral median tract
VNC	ventral nerve cord
vVCL II	ventral part of ventral commissure loop II

then does the auditory pathway acquire sufficient sensitivity to be of use in behavior. In the central nervous system the receptors enter a neuropil containing the dendritic arborizations of numerous interneurons, some of which are their targets (Section 2.2.4). Comparisons of the neuro-architecture of this neuropil in different insects reveal it to be organized according to a common ground plan. The development of some sound-sensitive central neurons has been followed from the differentiation of the stem cell (neuroblast; Section 2.3.1) to the birth of the neurons themselves to their subsequent growth through embryogenesis to maturity (Sections 2.3.2 and 2.3.3).

The study of the development of homologous groups of receptors has provided the fundamental and exciting discovery that the tympanal receptors are part of the more extensive pleural chordotonal system but have specialized to become associated with an external cuticular apparatus (the ear) in the first abdominal segment. The projections of pleural and tympanal receptors occupy the same general neuropilar space and form a serially organized system of projections in the central nervous system.

Because the chordotonal system is a fundamental component of all insect nervous systems, development of the equivalent system can be studied in both tympanate and atympanate insects (Section 2.4; see also Hoy, Chapter 1). Furthermore, the presence of the homologous chordotonal system in *Drosophila* allows a genetic analysis of the regulatory mechanisms governing, for example, the number of tympanal receptors and their identity (Section 2.5).

A number of the regulatory mechanisms that direct the development of the auditory system in insects can also be investigated by experimental manipulation of the developmental program itself. Plasticity among neurons of the nervous system has been revealed by studying the effects of regeneration following the deletion of auditory receptors or the pioneer cells for the auditory nerve during critical phases of development (see Section 2.6).

No study of the development of the auditory system can be complete without a consideration of the development of behavior (see Section 2.7). Regrettably, this is one area about which we have very little information. However, because the auditory pathway is ontogenetically part of the chordotonal system, about which we do have some information, comparisons between the two systems allow some inferences to be made.

2. Discussion

2.1. Evolutionary Strategies in Insect Development

Insects can be broadly divided into two groups according to the developmental strategy employed in order to reach adulthood (see also Hoy, Chapter 1). Insects such as the grasshopper are deemed hemimetabolous because they do not undergo a full metamorphosis during development. These insects hatch from the egg as a simplified version of the adult. Development is divided into two major phases: an embryonic phase, which is completed in about 20 days, depending on species and temperature, and a postembryonic phase, occupying a slightly longer period, during which the insect undergoes a series of molts (Fig. 4.1A). By contrast, insects such as *Drosophila* are deemed to be holometabolous because the embryonic phase is followed by a crawling larval stage, which metamorphoses into a quiescent pupal stage, and from which the flying adult emerges (Fig. 4.1B). Whereas the pattern of development in the very early embryo differs in the two groups of insects, both eventually reach a segmented germ band stage, after which embryonic development of the nervous system continues in a remarkably similar fashion until hatching. Thereafter, the two developmental plans diverge again.

The insect body plan is segmental in its organization. Where fusion of body segments has occurred during evolution, this is generally accompanied by fusion in the nervous system, though importantly, nervous system

A   Hemimetabolous

Embryonic

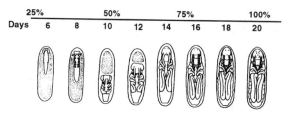

Postembryonic (Instars)

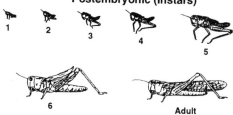

B   Holometabolous

Embryonic

Postembryonic

Larva

Adult

segmentation need not conform to the body segmentation seen externally. In insects, the nervous system consists of a central chain of ganglia, lying ventrally along the longitudinal body axis in a ladder-like array, the ventral nerve cord (VNC; see Fig. 4.12). The ganglia are collections of nerve cells linked by connectives consisting of the axonal processes of intersegmental neurons. As in the vertebrate nervous system, there are three classes of neurons in the insect nervous system: sensory afferents, interneurons, and motor neurons.

During evolution varying degrees of ganglionic fusion have resulted in central nervous systems with very different external organizations among insect species. Despite this diversity, comparisons of transverse sections of equivalent ganglia in the ventral nerve cord show that there is a considerable conservatism in their neuroarchitecture (Boyan, 1993).

The diversity of insect form is also reflected in the range of locations on the body in which an ear is found (Fig. 4.2; see also Hoy, Chapter 1). In some insects the ears are located bilaterally and symmetrically on the first abdominal segment, while in others an ear is found on the tibia of each prothoracic leg or in the ventral midline of the metathoracic segment (see Fig. 4.2). A different segment from within the serially continuous peripheral chordotonal system has therefore been selected for specialization as an ear in different species (see Michelsen, Chapter 2; Robert and Hoy, Chapter 6).

The structure of the ear itself also varies considerably between species: in the number and arrangement of sensory cells, in the mode of attachment of sensory cells to vibrating structures such as the external cuticle (tympanum), and in the position of the tympanal organ with respect to leg or body tracheae (see Michelsen, Chapter 2). As might be expected, the developmental strategies used to acquire the mature number and organization of tympanal receptor cells within the ear also varies among the various insect groups. These strategies can best be illustrated if we follow the development of the auditory system in three insect groups (the grasshopper, the bushcricket, and the cricket) for which the developmental program from

◄──

FIGURE 4.1. The two main types of developmental program among insects are summarized here. A: In hemimetabolous insects such as the grasshopper, an embryonic phase in the egg is followed after hatching by a series of postembryonic molts to reach adulthood. B: In the holometabolous fruitfly *Drosophila*, the embryonic phase is followed by a postembryonic crawling larval stage, which then pupates into the flighted adult. The stage of embryonic development in the grasshopper is normally given as a percentage of the time between egg laying and hatching (Bentley et al. 1979), whereas in *Drosophila* it is given either as a stage (based on recognizable morphological characteristics) or as hours of development at 23° to 25°C (Wright 1974). Postembryonic development, by contrast, is quantized by molts into larval stages (e.g., *Drosophila*) or instars (e.g., grasshopper). (A modified from Reichert, 1990, with permission of Thieme Verlag. B modified from Lawrence, 1993, with permission of Blackwell Scientific Publications.)

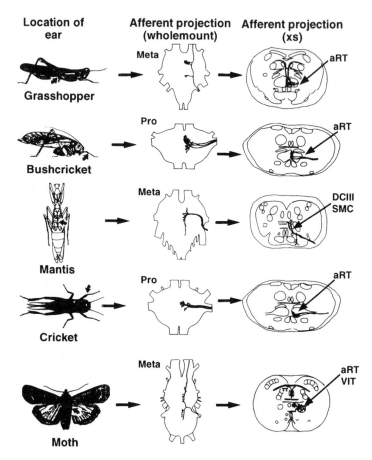

FIGURE 4.2. Locations of the auditory apparatus (ear) on the body of various insects (curved arrows) and the afferent projection of auditory receptors into the ventral nerve cord as seen in wholemount (center) and cross section (xs, right). Note that regardless of where the ear is located (grasshopper, first abdominal segment; bushcricket, prothoracic tibia; mantis, ventral metathoracic midline; cricket, prothoracic tibia; moth, first abdominal segment), the tympanal afferents always project into the same areas of neuropil — the various parts of the Ring Tract (aRT, RT, SMC, dVCL II) — in their respective ganglia. In some species (grasshopper, moth, and mantis) the afferents also project via the VIT into other neuromeres along the VNC. Drawings of insects are for illustrative purposes only and do not necessarily depict the exact species used to obtain the anatomical data shown. (Modified from Boyan, 1993, with permission of Elsevier Science.)

embryo to adult is available. One common pattern that emerges is that the tympanal receptors differentiate and develop embryonically, whereas the cuticular structures of the ear, which mediate sound transduction to the receptors, develop much later, postembryonically.

2.2. Development of the Peripheral Auditory System

2.2.1. Embryonic Differentiation of Tympanal Receptors

2.2.1.1. Grasshopper

Using immunohistochemical techniques, Meier and Reichert (1990) demonstrated that all 100 tympanal receptor cells differentiate as a lateral cluster from the body wall epithelium of the first abdominal segment (A1) during embryogenesis (Fig. 4.3). At 40% of embryogenesis, the auditory receptor cells are formed by an epithelial invagination at the posterior segment border. After completion of the invagination process, the developing sensory neurons are located in a compactly arranged oval configuration in the posteriolateral part of the embryonic body wall, with most dendrites terminating between adjacent cells of the outermost embryonic epithelium. Within the next 5% of embryonic development, a phase of cell translocation begins. A subset of rostroventral receptor cells initiates a migration towards a more anterior and proximal position. Other cells follow this pioneering group.

During this migration the dendrites lose their contact with the body wall epithelium and come to lie inside the body cavity. At 45%, cell movement has generated a morphology that is already typical for the auditory organ. A proximal smaller group of sensory cells has moved into a position next to the intersegmental nerve and forms the anterior part of the auditory organ. The more distal larger group remains in the original position at the site of invagination. Axogenesis commences between the 45% and 50% stages. During axogenesis the outgrowing axons and their growth cones are directed anteriorly towards the intersegmental nerve. At the 50% stage the axons have made contact with the intersegmental nerve. They then fasciculate with the nerve and begin to grow along it toward the central nervous system.

Comparisons of cellular differentiation sites, axogenesis, and cell migration described earlier in the grasshopper with the developmental plan in the auditory system of the mouse are revealing. During embryonic development of the mouse (Carney and Silver 1983) epitheloid cells differentiating from the rostrolateral wall of the otic cup migrate outward and condense (at E 10.5) to form a funnel-shaped configuration. During E 11.5 through E 12.5, pioneer distal auditory axons extend toward the otocyst, moving along cells of the preformed funnel. Axon growth cones invade the wall of the otocyst, moving tangentially along radially arranged cells that bridge the otocyst and funnel. The epitheloid cells, therefore, provide a preformed scaffold for outgrowing auditory axons into the otocyst (Carney and Silver 1983).

The similarity of the basic developmental events in the formation of the embryonic auditory pathway of insect and mammal is very likely a reflection of common molecular mechanisms, such as those that produce the

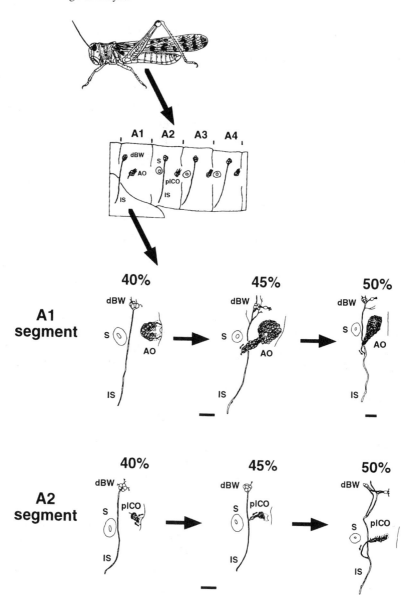

segmentation of the nervous system (Holland, Ingham, and Krauss 1992). Indeed, a feature of the body plan not only of insects but also of vertebrates is its segmentation (Leise 1990). This modular plan is reflected in the organization of the peripheral nervous system (see Fig. 4.3). Meier and Reichert (1990) showed that in abdominal segments A2 to A8, receptor groups differentiate from the body wall at the same time and in an equivalent location as the auditory organ in the first abdominal segment. These other receptor cells form the pleural chordotonal organs of the adult (Hustert 1975, 1978). Like the tympanal receptors, they are generated at the 40% stage by epithelial invagination of the body wall ectoderm into the body cavity near the posterior segment boundary (see Fig. 4.3). The invagination results in a cluster of cells that have their dendrites converging onto a small area of the epithelium.

During the next 5% of development, the pleural chordotonal receptor cells migrate to a more anterior position in each abdominal segment. This movement is similar to that undertaken by the tympanal receptors in segment A1. At the 45% stage the receptor cell dendrites lose contact with the original invagination site and reorient so that dendrites point toward the posterior segment boundary. This process leads (at 50%) to a row of cells oriented in the anterior-posterior axis of each abdominal segment. The major difference between an auditory organ and a pleural chordotonal organ lies in the respective number of receptor cells present: approximately 100 in the auditory organ and 10 to 15 in a pleural chordotonal organ. With this analysis Meier and Reichert (1990) elegantly demonstrated that the auditory organ is ontogenetically a segmental specialization of the more extensive pleural chordotonal system.

2.2.1.2. Bushcricket

Arguably the most exciting finding concerning sensory cell organization in an invertebrate ear to date is that of Oldfield (1982, 1985), who showed that

---

FIGURE 4.3. Embryonic development of the pleural chordotonal system in the grasshopper (Acrididae). The pleural chordotonal receptors form a serial array in the lateral body wall of the abdomen. In the first abdominal segment (A1), the receptors are associated with an auditory apparatus. Embryonic development of the auditory organ (AO) is shown at the 40%, 45%, and 50% stages. The tympanal receptors differentiate from the lateral body wall of the first abdominal segment, invaginate, and migrate anteriorly (small arrow). At 50% they direct axonal processes (small arrow) onto the preformed intersegmental nerve (IS). In other abdominal segments, such as A2, a pleural chordotonal organ is present instead of an auditory organ. The pleural chordotonal receptors differentiate in the same way and over the same time span as the tympanal receptors in segment A1. They migrate anteriorly (small arrow), and their outgrowing axons join the intersegmental nerve. dBW, dorsal body wall cells; S, spiracle. Scale bar, 50 µm. (Modified from Meier and Reichert, 1990, with permission of John Wiley and Sons, Inc.)

in the bushcricket the linear arrangement of the approximately 35 or so sensory cells along the crista acustica mirrors their best frequencies (Fig. 4.4A,B). Each sensory cell in the array is tuned to a different best frequency: cells with larger somata located proximately in the leg are tuned to lower frequencies, while cells with smaller somata located distally in the leg are tuned to higher frequencies. This linear array of tympanal receptors, scolopales, and attachment cells presents an ideal model system for studying the relationship between morphogenesis and physiological function in the developing ear.

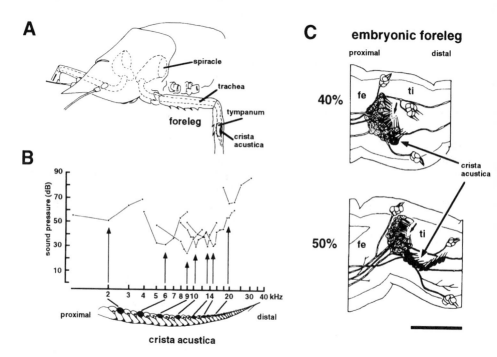

FIGURE 4.4. Structure and embryonic development of the ear in the bushcricket (Tettigoniidae). A: The tympanal receptors of the tettigoniid ear are organized into an elongated linear array called the *crista acustica*, located behind the tympanum of each foreleg. B: The organization of the receptors reflects their tuning properties: receptors located proximally are tuned to low frequencies and those located distally are tuned to higher frequencies. (Modified from Oldfield, 1982, 1985, with permission of Elsevier Science and Springer-Verlag.) C: The tympanal receptors of the crista acustica differentiate from the leg epithelium at the femur/tibia boundary early in embryonic development (40%) along with the receptors of the complex tibial organ (arrow). The tympanal receptors invaginate, then migrate distally in the tibia, and the crista acustica elongates as more cells are added during subsequent embryonic stages (50%). Scale bar, 100 μm. fe, femur; ti, tibia. (Modified from Meier and Reichert, 1990, with permission of John Wiley and Sons, Inc.)

The embryonic origins of the crista acustica in the bushcricket *Poecilimon affinis* have been elegantly demonstrated by Meier and Reichert (1990). Their study shows that a complex tibial organ is formed, which in turn gives rise to the subgenual organ and to a group of associated auditory receptor cells (Fig. 4.4C). At the early 40% stage, a large group of ectodermal cells invaginates from the limb bud epithelium into the lumen of the proximal part of the prothoracic tibia. These cells will form the complex tibial organ of the prothoracic leg. The cells within this invaginating cluster have apical dendrites oriented toward the center of invagination and in contact with the outermost epithelial layer.

After 40%, several receptor cells segregate at the distal edge of the complex tibial organ, lose their dendritic contact to the limb epithelium, reorient their dendrites posteriorly, and so form the beginning of the auditory organ. While initially only three to four receptor cells occupy this position, their number increases during subsequent development so that by the 50% stage, 13 auditory receptor cells are present in the crista acustica. These cells are already arranged in the one-dimensional linear array so characteristic of the adult crista acustica (see Fig. 4.4B; Schumacher 1973; Oldfield 1982, 1985).

2.2.1.3. Cricket

Embryonic development of the tympanal receptors in the cricket has been described by Klose (1991) and has interesting parallels to the developmental pattern occurring in both the bushcricket and grasshopper limb bud. Initially, it was believed that the tympanal organ arises postembryonically (Ball and Young 1974), but Klose showed (Fig. 4.5) that the tympanal organ originates from the same complex of sensory cells as the subgenual organ near the Til pioneers, cells equivalent to those previously described for the grasshopper limb bud (Bentley and Keshishian 1982; Ho and Goodman 1982; Bentley and Caudy 1983).

The precursors for the tympanal organ receptor cells differentiate from the leg epithelium at around 35% of embryonic development. The cells at first form a fan-shaped structure, but this changes after the 50% stage as the tympanal cells increase in number (to eight) and migrate in the distal direction, forming an angle of 120° with the remaining subgenual cells, as in the adult. Elongation of the developing tympanal organ continues over the remaining 50% of embryogenesis as more and more cells differentiate. At the 60% stage there are about 12 cells in the tympanal organ, at 70% there are around 20 cells, at 80% 35 cells, at 90% almost 50, and at hatching 53 cells (see Fig. 4.5).

In the mesothoracic and metathoracic legs, the early developmental plan is identical to that in the foreleg, but subsequent growth of the homologous structures stops during embryogenesis, which means that in the adult fewer cells are present in these organs than in the prothoracic leg.

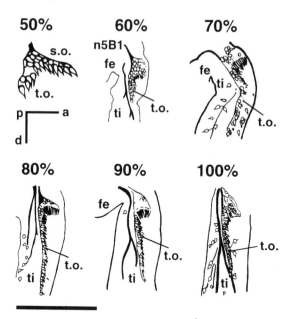

FIGURE 4.5. Embryonic development of the tympanal organ (t.o.) within the tibial sense organ complex (s.o.) in the foreleg of the cricket (Gryllidae). Receptors differentiate from the epithelium of the leg near the femur/tibia border. As new receptors form, older ones displace distally, forming an elongated crista acustica in the foreleg. Receptor axons from the t.o. project into the central nervous system via leg nerve 5B1. p, proximal; d, distal; a, anterior; fe, femur; ti, tibia. Scale bar, 100 μm at 50%, 200 μm for all other stages. (Modified from Klose, 1991, with permission of Prof. Dr. K. Schildberger.)

2.2.2. Embryonic Development of Tympanal Projections into the Central Nervous System

Once differentiated, the next phase in the development of the auditory pathway involves the navigation of receptor axons from the periphery toward the central nervous system. Cellular and molecular data from the grasshopper, cricket, and the fruitfly *Drosophila* are providing an insight into the developmental mechanisms that regulate this ingrowth of chordotonal (tympanal) receptor projections into the central nervous system (Meier and Reichert 1995).

We have seen that in the grasshopper the differentiating auditory receptor cells of the first abdominal segment initiate axogenesis at around 50% of embryonic development and direct their growth cones toward the fibers of the intersegmental nerve (see Fig. 4.3). The intersegmental nerve is already in place prior to 40% of embryonic development. It will become the tympanal nerve and is formed by a two-stage process: central fibers (i.e., motor neurons) from the "U" fascicle extend growth cones into the periphery, while dorsal body wall cells (the dorsal chordotonal organ cells, dch3) in the periphery direct their processes centrally. During normal development, growth cones from peripheral and central cell groups meet approximately halfway between the central nervous system and body wall (Fig. 4.6A). They then fasciculate, presumably as a result of the mutual expression of cell–cell adhesion molecules (such as fasciclin I), and grow over each

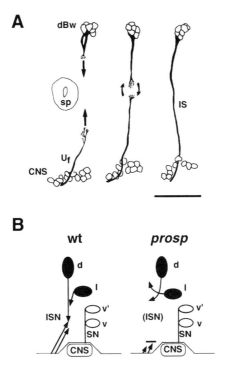

FIGURE 4.6. A: Formation of the auditory nerve during embryonic development in the grasshopper. Three consecutive stages (each separated by approximately 1% of developmental time) from adjacent abdominal segments are shown. The axons of the dorsal body wall (dBw) neurons and those of the central neurons pioneering the U fascicle (U$_f$) extend in opposite directions. The auditory (intersegmental) nerve in the thoracic and abdominal segments is formed by fasciculation of both axon bundles near the body wall spiracle (sp). Scale bar, 100 μm. (Modified from Meier and Reichert, 1995, with permission of Birkhäuser Verlag.) B: Effects of the *prospero* mutation (*prosp*) on the formation of the auditory (intersegmental) nerve in *Drosophila* embryos. The position of the dorsal (d), lateral (l), and ventral (v, v') neuronal clusters along the intersegmental (ISN) and segmental (SN) nerves is shown for wild-type (wt) and *prosp* flies. In the *prospero* mutation, the ISN fails to form because there is a misguiding both of peripheral axons from the dorsal and lateral body wall receptors, and of central axons from the central nervous system. The SN is not affected by the mutation. (Modified from Meier and Reichert, 1995, with permission of Birkhäuser Verlag.)

other in opposite directions to create the nerve. This stereotypic mode of intersegmental nerve formation can be seen in all thoracic and pregenital abdominal segments.

In the cricket the receptor cells of the femoral chordotonal organ, the subgenual organ, and the tympanal organ all fasciculate with nerve 5B1, which begins growth at the 30% stage and is pioneered by the so-called Til cells. If the pioneer cells are killed by heat shock, then outgrowing cells of the tympanal organ fail to reach the central nervous system because of difficulties in crossing the tibia-femur segment boundary (Klose 1991), a similar problem as that described by Klose and Bentley (1989) for the grasshopper limb bud.

The mechanisms determining these first navigational decisions are now becoming understood. Axonal pathfinding in the homologous system in *Drosophila* is regulated by the *prospero* gene (Doe et al. 1991; Vaessin et al. 1991). In the periphery, loss of function in this gene leads to a reversal in axon outgrowth polarity for the developing cells in the dorsal and lateral clusters (Fig. 4.6B). The affected sensory axons extend dorsally instead of ventrally. Centrally, the motoneurons also grow in an aberrant manner and do not pioneer the correct nerve. The intersegmental nerve does not form.

2.2.3. Postembryonic Development of the Ear

2.2.3.1. Grasshopper

While the auditory organ, consisting of sensory units, each comprising a receptor cell (scolopidium) and its associated accessory cells, is fully present at the end of embryogenesis, the auditory structures with which it is associated in the mature ear (the various parts of Müller's organ, Fig. 4.7A) are not all fully formed at hatching, nor is the orientation of the various receptor cell body groups the same as in the adult (Michel and Petersen 1982). The tympanal receptors of the grasshopper have been classified into four groups (a, b, c, d: Gray 1960; Michel and Petersen 1982; I, II, III, IV: Römer 1976). The a, b, and c type receptors are most sensitive to frequencies at around 3 to 5 kHz (low-frequency receptors), whereas the d cells are most sensitive to frequencies at around 12 to 20 kHz (high-frequency receptors; Michelsen 1971; Römer 1976). The cuticular structures with which the various receptors are associated develop in a stepwise manner with each instar (Table 4.2; Michel and Petersen, 1982). The a cells are attached to the elevated process, which is first clearly distinguishable in the second instar. The b cells are attached to the styliform body, and this appears only at the final molt. The c cells attach to the folded body, which only appears at the fourth instar. The high-frequency d cells ultimately insert near the pyriform vesicle; this structure appears in the fourth instar. Prior to this, the d cells insert in the immediate vicinity of the posterior edge of the elevated process, and their attachment site then moves posteriorly in a stepwise manner with each successive molt.

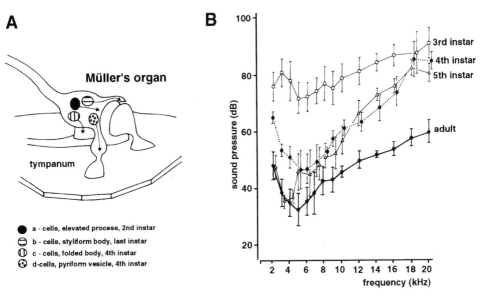

A

Müller's organ

tympanum

● a - cells, elevated process, 2nd instar
⊖ b - cells, styliform body, last instar
Ⓘ c - cells, folded body, 4th instar
☉ d-cells, pyriform vesicle, 4th instar

B

FIGURE 4.7. Postembryonic development of the grasshopper ear. A: The tympanal receptors are organized into four groups (represented by the symbols shown) within Müller's organ, which is situated on the internal side of the tympanum. Although all receptors are present at hatching, each receptor group becomes associated with its correct structure (dendrite locations are indicated by arrows) postembryonically at the time indicated. (Modified from Halex, Kaiser, and Kalmring 1988, with permission of Springer-Verlag.) B: Summed auditory threshold curves from the tympanal nerve at different postembryonic instars reveal a progressive increase in overall sensitivity as the ear matures. (Modified from Petersen, Kalmring, and Cokl, 1982, with permission of Blackwell Scientific Publications.)

TABLE 4.2. Postembryonic development of receptor attachment sites in the ear of the grasshopper.

Receptor type	Tuning (kHz)	Insertion	Appearance
Type I, a	3–5	Elevated process	2nd instar
Type II, c	3–5	Folded body	4th instar
Type IV, d	12–20	Pyriform vesicle	4th instar
Type III, b	3–5	Styliform body	5th–6th instar

Each of the four receptor types inserts into a different part of Müller's organ, which develops in a stepwise manner during postembryonic development.

Not only the inner cuticular structures, but also the tympanum of the ear itself, changes during postembryonic development. The layers of epidermal cells behind the tympanum disappear progressively, with the thin cuticle of the mature tympanum only appearing at the imaginal molt (Michel and Petersen 1982). Is such a developmental plan reflected in other auditory systems? In mammals, the area of the tympanic membrane, the length of the lever arms of the malleus and incus, the surface area of the oval window, and the volume of the bulla all show systematic changes during neonatal life (Huangfu and Saunders 1983). Appropriately, over postnatal days 10 to 20 there is a 70% improvement in threshold for physiological responses to a 20-kHz tone recorded from the cochlear nucleus and round window (Huangfu and Saunders 1983).

Not surprisingly, the stepwise development of the grasshopper ear also affects the energy transduction to the receptor cells and consequently the physiological properties of the receptor cells themselves. Early in postembryonic development all the receptors are relatively insensitive to sound because their attachment sites within Müller's organ are not yet formed. The majority of receptors responding to sound in early instars are tuned to low frequencies (Fig. 4.7B). The high-frequency d cells acquire their mature frequency-response characteritics only at the last molt as they attach in the correct manner to the elevated process. The fact that the low-frequency receptors, though less sensitive, are still tuned earlier in postembryonic development to the same frequencies as in the adult suggests that intrinsic properties of the receptor cells themselves, rather than the attachment site on the tympanum, may be the final determinant of frequency sensitivity (see Ball, Oldfield, and Rudolph 1989).

This stepwise development of auditory frequency sensitivity is, in principle, very similar to that described for the avian brain stem auditory nuclei (Lippe 1987). In the chicken, the characteristic frequency measured at four sites in the nucleus magnocellularis and nucleus laminaris increases in a stepwise manner among embryonic E17, E19–20, postembryonic, and adult stages. The tonotopic organization of these nuclei, therefore, shifts during late embryonic development, possibly due to structural changes within the basilar papilla (Lippe 1987).

2.2.3.2. Bushcricket

While the differentiation of receptor cells from the leg epithelium in bushcrickets occurs during embryogenesis, the final differentiation of the scolopidia only occurs during postembryonic development (Rössler 1992a). Rössler demonstrated (Fig. 4.8) that in the first instar of *Ephippiger ephippiger*, the dendrites and scolopales of the receptor cells within the crista acustica are oriented horizontally. In the second instar, the dendrites are bent towards the hemolymph channel. The dendrites, cap cells, and scolopale caps and rods enlarge in subsequent instars, with the dendrite achieving its final length only in the fifth instar. The linear row of cap cells to which the dendrites are attached increases in size (but not number) (Fig. 4.9A) and shifts distally during development, resulting in an increase in the overall length of the crista acustica.

The associated cuticular structures of the ear, such as the acoustic trachea (Fig. 4.9B), the tympana, the tympanal covers, the acoustic spiracle (Fig.

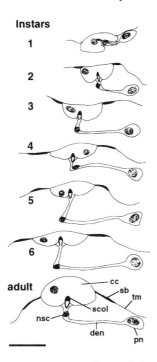

FIGURE 4.8. Drawings from transverse sections of the foretibia reveal the post-embryonic development of the third scolopidium in the crista acustica of the bushcricket ear through six instars to the adult stage. cc, cap cell; den, dendrite; nsc, nucleus of the scolopale cell; pn, perikaryon of the sensory neuron; sb, supporting bands; scol, scolopale cap and rods; tm, tectorial membrane. Scale bar, 50 μm. (Modified from Rössler, 1992a, with permission of Springer-Verlag.)

4.9C), and the tentorial membrane, all develop in a step-by-step manner during consecutive larval instars (Rössler 1992a). The state of these structures determines the efficiency of energy transmission to the receptor dendrites. Threshold curves, therefore, show the same frequency maxima as in the adult, only in the fourth instar, and the same absolute sensitivities only after the final molt (Fig. 4.9D).

As part of their landmark study on the embryonic development of the crista acustica in bushcrickets, Meier and Reichert (1990) showed that the same basic developmental plan seen in the foreleg is followed in each of the mesothoracic and metathoracic legs. They concluded that the linear arrays of mechanosensory cells that segregate out of the complex tibial organs in these other legs are segmentally homologous to the auditory crista acustica in the pothoracic leg. Despite homology, however, the number of receptors within the crista acustica in each leg differs: in the prothoracic leg of *E. ephippiger* there are 28 scolopidia, in the mesothoracic leg 11, and in the metathoracic leg 7 (Rössler 1992b). Another major difference involves

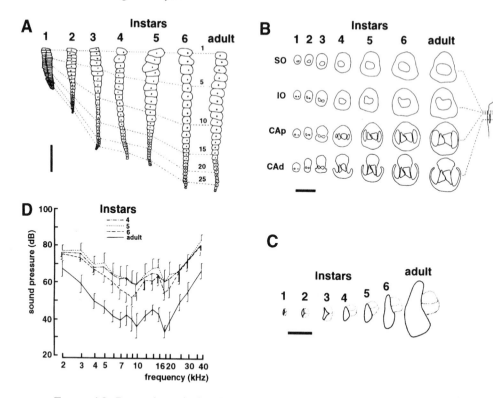

Figure 4.9. Postembryonic development of the buschcricket ear. A: Development of the row of cap cells in the crista acustica through six instars to the adult stage. The location of the same cell (indicated by its number in the array) at different ages is indicated by dashed lines. Scale bar, 100 μm. B: Development of the proximal tibia and the acoustic trachea in the foreleg through six instars to the adult stage. The plane of the section in the tibia is indicated by dashed lines. SO, subgenual organ; IO, intermediate organ; CAp, proximal part of crista acustica; CAd, distal part of crista acustica. Scale bar, 500 μm. C: Development of the prothoracic acoustic spiracle and its associated respiratory spiracle (dotted outline) through six instars to the adult stage. Scale bar, 500 μm. D: Change in summed auditory threshold of receptors in the tympanal nerve through the final four instars to the adult stage. (Modified from Rössler, 1992a, with permission of Springer-Verlag.)

the organization of scolopidia and accessory structures in the crista acustica of the three legs. Because there are also differences in leg and tracheal morphology, it is not surprising that auditory threshold curves of relevance for airborne acoustic communication are only obtained in the prothoracic leg. Different, not yet defined, mechanisms within each leg, therefore, modify a common plan for constructing a crista acustica and result in the segmental specializations we see at the end of postembryonic development.

2.2.3.3. Cricket

Because differentiated scolopales appear before the final cuticular structure of the ear develops, Ball and Hill (1978) decided to test whether the tympanal receptors of the cricket have an innate tuning mechanism or are dependent on their attachment sites for their physiology. Ball and Hill first measured tuning curves from physiologically identified central neurons in postembryonic (penultimate, ultimate) and adult stages (Fig. 4.10A). The tuning curves show identical best frequencies at all stages, but a dramatic increase in sensitivity between the ultimate and adult stages. These authors then attached a piezoelectric crystal to the foreleg of a late-instar cricket in order to activate the tympanal receptors directly with vibratory stimuli. The resulting tuning curves (Fig. 4.10B) were shown to have the same best frequencies as for sound activation (Fig. 4.10A) and the same stepwise increase in sensitivity during development, suggesting intrinsic tuning mechanisms. The auditory system, although physiologically functional at this immature level, is much less sensitive than in the adult due to inefficient energy transmission across the tympanum. As in the grasshopper, the thin silver-colored auditory tympanum, largely free of epithelial cells on its inner surface, only appears at the final molt (Ball and Young 1974).

Similar processes may be responsible for the fact that local mechanical responses to high-frequency stimuli also change during ontogeny in the gerbil (Harris and Dallos 1984). Using the isoresponse functions of the

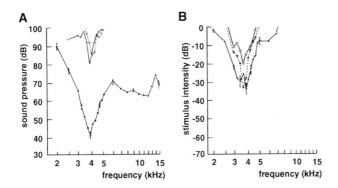

FIGURE 4.10. Postembryonic development of physiological properties in the auditory system of the cricket. A: Auditory thresholds of an ascending interneuron recorded in the cervical connective in response to pure tones in penultimate (◇) and ultimate (▽) postembryonic instars, and in the adult (◆). Whereas absolute sensitivity increases during development, the tuning properties of the adult ear are already present at the penultimate instar. B: Thresholds recorded from the same neuron type as in A but in response to vibration of the leg via a piezoelectric crystal in penultimate (□) and ultimate (◆) postembryonic instars and in the adult (▼). Note that the neuron is tuned to the same frequency in both stimulus situations. (Modified from Ball, Oldfield, and Rudolph 1989, with permission of Cornell University Press.)

cochlear microphonic, these authors showed that frequency sensitivity changes in a stepped manner during late postnatal development, possibly due to changes in epithelial cell layers on the scala tympani side and/or to changes in the cellular structure of the basilar membrane. According to Harris and Dallos (1984), these morphological changes indicate a progressive shift in mass and stiffness that is likely to affect local mechanical resonances.

2.2.4. Development of Central Projections

The actual ingrowth of the auditory receptor axons into the central nervous system has not been documented for any insect. In the grasshopper, the central projections of receptors in the thoracic ganglia of the third postembryonic instar already bear a close resemblance to those of the adult (Fig. 4.11; Petersen, Kalmring, and Cokl 1982; Halex, Kaiser, and Kalmring 1988). We assume that this projection is established late in embryogenesis, as demonstrated for the filiform hair projections of the cercus-to-giant interneuron system (Shankland, Bentley, and Goodman 1982). Within the auditory association area in the prothoracic ganglion of the bushcricket, and that within the metathoracic ganglion of the grasshopper (see Fig. 4.11), there is a clear tonotopic organization of afferent projections.

The distribution of terminals reflects the best frequency of the receptor in the periphery (Römer 1983; Römer, Marquart, and Hardt 1988; Ebendt, Friedel, and Kalmring 1994). The developmental rules that regulate this distribution are not clear but may be similar to that described for the cercal

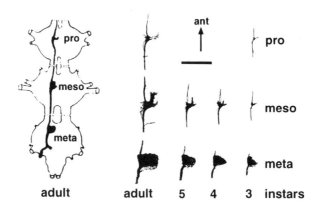

FIGURE 4.11. Postembryonic development of tympanal afferent projections in the prothoracic (pro), mesothoracic (meso), and metathoracic (meta) ganglia of the ventral nerve cord in the grasshopper. The basic projection is already well established in early instars. Note that not all afferents entering the central nervous system in the metathoracic ganglion extend processes to more anterior ganglia. ant, anterior. Scale bar, 500 µm. (Modified from Halex, Kaiser, and Kalmring, 1988, with permission of Springer-Verlag.)

system of the cricket (Bacon and Murphey 1984; Murphey 1986) and the axonal projections of retinal ganglion cells to the tectum in fish and frogs (Gilbert 1991): a hierarchy of factors, such as coincident activity in afferents, molecular labeling of target areas, and mutual inhibition between afferent terminals, generates a topographical mapping of the periphery.

Whatever the rules, they are likely to be similar in all insects. Evidence for this comes from a comparison of wholemount projections of tympanal (auditory) afferents from the ear into the central nervous system in various insects (see Fig. 4.2). This shows that although the receptors enter different ganglia of the central nervous system via different nerve roots, all projections in the central nervous system remain ipsilateral to their ear of origin. Further, transverse sections of the ganglia reveal that in each case the tympanal receptors project to the equivalent fiber tracts and commissures of the respective ganglia of the ventral nerve cord. The target neuropilar areas are the various parts of the ring tract (dVCLII, aRT, SMC), and the ventral intermediate tract (VIT). These regions appear to represent the auditory association areas in tympanate insects with considerably different lifestyles.

A serial homology among peripherally located chordotonal receptors might also be reflected in the projections of these afferents within the central nervous system. The central projections of tympanal afferents and pleural chordotonal organs are found in serially repeated neuropilar regions of the thoracic ganglia and in the same parts of the ring tract (dVCLII, aRT, SMC), and in the ventral intermediate tract (VIT) throughout the ventral nerve cord (Hustert 1978; Tyrer and Gregory 1982; Römer and Marquart 1984; Pearson et al. 1985; Pflüger, Bräunig, and Hustert 1988; Prier and Boyan 1993). This makes clear that afferents serving what might be regarded as two different modalities, hearing and stretch, are ontogenetically part of the same sensory system and that the tympanal organ is a segmental specialization within the chordotonal system.

2.3. Development of the Central Auditory System

2.3.1. Differentiation of Central Cells During Embryogenesis

A large number of central neurons that receive synaptic input from the chordotonal (tympanal) system have now been identified in various species of insects (Boyan 1984, 1993; Boyan and Williams 1995). Among this group, several have been studied developmentally, mostly in the grasshopper and *Drosophila* chordotonal systems. The most completely studied insect neuron of all is probably neuron 714 (also known as the G neuron) of the grasshopper (Bastiani, Pearson, and Goodman 1984; Pearson et al. 1985). This particular neuron has proven to be a model system for revealing transmitter identities and cell-surface recognition molecules using immunohistochemical techniques (Doe 1992); for examining the electrical properties (Abrams and Pearson 1982; Boyan 1986) and synaptic connectivities (Pearson et al. 1985; Boyan 1992) of lineally related cells using intracellular recording techniques; for studying the activity of neurons during normal behavior (Wolf 1984); for axogenesis and pathway formation (Raper, Bastiani, and Goodman 1983a,b); and for following the developmental history of a single neuron from embryo to adult (Boyan 1983; Raper, Bastiani, and Goodman 1983a,b; Bastiani, Pearson, and Goodman 1984; Bastiani, Raper, and Goodman 1984).

All central neurons in insects derive from precursor cells called *neuroblasts* (Bate 1976). These neuroblasts differentiate from the neuroepithelium during early embryogenesis and form aggregates of 61 in each segment of the ventral nerve cord (Fig. 4.12). There are 30 neuroblasts in each hemiganglion and one in the midline of each neuromere. The neuroblasts are organized into rows and columns and are numbered accordingly. The pattern is stereotypic not only for the grasshopper but for all insects studied. During embryogenesis each neuroblast gives rise to a stereotypic set of progeny, and the lineage of every neuron in the central nervous system can theoretically be traced back to one of these neuroblasts. Neuroblast 7-4, for example, is always found in row 7 and column 4 of the

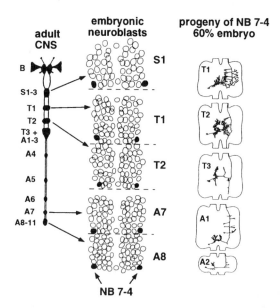

FIGURE 4.12. Embryonic morphologies of interneuron 714 and its serial homologs in the grasshopper. Camera lucida drawings of the pattern of neuroblasts in representative segments (S1, T1, T2, A7, A8) of the central nervous system of the early embryo. Neuroblast 7-4 (row 7 and column 4), which produces the lineage including neuron 714, is indicated in black and is present symmetrically in each hemiganglion. Morphologies of the 714 serial homologs in segments T1, T2, T3, A1, and A2 at 60% to 70% of embryonic development. (Modified from Boyan, 1993, with permission of Elsevier Science.)

plate of neuroblasts in any given segment of the central nervous system (Fig. 4.12). Because neuroblast 7-4 is present as a serial array in the central nervous system, then the equivalent lineage of progeny in present also, and individual progeny have been identified in all neuromeres between the second abdominal and suboesophageal ganglia (see Fig. 4.12). Interestingly, all the neurons in the array receive input from the chordotonal system, either from the pleural chordotonal receptors and/or from the serially equivalent auditory receptors, depending on where they are located. Functionally, each neuron represents a segmentally specialized unit, with mesothoracic progeny outputing to leg motoneurons, and abdominal ones to flight motoneurons or the respiratory system (Boyan 1993; Boyan and Ball 1993).

2.3.2. Axogenesis in the Embryonic Chordotonal System

If we consider neuron 714 as an example (Fig. 4.13), then the cell first directs an axon bearing a growth cone toward the ganglionic midline at about 35% to 40% of embryogenesis. The growth cone initially follows a path pioneered by cells that were born earlier from the same neuroblast as neuron 714. The growth cone grows across the midline to fasciculate with a bundle of longitudinally organized axons from the so-called A/P fascicle, the future adult lateral dorsal tract (LDT). Selective ablation of A and P cells shows that the growth cone of the 714 neuron, in fact, recognizes only the P axons in this fascicle and not the A axons (Raper, Bastiani, and Goodman 1984b; Bastiani, Raper, and Goodman 1984; Bastiani and Goodman 1986).

Having recognized the A/P fascicle, the primary axonal growth cone of the 714 neuron turns anteriorly, whereas the secondary growth cone turns posteriorly onto a different fascicle (Raper, Bastiani, and Goodman 1983a). The anteriorly growing axon of the 714 neuron requires the presence of anteriorly situated P neurons to continue directed growth. Ablation of these cells leaves the 714 neuron growth cone "confused" and it ceases growth (Raper, Bastiani, and Goodman 1984a,b). A segmentally organized set of pioneer neurons is therefore vital to the establishment of the correct pattern of growth in neuron 714.

In the adult, the 7-4 progeny have dendritic arborizations that extend anteriorly in the ventral intermediate tract (VIT) in their respective neuromeres (see Fig. 4.13B). These dendritic arborizations arise as a result of dendritic growth cones that form on each side of the ganglion at around 52% of embryogenesis (see Fig. 4.13A). The temporal and spatial diversity of choices made by the growth cones associated with the various branches of IN 714 may reflect the selective expression of cell-surface antigens on various parts of the neuron during certain specific periods of embryonic development. Neurons whose axons grow in contact with each other, or fasciculate, have in several cases been demonstrated to share common surface antigens in both the grasshopper and *Drosophila* (Grenningloh et

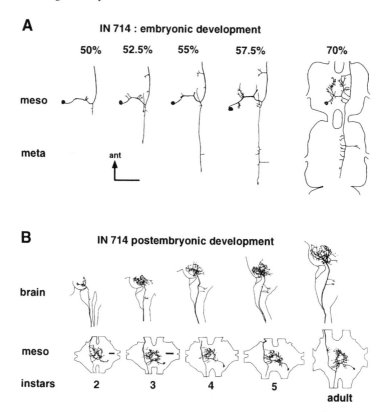

A   **IN 714 : embryonic development**

50%   52.5%   55%   57.5%   70%

meso

meta   ant

B   **IN 714 postembryonic development**

brain

meso

instars   2   3   4   5

adult

FIGURE 4.13. Development of the branching pattern of interneuron 714 of the grasshopper. A: Embryonic development between 50% and 70% of embryogenesis in the mesothoracic (meso) and metathoracic (meta) neuromeres. B: Postembryonic development of the arborizations in the mesothoracic ganglion (bottom) and brain (top) from instar 2 to adulthood. The ascending axon of neuron 714 terminates in the lower lateral protocerebrum of the brain. Only this region of the brain is shown each time. ant, anterior. Scale bars, 100 μm in A and B,2; 200 μm in B,3. (Modified from Boyan and Ball, 1993, with permission of John Wiley and Sons, Inc.)

al. 1990). Early in development, while still located in the commissure, the axon of 714 expresses fasciclin I; later, after recognizing the longitudinal A/P fascicle, the axon expresses fasciclin II (Bastiani et al. 1987).

By 50% of embryonic development, the axonal growth cone of the 714 neuron has reached the next anterior segment on its way to the brain and 50% to 60% of most of the major intraganglionic branches develop (see Fig. 4.13A). At 60% the segmental homologs clearly already have the morphology of the adult neurons (see Fig. 4.13A,B). This suggests that these neurons may not initially send out supernumerary branches that are later withdrawn but rather that development is very specifically directed into discrete areas of the nervous system.

2.3.3. Postembryonic Development in the Central Chordotonal System

Here again the grasshopper provides us with our most detailed description of the postembryonic development at the single cell level in an insect auditory pathway. By the 70% stage of embryogenesis, the dendritic arborizations of neuron 714 are already strongly reminiscent of the adult morphology (see Fig. 4.13; Raper, Bastiani, and Goodman 1983a,b). It is not known when the terminal arborizations of ascending acoustic neurons arrive in the brain, but if they arrive after 60% of embryogenesis (which is highly likely), then they grow into an already preformed neuropil in the lateral protocerebrum (Boyan et al. 1995). Furthermore, such ascending neurons do not pioneer the pathway to the brain; this pathway has already been pioneered much earlier by descending brain cells (Boyan et al. 1995).

Given this rapid embryonic growth, it is not surprising that postembryonic development of terminal arborizations in the brain, and postsynaptic arborizations in the mesothoracic ganglion, involve largely the acquisition of additional higher order branches — the gross morphology of the neuron changes little (see Fig. 4.13B; Boyan 1983). The additional branching may represent a response to afferent innervation, which is known to shape postsynaptic branching patterns in the cercal system (Shankland, Bentley, and Goodman 1982).

A similar sequence of developmental events has been demonstrated in central auditory pathways in vertebrates. In mammals, neocortical organization, particularly the formation of the corpus callosum, changes rapidly during the latter half of embryogenesis. The projection pattern of auditory callosal afferents into the gray matter of the opposite cerebral hemisphere, for example, forms during very late embryogenesis and early postnatal life (Feng and Brugge 1983). Subsequent to the arrival of the auditory callosal axons at their final destination by about the third postnatal day, cellular interactions over the next 3 weeks establish the innervation patterns characteristic of the adult cortex (Feng and Brugge 1983).

Physiological responses in afferents and interneurons of the grasshopper correlate with the morphological changes in the ear itself, as described earlier. Central neurons are tuned to low frequencies (5 kHz) in the early developmental stages (instars), and this is retained in the adult. The response to high frequencies (<12 kHz) is initially low, before Müller's organ is complete, but increases progressively (along with the absolute sensitivity) at each instar as the high-frequency d cells acquire their correct insertion site (Michel and Petersen 1982; Boyan 1983).

The morphological and sensitivity changes in interneurons during postembryonic development may also reflect several other developmental processes: (1) Axons belonging to afferents tuned to different frequencies may reach the central nervous system at different times. Synaptic connections with central neurons would therefore depend on the age and relative competitive ability of the different receptors. In the cercal system of a number of insects, new afferents and their sensory structures are added at each instar (Murphey, Jacklet, and Schuster 1980; Blagburn 1989; Lnenicka and Murphey 1989), and there is a competition between receptors for arborization space within the central nervous system and for synaptic connections with interneurons (Murphey 1986).

(2) Ingrowing afferent terminals have themselves been shown to influence the distribution of dendrites of giant neurons of the cricket and grasshopper cercal systems (Shankland, Bentley, and Goodman 1982; Shepherd and Murphey 1986; Shepherd, Kämper, and Murphey 1988) in much the same way as synaptic rearrangement produces changing branching patterns in neonatal rabbit ciliary neurons (Purves and Lichtman 1985). In the cercal system of the cricket, for example, synaptic connections established between filiform afferents and interneurons early in development are progressively remodeled. This occurs systematically according to the age of the afferent, and hence the information it carries, and takes the form of a change in the weighting of the synaptic connection a specific type of afferent makes with two identified interneurons (Chiba, Shepherd, and Murphey 1988; Lnenicka and Murphey 1989). Changes in the weighting of synaptic input are likely to occur during postembryonic development in the chordotonal pathway, but here it is a shift between two sensory modalities — sound and vibration.

(3) Even though tympanal receptor projections are present in early instars, they do not deliver auditory-based activity to postsynaptic cells because the ear itself is not sufficiently developed. Overall sensitivity increases as the cuticle comprising the tympanal membrane, which transmits acoustic energy to Müller's organ, becomes thinner with each instar (Ball and Hill 1978). As a working hypothesis, it is likely that the central chordotonal pathway effectively functions as a vibratory one early in larval development and becomes multimodal later as synaptic input from auditory receptors arrives.

Within the central nervous system, the equivalent lineage to that produced by neuroblast 7-4 in the grasshopper has also been described in the embryonic thoracic segments of the fly (*Drosophila*) and the moth (*Manduca*; Fig. 4.14; Thomas et al. 1984). Descriptions of axon outgrowth reveal that initially the same series of events occurs in *Drosophila* as in the grasshopper. Interneurons that are the progeny of the equivalent neuroblast (NB 7-4) and have the same growth pattern are present at equivalent locations in the embryonic neuropils of all three species and are therefore putative homologs. In the adult, neurons that have a strong morphological similarity to the progeny of neuroblast 7-4 have also been described in mantids and cockroaches. Furthermore, in each case these neurons respond to sound stimuli (Pearson et al. 1985; Yager and Hoy 1989; Ritzmann et al. 1991). Any convergence of functional properties is by no means a necessary correlate of developmental relatedness because even within the central nervous system of the grasshopper, serially homologous

embryonic IN 714

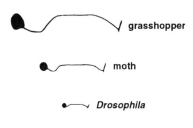

FIGURE 4.14. Drawings at the same scale of putative homologous neurons in grasshopper, moth, and *Drosophila*, illustrating a basically similar morphology but a vast difference in size. The cell bodies, crossing (commissural) segments, and the initial axon segment of INs 714 in each species are shown. The cell bodies of each neuron belong to the equivalent cell cluster, the crossing segments are located in the equivalent commissures, and the axons in the equivalent longitudinal fiber bundles. Median dendritic projections into the VIT have not yet formed. Scale bar, 50 μm. (Modified from Boyan and Ball, 1993, with permission of Elsevier Science.)

neurons from the 7-4 lineage have been shown to possess heterogeneous physiological properties (Pearson et al. 1985). Nevertheless, if, as proposed earlier, the chordotonal system is common to all these insects, then aspects of sensory processing in ontogenetically equivalent neuropilar regions of different insects may be very similar, even when the modality of the information being processed is different.

2.4. Development of the Chordotonal Pathway in Atympanate Insects

In tympanate, stridulating grasshoppers (Acridids), such as *Locusta* and *Schistocerca*, the tympanal afferents project into specific areas of the neuropil, such as the ring tract and VIT, where they synapse with identified interneurons, a subset of which are the progeny of neuroblast 7-4. However, different selective pressures acting on the tympanal pathways of various acridids have produced a spectrum of modern phenotypes, among them species that possess normal tympanal structures but are silent, and others that have considerably reduced auditory organs and are silent (Mason 1969; Riede, Kämper, and Höfler 1990). The same basic projection pattern of tympanal afferents into the ring tract reported for silent species is conserved in tympanate, stridulating acridids and even in those without fully developed auditory organs. Riede, Kämper, and Höfler (1990) go so far as to suggest that tympanal receptors predate acoustic communication.

This hypothesis can be tested in primitive grasshoppers such as the morabines, proscopids, and gumastacids, all of which never develop external auditory structures (ears) at all (see also Hoy, Chapter 1). From the work of

Meier and Reichert (1990), we know that the pleural chordotonal receptors of the acridid grasshopper differentiate from the body wall in the equivalent position to the tympanal afferents, and project to the equivalent part of the central nervous system. Meier and Reichert (1990) then examined the developing peripheral nervous system of the primitive morabine grasshopper *Heide amiculi* and found that at 40% and 45% of embryonic development, the pleural chordotonal organs in the abdominal segments are generated by epithelial invagination of the pleural ectoderm near the posterior segment boundary in the same manner as in *Schistocerca* (Fig. 4.15).

There follows a phase of cell migration toward the anterior part of the segment and a phase of axogenesis during which the axons project onto the

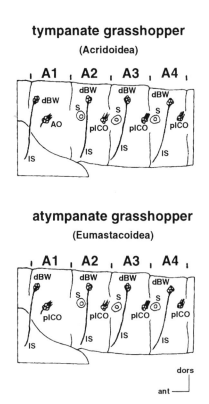

FIGURE 4.15. The pleural chordotonal organs in the first four abdominal segments of the tympanate grasshopper *Locusta migratoria* and the atympanate *Heide amiculi* shown in lateral view at 40% of embryogenesis. The pattern of development is the same in both groups of insects. Whereas the A1 pleural chordotonal receptors (plCO) of *Heide amiculi* form in the same position as those (AO) in *Locusta migratoria*, the atympanate grasshopper is deaf because its receptors do not become associated with an auditory apparatus. ant, anterior; dors, dorsal; IS, intersegmental nerve; dBW, dorsal body wall. (Modified from Meier and Reichert, 1990, with permission of John Wiley and Sons, Inc.)

intersegmental nerve. Because a pleural chordotonal organ also forms in the first abdominal segment at the site where an auditory organ develops in the more advanced acridid grasshoppers, Meier and Reichert (1990) concluded that the pleural chordotonal organ represents the more primitive ancestral condition, supporting the hypothesis of Riede, Kämper, and Höfler (1990).

2.5. Genetic Regulation of Chordotonal Receptor Identity

The fly *Drosophila* and the grasshopper are separated by even greater evolutionary time than the various grasshopper species discussed earlier. Despite this, the dorsal, lateral, and ventral clusters of chordotonal receptors are also present in embryonic *Drosophila* (Fig. 4.16) and maintain the anatomical organization described for the grasshopper with respect to each other and to the segmental nerve (Meier, Chabaud, and Reichert 1991). Furthermore, the lateral cluster develops in a similar fashion to that in the grasshopper, except that the receptors do not become associated with an external ear. Functionally, nothing is known about these receptors in *Drosophila*, but the anatomical evidence provided by Meier, Chabaud, and Reichert (1991) suggests that one might expect such receptors to play a similar role to that proposed for their putative homologs in the grasshopper (Hustert 1978), namely, in stretch reception, possibly monitoring abdominal pumping during respiration or egg laying.

If ontogenetically equivalent chordotonal systems are present in the dipteran and orthopteroid insects, then genetic approaches can be employed to explore the developmental mechanisms responsible for determining chordotonal receptor identity. It is known that in *Drosophila* the genes *cut* (Bodmer and Jan 1987), *numb* (Uemura et al. 1989), and *pox neuro* (Dambly-Chaudiere et al. 1992) all change the fate of peripheral neurons when deleted. Within the chordotonal system, mutations are known that affect the number and location of clusters and of cells within clusters. Some mutations affect all of the segmentally repeated body wall receptors in the same way, whereas others convert abdominal patterns of receptors to thoracic patterns (see Fig. 4.16). In the wild-type fly, the lateral cluster contains three cells in a thoracic neuromere and five in an abdominal neuromere. In flies containing mutations in the *engrailed* gene, part of the lateral cluster is eliminated in both the thorax and abdomen, demonstrating that the chordotonal organs occupy a comparable (posterior) position along the anterior-posterior axis in each segment. In flies containing mutations in the *rhomboid* gene, the number of cells in the abdominal lateral cluster is reduced to three cells (the thoracic condition) while the positional relationships of the two clusters are preserved with respect to the segmental nerve. In certain *abdominal-A* mutants, the abdominal pattern of receptors is

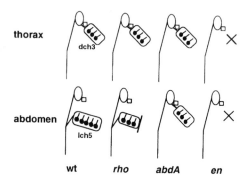

FIGURE 4.16. The alterations in number and pattern of the body wall receptors in various mutants of *Drosophila* lend support to the idea that these receptors are serially homologous. wt indicates the wild-type pattern. In some *engrailed* mutants (*en*), the lateral cluster is missing entirely in both the thorax and abdomen, indicating that the cluster lies in the same (posterior) compartment of the body. In some *rhomboid* mutants (*rho*), the number of cells in the lateral cluster is reduced to three (the same number as in the thorax) while the position of lateral and dorsal clusters with respect to the segmental nerve is preserved. In certain *abdominal A* mutants (*abdA*), the abdominal receptors take on a thoracic number and arrangement. (Modified from Meier, Chabaud, and Reichert, 1991, with permission of Birkhäuser Verlag.)

modified to a thoracic one by a reduction in receptor cell number (to three) in the abdominal lateral cluster, a change in cluster position, and a switch in the projection of receptor axons to the position normal for the thoracic cluster. Single genes can thus switch the identity of a cluster of sensory cells and mimic within a single segment the changes normally observed serially along the segmented body.

2.6. Regeneration in the Chordotonal System

Regeneration experiments can reveal much about the regulatory mechanisms that shape the normal development of the chordotonal (auditory) pathway. Such experiments can be designed according to several scenarios. The experiment can be carried out embryonically, in which case the regenerating axons grow into a tissue that is not too dissimilar in age from themselves; that is, gradients of extracellular molecules in the central nervous system are likely to be unchanged, and pioneer cells and guidepost cells may well still be present. These types of experiments have been performed in the motor system (Whitington and Seifert 1981) and in the cercal sensory system (Shankland and Goodman 1982) of embryonic grasshoppers.

In the chordotonal sensory system, killing by heat shock of pioneer cells in the foreleg of the cricket does not affect normal development of the tympanal organ but does prevent the outgrowing tympanal afferents from crossing the femur-tibia segment boundary on their way into the central nervous system (Klose 1991). If the chordotonal system is anything like the cercal sensory, then (1) there is a competition between receptors for arborization space within the central nervous system and for synaptic connections with interneurons (Murphey 1986); and (2) ingrowing afferent terminals influence the distribution of dendrites of central neurons (Shankland, Bentley, and Goodman 1982). Interestingly, cell ablation does not appear to have been carried out in the ear itself in embryonic insects to see if the receptors themselves are regulated in a manner similar to neuroblasts in the central nervous system (Doe and Goodman 1985).

If the regeneration experiment is carried out postembryonically, the regenerating axons grow into a tissue that, though still developing, is structurally and molecularly quite different from that into which they originally grew in the embryo. Pioneer cells are no longer present, and the target cells themselves may have changed their structure in response to denervation. Mechanisms of axon navigation, and interactions between arriving afferents and central targets can be examined in this type of experiment.

Such regeneration experiments performed on the auditory pathway of larval crickets have broadly followed two strategies. On the one hand, the whole foreleg bearing the ear has been removed and the effects of denervation studied in an animal in which the foreleg, but not necessarily the ear or its afferents, regenerates (e.g., Schildberger et al. 1986). On the other hand, the auditory nerve in one foreleg can be crushed during early larval instars without affecting the leg itself (Hoy, Nolen, and Casaday 1985; Pallas and Hoy 1986; Brodfuehrer and Hoy 1988). Foreleg amputations and transplantations in larval crickets were performed by Ball (1979) and Biggin (1981). Forelegs transplanted to the site of the mesothoracic leg in third to sixth instar animals (before a tympanum is present) survived subsequent molts and formed a tympanum. Thus the prothoracic (ear-bearing) nature of the leg was already determined prior to the third instar.

Regeneration of the prothoracic leg following amputation resulted in the formation of tympanal cuticle only when performed sufficiently early (prior to the third instar; Ball 1979). Regeneration of a foreleg following amputation between the coxa and trochanter in early instars did not result in the tympanal receptors being regenerated (Huber et al. 1984; Huber 1987). However, when the amputation was carried out between the femur and tibia in middle instar crickets, in which the tympanal organ had already formed, then the tympanal organ was morphologically and functionally regenerated (Schildberger and Huber 1988).

Schildberger et al. (1986) used the unilateral preparation created by amputation of the leg in the third/fourth instar, in which no functional ear formed on the regenerated leg, to look for subsequent structural and functional changes in identified central auditory neurons. Identified interneurons were shown to be plastic in that they sprouted dendritic processes following deafferentation during a "critical phase" early in larval development. The interneurons then directed their newly formed branches toward the auditory neuropil containing afferents from the intact side (Fig. 4.17A). Novel functional synaptic connections formed between the afferents and the "restructured" interneurons on the still intact contralateral side. Such critical phases in development during which the central components of sensory systems respond to changes in the periphery in a highly plastic manner have been reported for a range of preparations, including the cercal system of the cricket (Matsumoto and Murphey 1978; Murphey and Levine 1980) and the auditory system of the owl (Knudsen 1985).

In a parallel study on another species of cricket, Hoy, Nolen, and Casaday (1985) and Pallas and Hoy (1986) showed that auditory deprivation induced by unilateral nerve crush in one foreleg also evoked dendritic sprouting into the still intact contralateral neuropil by identified interneurons. These interneurons were also able to form functional synaptic connections with the remaining afferents.

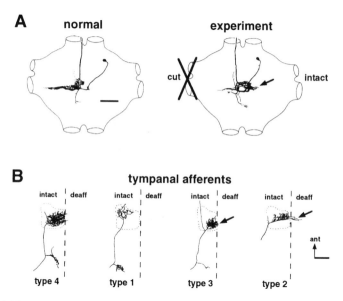

FIGURE 4.17. Regeneration and plasticity in the auditory pathway of the cricket (A) and grasshopper (B). A: Transection of the auditory nerve in the left foreleg during early postembryonic development in the cricket causes interneuron 1 in the prothoracic ganglion to redirect its dendritic processes from the lesioned side across the midline and to the intact side, where it forms novel synapses with tympanal afferents. Scale bar, 200 µm. (Modified from Schildberger et al., 1986, with permission of Springer-Verlag.) B: In the grasshopper, lesioning of the tympanal nerve on the right side of the body (deafferented) during early postembryonic development causes type 3 and type 2 tympanal afferents from the intact side to extend their axonal processes across the midline of the metathoracic ganglion and to invade the neuropil on the deafferented side, where the intrinsic afferents are now missing. Scale bar, 100 µm. (Modified from Lakes, Kalmring, and Engelhard, 1990, with permission of Springer-Verlag.)

Because the auditory pathway of the cricket is symmetrically organized, most interneurons arise from bilaterally paired neuroblasts in the embryo (see Fig. 4.12). The contralateral auditory neuropil into which the test interneuron sprouts following unilateral denervation is also normally innervated by its bilateral "control" homolog. In an elaboration of the deafferentation approach described earlier, Brodfuehrer and Hoy (1988) reported that the sprouting dendrites of the "deafferented" interneuron and those of its "intact" bilateral homolog do not functionally compete for synaptic connections with the remaining afferents. This is surprising and contrasts with studies in the cercal sensory system of the cricket, in which an examination of evoked excitatory postsynoptic potentials (EPSPs), the density of dendritic branching, and the actual distribution of synapses all revealed strong competition between normal afferents, as well as between normal and ectopic afferents, for synaptic sites on identified postsynaptic cells (Shepherd and Murphey 1986; Bacon and Blagburn 1992). One of these interneurons has also been shown to presynaptically modulate the efficacy of the synapse formed with it by a particular cercal afferent at the expense of another (Davis and Murphey 1994).

In early larval grasshoppers, Lakes, Kalmring, and Engelhard (1990) extirpated the receptors themselves after cutting the tympanal nerve at the ear. Because sensory receptors do not regenerate in the postembryonic grasshopper, this procedure effectively unilaterally denervated the auditory pathway. In one species of grasshopper, the response of three of the four types of afferents from the intact side was to sprout and grow into the anterior ring tract on the contralateral "degenerated" side of the metathoracic ganglion (Fig. 4.17B). Morphological changes were also evoked by denervation in some, but not all, interneurons tested, and some interneurons regained their "normal" physiological response properties as a result of synaptic remodeling (Lakes, Kalmring, and Engelhard 1990; Lakes and Kalmring 1991).

Regeneration and transplantation experiments performed in vertebrate auditory/vestibular pathways reveal a number of developmental mechanisms that appear to be fundamentally similar to those described earlier for the invertebrate auditory system. In adult chicks, for example, new hair cells replace those damaged in the basilar papilla by sound or drugs (Swanson 1988). Recovery of sensory function appears to depend on the amount of basilar epithelium remaining after trauma. This suggests a subsequent differentiation and proliferation of epithelial cells like that demonstrated for the central nervous system of insects (Doe and Goodman 1985) but not yet in the periphery. In the mammalian vestibular system, hair cells killed by drugs may indeed be regulated by cellular mechanisms in which another sensory epithelial cell type is transformed into a hair cell (Rubel, Dew, and Roberson 1995). The cues for functional regeneration of hair cell stereocilia in the embryonic chick appear to reside in the ear itself. Corwin and Cotanche (1989) performed experiments in which they denervated embryonic chicken ears before the time of hair cell production. Following transplantation of these ears to host embryos, Corwin and Cotanche report that vestibular and auditory hair cell phenotypes differentiated appropriately, with the correct gradients of hair cell structural phenotypes. This suggests that the normal development of gradients in hair cell stereocilia properties is controlled by location-specific cues originating in the ear itself. Such local cues in the epithelium may involve a family of molecules called *transforming growth factor β* (TGF-β; Mahanthappa 1994). Transforming growth factors may function via a ligand and receptor mechanism similar to interactions involving proteins of the *Notch* gene that regulate cell fate in the grasshopper and *Drosophila* nervous system (Menne and Klämbt 1994).

2.7. Development of Auditory Behavior

Studies correlating the behavior of the insect to the developing auditory pathway are rare. Prior to the imaginal molt, the cuticular tympanum of crickets is so thick that only the most intense sounds are transmitted and the insect may, to all intents and purposes, be considered deaf to airborne sound (Ball and Young 1974; Ball and Hill 1978). The experiments with foreleg vibration described early (see Section 2.2.3.3) show that the tympanal receptors in late instars of crickets are still tuned to the same frequencies as in adults (Ball and Hill 1978). Shuvalov and Popov (1971) showed that sounds in the frequency range of 2.5 to 45 kHz evoke escape responses in young female crickets up to 5 days after the imaginal molt. Frequencies of 4 kHz and 12 to 15 kHz were particularly effective and represent frequencies that correspond to the ranges of optimal sensitivity of groups of tympanal afferents. The sound frequencies that can evoke these escape responses change during ontogeny. In sexually mature female crickets, only the high frequencies remain effective. This suggests that either a remodeling of synaptic weighting in the tympanal pathway occurs, perhaps along the lines described for the cercal receptor/giant interneuron pathway of the cricket (Llenicka and Murphey 1989), or hormonal influences change interneuron sensitivities to particular afferent input, as demonstrated for juvenile hormone in the cricket auditory pathway (Atkins, Henley, and Stout 1990).

It is also possible that ground vibration plays a role in activating central chordotonal neurons during postembryonic development. From an evolutionary viewpoint, vibration reception is likely to be more primitive than auditory, as evidenced by the presence of a chordotonal pathway with the same morphology in both singing and nonsinging grasshoppers (Riede, Kämper, and Höfler 1990). Even though tympanal receptor projections are present, the central chordotonal pathway may function as a vibratory pathway early in larval development and become multimodal later with the appearance of synaptic input from the ear. This has been demonstrated for adult grasshoppers and bushcrickets in which receptors in all six legs respond to vibration stimuli and provide synaptic input to a number of interneurons that also respond to airborne sound (Cokl, Kalmring, and Wittig 1977; Kühne, Lewis, and Kalmring 1980; Silver, Kalmring, and Kühne 1980).

Petersen, Kalmring, and Cokl (1982) showed that this convergence of vibratory and auditory inputs is also present in later larval instars. If there is a behavioral significance for such a sensory neural network in immature insects, it may relate to the fact that the auditory pathway subserves not only reproductive behavior but also has a broader communicative function (Alexander 1967). Aggregation of larval grasshoppers, for example, may be promoted by vibratory communication. Central interneurons of the chordotonal pathway, such as INs 714, 314, which receive input from both tympanal and vibratory receptors, have been shown to provide input at the co-contraction phase to the neural network responsible for the jump (Pearson, Heitler, and Steeves 1980; Pearson and Robertson 1981). The locust jump is an escape behavior and as such is present in early larval instars (Gabriel 1985; Queathem 1991).

3. Conclusions

The aim of this review was to describe the development of the auditory pathway in a range of insect species and thus to uncover some general principles governing the ontogeny of the ear, of auditory receptors, and of auditory interneurons. A good whereas is that whereas the sensory receptor cells that make up the adult ear all differentiate from the epithelium during embryogenesis, only during postembryonic development do these receptors become correctly oriented with respect to the developing cuticular structures of the ear, and only at the adult stage is the insect ear fully functional. This developmental plan has some evolutionary echoes in vertebrate auditory development. In birds and mammals, for example, proliferation of inner ear sensory hair cells also only occurs during embryonic development, whereas in amphibians and fish, proliferation continues well into postembryonic development (Corwin 1983; Popper and Hoxter 1984). In mammals, epitheloid cells differentiate from the rostrolateral wall of the otic cup, migrate outward, and condense to form a funnel-shaped scaffold for outgrowing auditory axons into the otocyst (Carney and Silver 1983) — a process strongly reminiscent of the events desribed for the embryonic grasshopper chordotonal system by Meier and Reichert (1990).

Another feature of insect auditory system development that becomes immediately obvious from the data reviewed is its conservative nature. At the molecular/genetic level, homologous precursors and neurons in both the grasshopper and *Drosophila* express homologous proteins, during both differentiation and axogenesis (Bastiani et al. 1987; Patel et al. 1989; Ball et al. 1991; Siegler, Manley, and Thompson 1991). Furthermore, common mechanisms are likely to regulate cell number in the serially arranged organs of the pleural chordotonal system in all species (see Fig. 4.17). Cell fate is genetically regulated, and it is at this molecular level of cell identity that the greatest advances are likely to come in our understanding of the development of insect auditory systems.

At the cellular level, pathway formation in the chordotonal system is via equivalent pioneer cells using mechanisms generally applicable to both vertebrate and invertebrate nervous system development. Current research shows the insect nervous system to be a serial repetition of anatomical and functional compartments or modules, an organizational plan well known from the vertebrate nervous system (Leise 1990). A serial array of auditory and pleural chordotonal sensory receptors differentiates at equivalent sites from the dorsal body wall of both tympanate and atympanate insects (see Fig. 4.15). The receptors in all the homologous sense organs in the array develop in a similar manner, even when the numbers of sensory cells between segments or insects differ.

Superimposed on this serial plan is the requirement for specialization in certain segments, for example, those involved in audition. Although segmentally homologous structures (such as the ear and pleural chordotonal organs) may differ strikingly in final form and function, lineage studies demonstrate how a basic organizational plan is modified during development to provide the diversity of structure and function seen along the length of the nervous system of the adult insect.

Acknowledgments. Eva Lodde is gratefully acknowledged for technical assistance.

References

Abrams TW, Pearson KG (1982) Effects of temperature on identified central neurons that control jumping in the grasshopper. J Neurosci 2:1538–1553.

Alexander RD (1967) Acoustical communication in arthropods. Ann Rev Entomol 12:495–526.

Atkins G, Henley J, Stout J (1990) A neural correlate for control of syllable period selective phonotaxis in the cricket *Acheta domestica* by juvenile hormone III. Soc Neurosci Abstr 20:758.

Bacon JP, Blagburn JM (1992) Ectopic sensory neurons in mutant cockroaches compete with normal cells for central targets. Development 115:773–784.

Bacon JP, Murphey RK (1984) Receptive fields of cricket (*Acheta domesticus*) interneurones are related to their dendritic structure. J Physiol (Lond) 352:601–623.

Ball EE (1979) Development of the auditory tympana in the cricket *Teleogryllus commodus* (Walker): experiments on regeneration and transplantation. Experientia 35:324–325.

Ball EE, Hill KG (1978) Functional development of the auditory system of the cricket, *Teleogryllus commodus*. J Comp Physiol A 127:131–138.

Ball E, Young D (1974) Structure and development of the auditory system in the prothoracic leg of the cricket *Teleogryllus commodus* (Walker). II. Postembryonic development. Z Zellforsch 147:313–324.

Ball EE, Oldfield BP, Rudolph KM (1989) Auditory organ structure, development, and function. In: Huber F, Moore TE, Loher W (eds) Cricket Behavior and Neurobiology. Ithaca, NY: Cornell University Press, pp. 391–422.

Ball EE, Rehm EJ, Patel NH, Goodman CS (1991) Evolution of insect segmentation and homeotic genes and their function during neurogenesis. Soc Neurosci Abstr 17:11.

Bastiani MJ, Goodman CS (1986) Guidance of neuronal growth cones in the grasshopper embryo. III. Recognition of specific glial pathways. J Neurosci 6:3541–3551.

Bastiani M, Pearson KG, Goodman CS (1984) From embryonic fascicles to adult tracts: organization of neuropile from a developmental perspective. J Exp Biol 112:45–64.

Bastiani MJ, Raper JA, Goodman CS (1984) Pathfinding by neuronal growth cones in grasshopper embryos. III. Selective affinity of the G growth cone for the P cells within the A/P fascicle. J Neurosci 4:2311–2328.

Bastiani MJ, Harrelson AL, Snow PM, Goodman CS (1987) Expression of fasciclin I and II glycoproteins on subsets of axon pathways during neuronal development in the grasshopper. Cell 48:745–755.

Bate CM (1976) Embryogenesis of an insect nervous system. I. A map of the thoracic and abdominal neuroblasts in *Locusta migratoria*. J Embryol Exp Morphol 35:107–123.

Bentley D, Caudy M (1983) Pioneer axons lose directed growth after selected killing of guidepost cells. Nature 304:62–65.

Bentley D, Keshishian H (1982) Pathfinding by peripheral pioneer neurons in grasshoppers. Science 218:1082–1088.

Bentley D, Keshishian H, Shankland M, Toroian-Raymond A (1979) Quantitative staging of embryonic development of the grasshopper *Schistocerca nitens*. J Embryol Exp Morphol 54:47–74.

Biggin RJ (1981) Pattern re-establishment — transplantation and regeneration of the leg in the cricket *Teleogryllus commodus* (Walker). J Embryol Exp Morphol 61:87–101.

Blagburn JM (1989) Synaptic specificity in the first instar cockroach: patterns of monosynaptic input from filiform hair afferents to giant interneurons. J Comp Physiol A 166:133–142.

Bodmer R, Jan YN (1987) Morphological differentiation of the embryonic peripheral neurons in *Drosophila*. Roux's Arch Dev Biol 196:69–77.

Boyan GS (1983) Postembryonic development in the auditory system of the locust. Anatomical and physiological characterization of interneurones ascending to the brain. J Comp Physiol 151:449–513.

Boyan GS (1984) Neural mechanisms of information processing by identified interneurones in Orthoptera. J Insect Physiol 30:27–41.

Boyan GS (1986) Modulation of auditory responsiveness in the locust. J Comp Physiol A 158:813–825.

Boyan GS (1992) Common synaptic drive to segmentally homologous interneurons in the locust. J Comp Neurol 321:544–554.

Boyan GS (1993) Another look at insect audition: the tympanic receptors as an evolutionary specialization of the chordotonal system. J Insect Physiol 39:187–200.

Boyan GS, Ball EE (1993) The grasshopper, *Drosophila*, and neuronal homology. Prog Neurobiol 41:657–682.

Boyan GS, Willams JLD (1995) Lineage analysis as an analytical tool in the insect central nervous system: bringing order to interneurons. In: Breidbach O, Kutsch W (eds) The Nervous Systems of Invertebrates: An Evolutionary and Comparative Approach. Basel: Birkhäuser Verlag, pp. 273–301.

Boyan GS, Therianos S, Williams JLD, Reichert H (1995) Axogenesis in the embryonic brain of the grasshopper *Schistocerca gregaria*. An identified cell analysis of early brain development. Development 121:75–86.

Brodfuehrer PD, Hoy RR (1988) Effect of auditory deprivation on the synaptic connectivity of a pair of identified interneurons in adult field crickets. J Neurobiol 19:17–38.

Campos-Ortega JA, Knust E (1990) Genetic mechanisms in early neurogenesis of *Drosophila melanogaster*. Ann Rev Genet 24:387–407.

Carney PR, Silver J (1983) Studies on cell migration and axon guidance in the developing distal auditory system of the mouse. J Comp Neurol 215:359–369.

Chiba A, Shepherd D, Murphey RK (1988) Synaptic rearrangement during postembryonic development in the cricket. Science 240:901–905.

Cokl A, Kalmring K, Wittig H (1977) The responses of auditory ventral cord neurons of *Locusta migratoria* to vibration stimuli. J Comp Physiol A 120:161–172.

Corwin JT (1983) Postembryonic growth of the macula neglecta auditory detector in the ray, *Raja clavata*: continual increases in hair cell number, neural convergence, and physiological sensitivity. J Comp Neurol 217:345–356.

Corwin JT, Cotanche DA (1989) Development of location-specific hair cell stereocilia in denervated embryonic ears. J Comp Neurol 288:529–537.

Dambly-Chaudiere C, Jamet E, Burri M, Bopp D, Basler K, Hafen E, Dumont N, Spielmann P, Ghysen A, Noll M (1992) The paired box gene *pox-neuro* — a determinant of poly-innervated sense organs in *Drosophila*. Cell 69:159–172.

Davis GW, Murphey RK (1994) Long-term regulation of short-term transmitter release properties: retrograde signalling and synaptic development. Trends Neurosci 17:9–13.

Doe CQ (1992) Molecular markers for identified neuroblasts and ganglion mother cells in the *Drosophila* central nervous system. Development 116:855–863.

Doe CQ, Goodman CS (1985) Early events in insect neurogenesis. I. Development and segmental differences in the pattern of neuronal precursor cells. Dev Biol 111:193–205.

Doe CQ, Chu-LaGraff Q, Wright DM, Scott MP (1991) The *prospero* gene specifies cell fates in the *Drosophila* central nervous system. Cell 65:451–464.

Ebendt R, Friedel J, Kalmring K (1994) Central projection of auditory receptors in the prothoracic ganglion of the bushcricket *Psorodonotus illyricus* (Tettigoniidae): computer-aided analysis of the end branch pattern. J Neurobiol 25:35–49.

Feng JZ, Brugge JF (1983) Postnatal development of auditory callosal connections in the kitten. J Comp Neurol 214:416–426.

Gabriel JM (1985) The development of the locust jumping mechanism. I. Allometric growth and its effect on jumping performance. J Exp Biol 118:313–326.

Gilbert S (1991) Developmental Biology. Sunderland, MA: Sinauer Associates.

Goodman CS, Doe CQ (1994) Embryonic development of the *Drosophila* central nervous system. In: Bate M, Martinez-Arias A (eds) The Development of *Drosophila*, Vol. I. Cold Spring Harbor, NY: Cold Spring Harbor Press, pp. 1131–1206.

Gray EG (1960) The fine structure of the insect ear. Philos Trans R Soc Lond B 243:75–94.

Grenningloh G, Bieber AJ, Rehm EJ, Snow PM, Traquina ZR, Hortsch M, Patel NH, Goodman CS (1990) Molecular genetics of neuronal recognition in *Drosophila*: evolution and function of immunoglobulin superfamily cell adhesion molecules. Cold Spring Harbor Symp Quant Biol 50:327–340.

Halex H, Kaiser W, Kalmring K (1988) Projection areas and branching patterns of the tympanal receptor cells in migratory locusts, *Locusta migratoria* and *Schistocerca gregaria*. Cell Tissue Res 253:517–528.

Harris DM, Dallos P (1984) Ontogenetic changes in frequency mapping of a mammalian ear. Science 225:741–743.

Ho RK, Goodman CS (1982) Peripheral pathways are pioneered by an array of central and peripheral neurons in grasshopper embryos. Nature 297:404–406.

Holland P, Ingham P, Krauss S (1992) Mice and flies head to head. Nature 358:627–628.

Hoy RR, Nolen TG, Casaday GC (1985) Dendritic sprouting and compensatory synaptogenesis in an identified interneuron follow auditory deprivation in a cricket. Proc Natl Acad Sci USA 82:7772–7776.

Huangfu M, Saunders JC (1983) Auditory development in the mouse: structural maturation of the middle ear. J Morphol 176:249–259.

Huber F (1987) Plasticity in the auditory system of crickets: phonotaxis with one ear and neuronal reorganization within the auditory pathway. J Comp Physiol A 161:583–604.

Huber F, Kleindienst HU, Weber T, Thorson J (1984) Auditory behavior of the cricket. III. Tracking of male calling song by surgically and developmentally one-eared females, and the curious role of the anterior tympanum. J Comp Physiol A 155:725–738.

Hustert R (1975) Neuromuscular coordination and proprioceptive control of rhythmical abdominal ventilation in intact *Locusta migratoria migratorioides*. J Comp Physiol A 97:159–179.

Hustert R (1978) Segmental and interganglionic projections from primary fibres of insect mechanoreceptors. Cell Tissue Res 194:337–351.

Klose M (1991) Aspekte der Entwicklung des peripheren sensorischen Nervensystems in den Beinanlagen von Grillen und Heuschrecken. Doctoral Thesis, Ludwig-Maximilians-Universität, München.

Klose M, Bentley D (1989) Transient pioneer neurons are essential for formation of an embryonic peripheral nerve. Science 245:982–984.

Knudsen EI (1985) Experience alters the spatial tuning of auditory units in the optic tectum during a sensitive period in the barn owl. J Neurosci 5:3094–3109.

Kühne R, Lewis B, Kalmring K (1980) The responses of ventral cord neurons of *Decticus verrucivorus* (L) to sound and vibration stimuli. Behav Proc 5:55–74.

Lakes R, Kalmring K (1991) Regeneration of the projection and synaptic connections of tympanic receptor fibers of *Locusta migratoria* (Orthoptera) after axotomy. J Neurobiol 22:169–181.

Lakes R, Kalmring K, Engelhard KH (1990) Changes in the auditory system of locusts (*Locusta migratoria* and *Schistocerca gregaria*) after deafferentation. J Comp Physiol A 166:553–563.

Lawrence PA (1993) The Making of a Fly. Oxford: Blackwell Science Ltd.

Leise EM (1990) Modular construction of nervous systems: a basic principle of design for invertebrates and vertebrates. Brain Res Rev 15:1–23.

Lippe WR (1987) Shift of tonotopic organization in brain stem auditory nuclei of the chicken during late embryonic development. Hear Res 25:205–208.

Lnenicka GA, Murphey RK (1989) The refinement of invertebrate synapses during development. J Neurobiol 20:339–355.

Mahanthappa NK (1994) Regeneration in the auditory system: lessons from other epithelia, and persisting puzzles. Trends Neurosci 17:357–359.

Mason JB (1969) The tympanal organ of Acridomorpha. Eos 44:267–355.

Matsumoto SG, Murphey RK (1978) Sensory deprivation in the cricket nervous system: evidence for a critical period. J Physiol (Lond) 285:159–170.

Meier T, Reichert H (1990) Embryonic development and evolutionary origin of the orthopteran auditory system. J Neurobiol 21:592–610.

Meier T, Reichert H (1995) Developmental mechanisms, homology and evolution of the insect peripheral nervous system. In: Breidbach O, Kutsch W (eds) The Nervous Systems of Invertebrates: An Evolutionary and Comparative Approach. Basel: Birkhäuser Verlag, pp. 249–271.

Meier T, Chabaud F, Reichert H (1991) Homologous patterns in the embryonic development of the peripheral nervous system in the grasshopper *Schistocerca gregaria* and the fly *Drosophila melanogaster*. Development 112:241–253.

Menne TV, Klämbt C (1994) The formation of commissures in the *Drosophila* CNS depends on the midline cells and on the *Notch* gene. Development 120:123–133.

Michel K, Petersen M (1982) Development of the tympanal organ in larvae of the migratory locust (*Locusta migratoria*). Cell Tissue Res 222:667–676.

Michelsen A (1971) The physiology of the locust ear. I. Frequency sensitivity of single cells in the isolated ear. Z Vergl Physiol 71:49–62.

Murphey RK (1986) Competition and the dynamics of axon arbor growth in the cricket. J Comp Neurol 251:100–110.

Murphey RK, Levine RB (1980) Mechanisms responsible for changes observed in response properties of partially deafferented insect interneurons. J Neurophysiol 43:367–382.

Murphey RK, Jacklet A, Schuster L (1980) A topographic map of sensory cell terminal arborizations in the cricket CNS: correlation with birthday and position in a sensory array. J Comp Neurol 191:53–64.

Oldfield BP (1982) Tonotopic organisation of auditory receptors in Tettigoniidae (Orthoptera: Ensifera). J Comp Physiol 147:461–469.

Oldfield BP (1985) The tuning of auditory receptors in bushcrickets. Hear Res 17:27–35.

Pallas SL, Hoy RR (1986) Regeneration of normal afferent input does not eliminate aberrant synaptic connections of an identified auditory interneuron in the cricket, *Teleogryllus oceanicus*. J Comp Neurol 248:348–359.

Patel NH, Martin-Blanco E, Coleman KG, Poole SJ, Ellis MC, Kornberg TB, Goodman CS (1989) Expression of engrailed proteins in arthropods, annelids, and chordates. Cell 58:955–968.

Pearson KG, Robertson RM (1981) Interneurons coactivating hindleg flexor and extensor motoneurons in the locust. J Comp Physiol A 144:391–400.

Pearson KG, Heitler WJ, Steeves JD (1980) Triggering of locust jump by multimodal inhibitory interneurons. J Neurophysiol 43:257–278.

Pearson KG, Boyan GS, Bastiani M, Goodman CS (1985) Heterogeneous properties of segmentally homologus interneurones in the ventral cord of locusts. J Comp Neurol 223:133–145.

Petersen M, Kalmring K, Cokl A (1982) The auditory system in larvae of the migratory locust. Physiol Entomol 7:43–54.

Pflüger HJ, Bräunig P, Hustert R (1988) The organization of mechanosensory neuropiles in locust thoracic ganglia. Philos Trans R Soc Lond B 321:1–26.

Popper AN, Hoxter B (1984) Growth of a fish ear: I. Quantitative analysis of hair cell and ganglion cell proliferation. Hear Res 15:133–142.

Prier KR, Boyan GS (1993) Chordotonal input onto an identified auditory interneuron (714) in the locust, *Schistocerca gregaria*. Proceedings of the 21st Göttingen Neurobiology Conference, Göttingen, West Germany, June 1993. Stuttgart: Georg Thieme Verlag, p. 223.

Purves D, Lichtman JW (1985) Principles of Neural Development. Sunderland, MA: Sinauer Associates.

Queathem E (1991) The ontogeny of grasshopper jumping performance. J Insect Physiol 37:129–138.

Raper JA, Bastiani M, Goodman CS (1983a) Pathfinding by neuronal growth cones in grasshopper embryos. I. Divergent choices made by the growth cones of sibling neurons. J Neurosci 3:20–30.

Raper JA, Bastiani M, Goodman CS (1983b) Pathfinding by neuronal growth cones in grasshopper embryos. II. Selective fasciculation onto specific axonal pathways. J Neurosci 3:31–41.

Raper JA, Bastiani MJ, Goodman CS (1984a) Guidance of neuronal growth cones: selective fasciculation in the grasshopper embryo. Cold Spring Harbor Symp Quant Biol 48:587–598.

Raper JA, Bastiani MJ, Goodman CS (1984b) Pathfinding by neuronal growth cones in grasshopper embryos. IV. The effects of ablating the A and P axons upon the behavior of the G growth cone. J Neurosci 4:2329–2345.

Reichert H (1990) Neurobiologie. Stuttgart: Thieme Verlag.

Riede K, Kämper G, Höfler I (1990) Tympana, auditory thresholds, and projection areas of tympanal nerves in singing and silent grasshoppers (Insecta, Acridoidea). Zoomorphology 109:223–230.

Ritzmann RE, Pollack AJ, Hudson SE, Hyvonen A (1991) Convergence of multimodal sensory signals at thoracic interneurons of the escape system of the cockroach, *Periplaneta americana*. Brain Res 563:175–183.

Römer H (1976) Die Informationsverarbeitung tympanaler Rezeptorelemente von *Locusta migratoria* (Acrididae, Orthoptera). J Comp Physiol A 109:101–122.

Römer H (1983) Tonotopic organization of the auditory neuropile in the buschcricket *Tettigonia viridissima*. Nature 306:60–62.

Römer H, Marquart V (1984) Morphology and physiology of auditory interneurons in the metathoracic ganglion of the locust. J Comp Physiol A 155:249–262.

Römer H, Marquart V, Hardt M (1988) Organization of a sensory neuropil in the auditory pathway of two groups of Orthoptera. J Comp Neurol 275:201–215.

Rössler W (1992a) Postembryonic development of the complex tibial organ in the foreleg of the bushcricket *Ephippiger ephippiger* (Orthoptera, Tettigoniidae). Cell Tissue Res 269:505–514.

Rössler W (1992b) Functional morphology and development of tibial organs in the legs I, II and III of the bushcricket *Ephippiger ephippiger* (Insecta, Ensifera). Zoomorphology 112:181–188.

Rubel EW, Dew LA, Roberson DW (1995) Mammalian vestibular hair cell regeneration. Science 267:701–707.

Schildberger K, Huber F (1988) Post-lesion plasticity in the auditory system of the cricket. In: Flohr H (ed) Post-lesion Neural Plasticity. Heidelberg: Springer-Verlag, pp. 564–575.

Schildberger K, Wohlers DW, Schmitz B, Kleindienst HU, Huber F (1986) Morphological and physiological changes in central auditory neurons following unilateral foreleg amputation in larval crickets. J Comp Physiol 158:291–300.

Schumacher R (1973) Morphologische Untersuchungen der tibialen Tympanalorgane von neun einheimsichen Laubheuschrecken-Arten (Orthoptera, Tettigonioidea). Z Morphol Tiere 75:267–282.

Shankland M, Goodman CS (1982) Development of the dendritic branching pattern of the medial giant interneuron in the grasshopper embryo. Dev Biol 92:489–506.

Shankland M, Bentley D, Goodman CS (1982) Afferent innervation shapes the dendritic branching pattern of the medial giant interneuron in grasshopper embryos raised in culture. Dev Biol 92:507–520.

Shepherd D, Murphey RK (1986) Competition regulates the efficacy of an identified synapse in crickets. J Neurosci 6:3152–3160.

Shepherd D, Kämper G, Murphey RK (1988) The synaptic origins of receptive field properties in the cricket cercal sensory system. J Comp Physiol A 162:7–11.

Shuvalov VF, Popov VA (1971) Reaction of females of the domestic cricket *Acheta domesticus* to sound signals and its changes in ontogensis. J Evol Biochem Physiol 7:612–616.

Siegler MVS, Manley PE Jr., Thompson KJ (1991) Sulphide silver staining for endogenous heavy metals reveals subsets of dorsal unpaired median (DUM) neurones in insects. J Exp Biol 157:565–571.

Silver S, Kalmring K, Kühne R (1980) The responses of central acoustic and vibratory interneurons in bush crickets and locusts to ultrasonic stimulation. Physiol Entomol 5:427–443.

Swanson GJ (1988) Regeneration of sensory hair cells in the vertebrate inner ear. Trends Neurosci 11:339–342.

Thomas JB, Bastiani MJ, Bate CM, Goodman CS (1984) From grasshopper to *Drosophila*: a common plan for neuronal development. Nature 310:203–207.

Tyrer NM, Gregory GE (1982) A guide to the anatomy of locust suboesophageal and thoracic ganglia. Philos Trans R Soc Lond B 297:91–123.

Uemura T, Shepherd S, Ackerman L, Jan LY, Jan YN (1989) *numb*, a gene required in determination of cell fate during sensory organ formation in *Drosophila* embryos. Cell 58:349–360.

Vaessin H, Grell E, Wolff E, Bier E, Jan LY, Jan YN (1991) *Prospero* is expressed in neuronal precursors and encodes a nuclear protein that is involved in the control of axonal outgrowth in *Drosophila*. Cell 67:941–953.

Whitington PM, Seifert E (1981) Identified neurons in an insect embryo: the pattern of neurons innervating the metathoracic leg of the locust. J Comp Neurol 200:203–212.

Wolf H (1984) Monitoring the activity of an auditory interneuron in a freemoving grasshopper. In: Kalmring K, Elsner N (eds) Acoustic and Vibrational Communication in Insects. Berlin: Paul Parey, pp. 51–60.

Yager DD, Hoy RR (1989) Audition in the praying mantis, *Mantis religiosa* L.: identification of an interneuron mediating ultrasonic hearing. J Comp Physiol A 165:471–493.

5
Neural Processing of Acoustic Signals

Gerald S. Pollack

1. Introduction

Like all sensory systems, auditory systems have been shaped by the stimuli that carry meaning for the animals they serve (see Hoy, Chapter 1; Michelsen, Chapter 2; Römer, Chapter 3; Robert and Hoy, Chapter 6; Barth, Chapter 7; Fullard, Chapter 8). It thus comes as no surprise that auditory neurons and neural circuits are specialized to detect and analyze those sounds that carry behaviorally important information. The strong effect of selective pressure is particularly evident among insects, where hearing has evolved independently many times (Fullard and Yack 1993; Hoy, Chapter 1), and often seems to be a "special-purpose" modality that serves restricted and obvious behavioral functions. Because of this close relationship between biological function and auditory neurophysiology, the first section of this chapter focuses on the behavioral functions of sound and on how biologically meaningful information is represented by the physical parameters of acoustic signals. Subsequent sections examine how this information is analyzed by the nervous system.

One common role for audition is predator detection, and some of the clearest examples of this occur in the bat-detection abilities of some nocturnal flying insects. Echolocating bats detect insects on the wing by emitting ultrasonic sounds and analyzing the echoes that are reflected by the bodies of their prey. Many nocturnally flying insects can hear these echolocation cries and, when they do, take evasive action (Hoy 1992; Fullard, Chapter 8). In many cases bat detection appears to be the main or exclusive function of hearing (Miller and Olsen 1979; Yager, May, and Fenton 1990; Fullard, Chapter 8), whereas in others the auditory system also serves for intraspecific communication (Moiseff, Pollack, and Hoy 1978; Spangler 1988a,b; Robert 1989; Libersat and Hoy 1991). Bat detection is considered in depth elsewhere in this volume (Fullard, Chapter 8) and will only be touched on in this chapter.

Another function of hearing is the detection and localization of sound-producing hosts. Several species of tachinid flies deposit eggs or larvae on or near singing insects, which they locate by homing in on their communication signals. Robert and Hoy discuss this fascinating example of interspecific eavesdropping in Chapter 6.

A third use of hearing, and the main focus of this chapter, is acoustic communication among conspecifics. Insects sing loud, conspicuous songs to attract mates from a distance (Regen 1913; Spooner 1968; Helversen and Helversen 1983; Doolan and Young 1989) and to proclaim, defend, and situate their territories (Cade 1979; Campbell and Shipp 1979; Thiele and Bailey 1980; Römer and Bailey 1986). Acoustic signals are also used in more intimate contexts, such as to induce copulation once a male and female have come together (Ewing 1983; Balakrishnan and Pollack 1996). Responses to conspecific signals are usually evoked only by stimuli that meet more-or-less stringent structural criteria. In other words, conspecific signals usually must be *recognized* as such before they elicit behavioral responses. The following sections summarize behavioral experiments that identify those features of sound signals that are most salient to insect listeners. Later sections consider how these features are detected and analyzed by the nervous system.

1.1. Signal Spectrum

The ears of most insects are sensitive to a wide range of frequencies, and often different auditory receptors within the ear are tuned to different frequencies within the audible range (Michelson 1966, 1971a; Nocke 1972; Oldfield 1985). The possibility thus exists of using different frequency channels for different purposes. One of the clearest examples of this is the phonotactic behavior of crickets. Crickets can hear sounds ranging in frequency from ca. 1 to 2 kHz to over 100 kHz. Intraspecific signals, however, are much narrower in bandwidth. The most conspicuous signal of crickets, the calling song, is dominated by a single frequency component, usually near 5 kHz, with higher harmonics present only at much lower intensity (Nolen and Hoy 1986a; Balakrishnan and Pollack 1996). Calling song (which, like all cricket songs, is produced only by males) serves to attract females for mating. Crickets walk or fly toward calling song models with carrier frequencies that are close to the dominant frequency of the natural song, but away from models with ultrasonic ($\geq$20 kHz) carrier frequencies (Moiseff, Pollack, and Hoy 1978; Pollack, Huber, and Weber 1984; Nolen and Hoy 1986a; Fig. 5.1). These two responses, referred to as positive and negative phonotaxis, respectively, serve the different behavioral functions of pair formation and predator (bat) evasion. They differ not only in frequency sensitivity and direction of locomotion but also in latency (Nolen and Hoy 1986a), in sensitivity to stimulus temporal pattern (Pollack, Huber, and Weber 1984; Nolen and Hoy 1986a; Pollack and El-Feghaly 1993), and in the basic "rule" that drives the reponse — turn toward the most strongly stimulated ear for positive phonotaxis and away from the most strongly

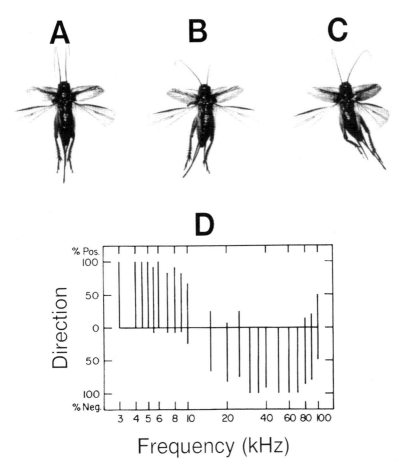

FIGURE 5.1. Positive and negative phonotaxis in flying *Teleogryllus oceanicus*. A–C: Photographs of crickets performing tethered flight (A) without sound stimulation, (B) with a 5-kHz model of conspecific calling song broadcast from the left, and (C) with a 40-kHz song model broadcast from the left. In (B) the abdomen, hind legs, and antennae are swung to the left, indicating an attempt to steer to the left, that is, toward the sound source. In C the cricket is steering to the right, that is, away from the sound source. D shows the direction of phonotaxis as a function of sound frequency. (Reprinted with permission from Moiseff, Pollack, and Hoy 1978; copyright 1978 National Academy of Sciences U.S.A.)

stimulated ear for negative phonotaxis (Moiseff, Pollack, and Hoy 1978; Pollack, Huber, and Weber 1984). Thus, these are qualitatively different behaviors rather than mere variants of a single phonotactic response.

Some crickets use different frequency bands for different intraspecific signals. In *Gryllus bimaculatus*, for example, the dominant frequency of the calling song is 4.5 kHz, whereas the courtship song, a distinct signal that induces the female to mount the male for copulation, has peak energy near 13.5 kHz. Libersat, Murray, and Hoy (1994) studied the role of sound frequency in responses to courtship song by muting males (by removing the sound-producing structures, the forewings) and substituting their songs with synthesized models. They found that energy near 13.5 kHz is required to induce females to mount courting males. In contrast, 4.5 kHz, which is the most effective frequency for eliciting positive phonotaxis, is ineffective in inducing the mounting response during courtship.

In the example, just given different signals are used for long-range (calling song) and short-range (courtship song) interactions. The cicada *Cystosoma saundersii* uses similar signals in these two contexts. Despite the similarity of the signals, long- and short-range behavioral responses show different selectivity for carrier frequency. Acoustic signaling occurs in a two-stage process. Females perform flight phonotaxis to arrive close to a singing male. After landing, they flick their wings in response to male song, which elicits courtship singing from the male. The courtship song is similar to the calling song, though emitted at lower intensity and less regularly. The threshold for flight phonotaxis is lowest at about 850 Hz, which is the dominant component of this species' rather broad-band signal (Young 1980). The wing-flick response, however, is most sensitive to about 1600 to 1700 Hz, the second most conspicuous component of the signal (Doolan and Young 1989).

It is thus clear that insects can discriminate among different frequency bands within their hearing range, but within a single band they do not appear to discriminate, at least at the behavioral level (Wyttenbach, May, and Hoy 1996). For example, the crickets *Gryllus campestris*, *Teleogryllus commodus*, and *T. oceanicus* perform phonotaxis toward calling-song models with frequencies ranging from 3 kHz up to at least 15 kHz (Hill 1974; Moiseff, Pollack, and Hoy 1978; Thorson, Weber, and Huber 1982; Pollack, Huber, and Weber 1984). Behavioral thresholds are lowest in response to the actual song frequency, reflecting the distribution of tuning characteristics among auditory receptor neurons (see Section 2.3). However, other frequencies within a broad low-frequency range are also able to elicit phonotaxis as long as they are of sufficient intensity. The generally good match between auditory tuning and the spectrum of the conspecific signal provides for sufficient sensitivity to detect signals over a relatively large distance while screening out noise (including signals of other species) at other frequencies.

1.2. Recognition of Temporal Pattern

Insect songs are not melodious; they lack the pronounced frequency modulation that characterizes many other animal communication sounds, such as bird song and human speech, and thus they are, quite literally, monotonous. Their information content thus resides not in their melodies but in their rhythms. The typical insect song consists of a series of brief sounds, usually referred to as *pulses* or *syllables*, which occur in repeated, stereotypic sequences. Temporal features such as pulse durations, interval between pulses, and pulse envelope shapes have been varied systematically in a number of behavioral experiments to determine their individual and collective importance for signal recognition (for reviews see Elsner and Popov 1978; Doherty and Hoy 1985; Helverson and Helversen 1994). The most extensive studies on temporal pattern as a cue to song identity have been done with orthopteran insects, that is, crickets (gryllids), grasshoppers (acridids), and bushcrickets (tettigoniids). Because these behavioral results have shaped many of the questions asked in neurophysiological experiments, it will be useful to review some of the major findings.

1.2.1. Temporal Pattern and Recognition of Cricket Songs

The temporal patterns of cricket songs are quite diverse, ranging from simple trills to complex patterns that include rhythmically different subsections. Songs can also have characteristic patterns of modulation of pulse amplitude (Fig. 5.2). Experiments with synthetic song models have shown that many of these features are important for determining behavioral responses.

The simplest cricket songs consist of regularly repeated sound pulses with constant intervals between the pulses; such songs are referred to as *trills*. The main temporal parameters of trills are the pulse period and pulse duration, both of which can influence phonotactic responses. For example, in many trilling species phonotactic responses are best elicited by song models with the species-typical value of the pulse period (Walker 1957; Popov and Shuvalov 1977; Doherty and Callos 1991).

Additional complexity arises when groups of sound pulses occur in short groupings, usually known as *chirps*, that are repeated regularly. This song structure introduces the new parameters of chirp period and chirp duration. In *G. campestris*, the organization of the song into chirps is relatively unimportant for its behavioral effectiveness; most females respond as readily to constant trills as to songs in which the pulses are grouped into separate chirps. The main cue for recognition is the pulse period (Thorsen, Weber, and Huber 1982; Fig. 5.3A).

In other chirping species, song recognition clearly depends on a number of temporal features in addition to pulse period. In *Acheta domesticus*, phonotactic responses are best elicited by the species-typical values of pulse period, pulse duration, number of pulses per chirp, and chirp period (Stout, DeHaan, and McGhee 1983). In *G. bimaculatus*, pulse period, chirp length, and chirp period are all important for phonotaxis. Interestingly, the ranges of acceptable values for the various parameters are dependent on one another. For example, if the pulse period is close to the species-typical value, then a wide range of chirp periods is behaviorally effective. If,

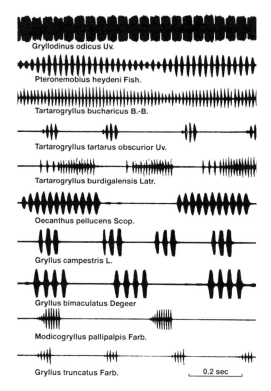

Gryllodinus odicus Uv.

Pteronemobius heydeni Fish.

Tartarogryllus bucharicus B.-B.

Tartarogryllus tartarus obscurior Uv.

Tartarogryllus burdigalensis Latr.

Oecanthus pellucens Scop.

Gryllus campestris L.

Gryllus bimaculatus Degeer

Modicogryllus pallipalpis Farb.

Gryllus truncatus Farb.   0.2 sec

FIGURE 5.2. Oscillograms of the songs of several European and Asian species of crickets, illustrating the diversity of temporal patterns. (Reprinted with permission from Popov et al. 1974; copyright 1974 Westdeutscher Verlag.)

FIGURE 5.3. Aspects of calling-song recognition in *Gryllus*. A: Selectivity for syllable rate in *G. campestris*. The graph shows the amount of time females spent "tracking" song models as they walked on a spherical treadmill. Performance was best for syllable intervals between 28 and 53 ms. Structures of the song models are shown above. (Reprinted with permission from Thorsen, Weber, and Huber 1982; copyright 1982 Springer-Verlag.) B: Selectivity for chirp period and syllable period depends on the values of these parameters (*G. bimaculatus*). Graphs illustrate the boundaries between acceptable song models that were tracked on a treadmill, indicated by filled circles, and unacceptable models that were not (open circles). Each graph summarizes the behavior of a different female. The range of acceptable chirp periods is greater if the syllable period is close to the species-typical value (indicated by the bar parallel to the x-axis), and, similarly, a wider range of syllable periods is accepted if the chirp period is "correct" (indicated by bar near *y*-axis). Definitions of chirp period and syllable period are indicated above the graphs. (Reprinted with permission from Doherty 1985a; copyright 1985 Springer-Verlag.)

A

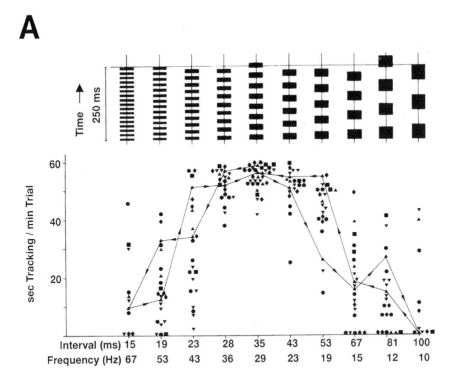

B

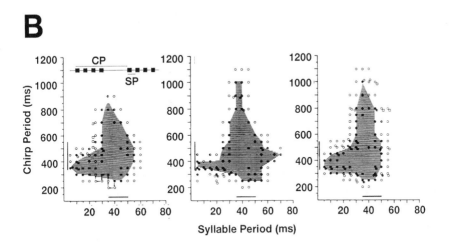

however, the pulse period deviates from the optimal value, then the range of effective chirp periods is narrower. Similarly, if the chirp period is optimal, then a wide range of syllable periods is effective, but if the chirp period deviates from the optimum, then the range of effective syllable periods is restricted (Doherty 1985a,b; Fig. 5.3B).

Some cricket songs include rhythmically distinct subsections. The *T. oceanicus* calling song consists of a brief series of regularly repeated sound pulses (the chirp) alternating with a longer series of pulse doublets (the trills; Fig. 5.4A). Surprisingly, this complexity of rhythm is not necessary for song recognition. This first became apparent when the stereotyped sequence of interpulse intervals that creates the normal temporal pattern was shuffled in order, producing a stimulus that retained the values and relative proportions of interpulse intervals of the normal song but obliterated its chirp/trill structure. This song was as attractive, in choice tests, as the natural pattern (Pollack and Hoy 1979), demonstrating that the syntactical information of the normal song is not required for recognition (Fig. 5.4A). Teasing the song's rhythm into its separate components demonstrated even more strikingly that not all of the features of the natural song are required. In choice tests, females were more strongly attracted to the chirp rhythm than to the trill rhythm, despite the fact that the chirp accounts for only about 16% of the normal song's duration (Pollack and Hoy 1981a; Fig. 5.4B). Males also respond phonotactically to their species' calling song. Males tend to occur in loose aggregates, and phonotaxis is an important mechanism for adjusting intermale spacing (Cade 1979; Campbell and Shipp 1979). In contrast to females, males are more strongly attracted by the trill pattern (Pollack 1982). Thus, the complexity of this species' song may in part reflect the partitioning of different messages into different song components. The congeneric species, *T. commodus*, also has a two-part song. In this case, however, both parts are required to elicit responses most effectively from females (selectivity of male *T. commodus* has not yet been studied) (Hennig and Weber 1997). Thus, song-recognition mechanisms can be quite diverse, even between closely related species.

T. oceanicus females evaluate several temporal parameters of the chirp rhythm, including pulse period, pulse duration, and, in some individuals, the relationship between these two, that is, duty cycle (Doolan and Pollack 1985). Selectivity for temporal pattern is poor near threshold and increases with intensity (Doolan and Pollack 1985; Pollack and El-Feghaly 1993; Fig. 5.5). This intensity dependence makes functional sense. As a sound signal travels from sender to receiver, it is reflected by vegetation, the ground, and any other obstacles and is further distorted by air turbulence and other environmental factors (Michelsen and Larsen 1983; Römer, Chapter 3). What starts out as a clear rhythmic pattern thus becomes increasingly obscure at greater distances from the source. Female *T. oceanicus* are not too choosy about the temporal structures of low-intensity signals, which in natural circumstances would have arisen from distant males; they defer

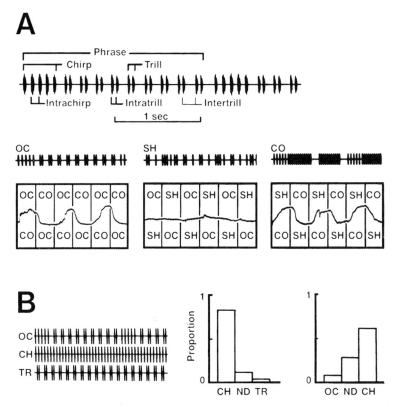

FIGURE 5.4. Selectivity for temporal pattern in *Teleogryllus oceanicus*. A: Oscillograms show the structure of the calling song (above) and of three synthesized models: OC models the normal song; in SH the values and relative proportions of the intrachirp, intratrill, and intertrill intervals are the same as in OC but their order is "shuffled" randomly; CO models the song of a related species, *T. commodus*. Tracings show the directions of abdominal steering movements (see Fig. 5.1) in choice tests. Upward deflections indicate steering movements to the left, and downward to the right; the lettering above and below each trace indicates which song model was played from the left and right, respectively. Left trace: when OC and CO are presented simultaneously, *T. oceanicus* females steer toward the OC (the conspecific song) and "track" switches in the song positions. Middle: Females fail to discriminate between OC and SH. Right: SH is preferred to the heterospecific model. (Reprinted with permission from Pollack and Hoy 1979; copyright 1979 American Association for the Advancement of Science.) B: CH and TR are models constructed from chirp and trill sections of the normal song. Females prefer CH both to TR and also to OC (ND = no discrimination). (Reprinted with permission from Pollack and Hoy 1981a; copyright 1981 Springer-Verlag.)

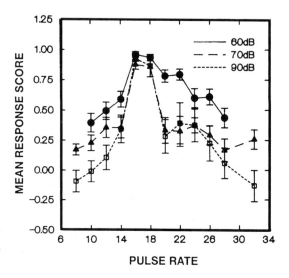

FIGURE 5.5. Intensity-dependent tuning for pulse rate. Graph shows response scores measured from flight steering movements. A score of +1 indicates strong positive phonotaxis, −1 indicates negative phonotaxis, and 0 indicates no response. Phonotaxis is tuned to the pulse rate of the chirp section of the song, ca. 16/s. Tuning is broad at low intensities and sharper at high intensities. (Reprinted with permission from Pollack and El-Feghaly 1993; copyright 1993 Springer-Verlag.)

making decisions based on temporal structure until they are sufficiently close to the source that the signal can be expected to retain its original rhythmic features in relatively pure form.

Male crickets begin to sing courtship song after making antennal contact with a female. The courtship song of *T. oceanicus* is, like the calling song, rhythmically complex, comprising a chirp that is similar to that of calling song and a trill section that consists of long trains of pulses repeated at a much higher rate (Balakrishnan and Pollack 1996). As was found for calling song (see earlier), the chirp pattern of the courtship song is more effective behaviorally than the trill, which can be omitted entirely with only a slight decrease in effectiveness (Fig. 5.6). Communication during courtship is multimodal (Balakrishnan and Pollack 1997), and Balakrishnan and Pollack (1996) speculated that the main signal provided by the trill may be vibration of the male's body as a result of the muscular activity underlying trilling, which might be sensed by the female either through the substrate or through direct contact with the male.

The diversity of structural features that are important for song recognition, and in particular the demonstration in *G. bimaculatus* that different parameters can be "traded off" against one another, suggests that "the neural circuit" for song recognition may actually comprise several circuits operating in parallel, each concerned with a different feature of the signal,

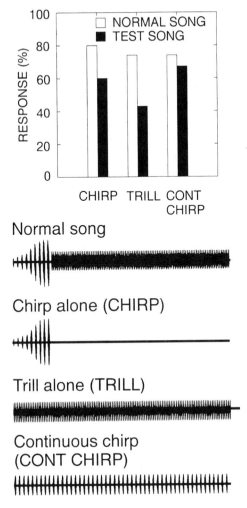

FIGURE 5.6. Selectivity for components of courtship song in *Teleogryllus oceanicus*. The normal courtship song consists of distinct chirp and trill sections. In behavioral tests of the ability to restore mounting of muted males by females, the chirp alone is nearly as effective as the entire song (response frequencies do not differ significantly), whereas the trill alone is significantly less effective. Increasing the length of the chirp pattern to match the duration of the normal song (CONT CHIRP) restores mounting to virtually normal levels. (Reprinted with permission from Balakrishnan and Pollack 1996; copyright 1996 Academic Press.)

with the final behavioral outcome determined by their pooled output. The diversity that is seen between species (or, in *G. bimaculatus*, between individuals) might then reflect differences in the weights given to the various signal features, rather than fundamental differences in the underlying recognition mechanisms.

1.2.2. Temporal Pattern and Recognition of Grasshopper Songs

Acoustic communication in grasshoppers is more complex than in crickets because it is generally bidirectional; males and females engage in a "conversation" that brings them together. The sequence is initiated by the male, who emits calling songs regularly as he moves through the environment. A receptive female responds with her own song, which causes the male to turn and jump toward her. The male continues to sing and continues to be guided by the replies of the stationary female until the two come together. Thus there are two distinct signals, those of the male and of the female, each of which must be recognized by the opposite sex.

This communication system has been studied most thoroughly in the European grasshopper, *Chorthippus biguttulus*. The male's song consists of repeated series of 20 to 60 syllables that are 50 to 100 ms in duration and are separated by pauses of 10 to 15 ms (Helversen 1972). The crucial parameter for recognition is the ratio of these two parameters; the longer the duration of a syllable, the longer the pause between syllables must be to evoke responses from females (Fig. 5.7). The use of the ratio of syllable length to syllable interval as the chief recognition cue, rather than the absolute values of these, may help to allow effective communication over a wide temperature range. Syllable length and interval both vary inversely with temperature and so also do the most effective values of these parameters for eliciting responses from females. However, because females will respond well even to nonoptimal lengths and intervals, provided their ratio is "correct," they can enter into a dialog even with a temperature-mismatched

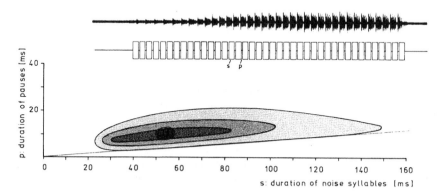

FIGURE 5.7. Temporal pattern recognition in *Chorthippus biggutulus*. Upper traces show the male's song and the structure of a synthetic model. The graph shows the relationship between syllable duration, the length of the pause between syllables, and the response frequency of females tested at 35°C; the contours indicate 20%, 50%, and 80% response rates. The small, dark area shows the range of durations and pauses in the male's natural song, also at 35°C. (Reprinted with permission from Helversen and Helversen 1983; copyright 1983 Springer-Verlag.)

male, which might occur, for example, if one of the two discussants was in full sunlight while the other was in shade (Helversen 1972).

Another song parameter important for females is the absence of long silent gaps in the syllables. Males sing by scraping a file on the inner side of each hind femur against the corresponding forewing. Sounds accompany each up-and-down movement of the leg. During a single syllable, each leg usually undergoes a series of three up-down cycles and produces six distinct sounds separated by silent gaps (at the top and bottom of each cycle). The two legs are slightly out of phase so that the gaps from one side are normally partly obscured by sound produced on the other side. Males who have autotomized one hind leg, however, presumably in an encounter with a predator, necessarily produce syllables with gaps. Females reject songs with gaps longer than 2 to 3 ms (Helversen 1972; Ronacher and Stumpner 1988; see Fig. 5.24). Presumably this selects against males who were unsuccessful in avoiding predators.

The female's response song is similar to the male's song but differs in several respects. The individual sound pulses have more gradual onsets and offsets than those of male song. Moreover, rather than the individual sounds resulting from each upstroke and downstroke of the leg fusing into nearly continuous syllables of relatively long duration, as occurs in males, in females the sound pulses remain distinct and are separated by gaps several milliseconds in length. The female's song is also less intense than the male's and, whereas both songs have broad spectra, the female's has relatively less energy in the high-frequency range (Helversen and Helversen 1997). These sex-specific differences in song structure are reflected in the selectivity of males and females for song parameters. Males respond best to songs with syllables that have gradual onsets (Helverson 1993), whereas females are relatively indifferent to the rate of onset (Helversen and Helversen 1997). Unlike females, which reject songs with gaps, males respond best to songs with gaps. Females prefer songs with syllables longer than 20 milliseconds (see Fig. 5.7), but males respond only if syllable duration is less than 20 milliseconds, corresponding to the distinct sound pulses of female song (Helversen and Helversen 1997). The optimum sound intensity for eliciting responses is lower for males than for females. Finally, females respond best to songs containing both low- and high-frequency components, whereas males respond best to low-frequency songs and, in fact, are inhibited by the addition of high-frequency components (Helversen and Helversen 1997).

1.2.3. Temporal Pattern and Recognition of Bushcricket Songs

Communication systems of bushcrickets are of two main types. The first and most common is similar to that of crickets; song is produced exclusively by males and elicits phonotactic responses from females. As in other orthoptera, behavioral responses show species-specific selectivity for temporal parameters such as syllable rate (Bailey and Robinson 1971).

The second type of bushcricket system, which is found most commonly in the subfamily Phaneropterinae, resembles that of grasshoppers in that females answer males with their own songs, which elicit both singing and phonotaxis from the male. In *Leptophyes punctatissima*, in which the male's song is very brief (<10 ms), females are relatively nonselective about the structural features of the male's signal and respond both to simple clicks and to abnormally long stimuli, provided they contain sufficient energy in the ultrasonic frequency band that this species uses for communication (Robinson 1980; Robinson, Rheinlaender, and Hartley 1986). In other species the songs of males are longer in duration and more complex, and their temporal structures are clearly analyzed by females. In *Ancistrura nigrovittata*, for example, syllable duration and syllable interval (Fig. 5.8) must be within acceptable ranges to elicit female response songs (Dobler, Heller, and Helversen 1994).

A crucial feature of this duetting system is the timing of the female's reply; her song must fall within a brief temporal window in order to elicit phonotaxis and further singing from the male (Heller and Helversen 1986; Robinson, Rheinlaender, and Hartley 1986). In species with short, simple songs, this time window may begin at the onset of the male's song. In other species, in which the male's song is longer and more complex, the song includes a conspicuous "trigger" feature, for example, an isolated sound pulse that follows the more complex portion after a silent interval and serves as a marker for the beginning of the female's response time window (Heller and Helversen 1986; see Fig. 5.8). The latency of the female's reply, which is remarkably constant and matched to the male's temporal acceptance window, serves as a species-identifying feature. The latency varies

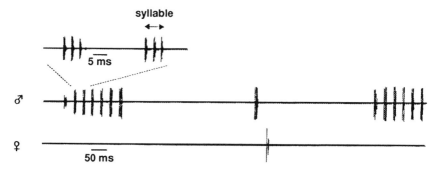

FIGURE 5.8. Acoustic signaling in phaneropterine tettigoniids. The oscillograms show an acoustic interaction in *Ancistrura nigrovittata*. The male's song consists of a series of three-pulse syllables, followed by a silent period, followed by a single "trigger" syllable. The female's response follows shortly after this trigger, but only if the durations and intervals between the earlier syllables were "correct". (Reprinted with permission from Dobler, Heller, and Helversen 1994; copyright 1994 Springer-Verlag.)

among species from the extremely short value of 16ms (*Ancistrura nuptialis*) to several hundred milliseconds (*Isophya leonorae*). The very short response latencies of some species suggests that the underlying neural circuitry in the female must be rather simple and is probably local to the thoracic ganglia. It seems unlikely that sophisticated analysis of the male's song structure can occur in this short interval. Instead, recognition of the temporal pattern of the more complex portion of the male's song, as described earlier for *A. nigrovittata*, probably serves to enable the female's response circuitry so that it can simply "fire" once the beginning of the species-specific time window is signaled by the song's trigger feature.

1.3. Analysis of Sound Direction

Most responses to sound signals require not only their identification but also their localization in space. Like other animals, insects determine sound direction by making binaural comparisons. This has been demonstrated by studying the orientation responses of unilaterally deafened animals. For example, many moths exhibit negative phonotaxis to batlike sounds. When moths in which one ear has been destroyed are stimulated with ultrasound, they steer away from the remaining ear, no matter what the direction of the sound source (Roeder 1967). Male grasshoppers, which make phonotactic turns towards the songs of females, turn only towards the intact side after one tympanal nerve has been severed (Ronacher, Helversen, and Helversen 1986). Crickets that have been unilaterally deafened turn toward the intact side in response to calling song (which evokes positive phonotaxis in intact crickets), and away from that side when stimulated with ultrasound (a negative phonotactic stimulus; Moiseff, Pollack, and Hoy 1978; Pollack, Huber, and Weber 1984). These experiments, taken together with the directional sensitivity of the auditory system (see Michelsen, Chapter 2), suggest that insects perform phonotaxis by using a simple strategy of turning toward (or, for negative phonotaxis, away from) the more strongly stimulated ear.

Some insects are capable not only of determining the laterality of a sound source but also of scaling their orientation responses according to its angular position. For example, for crickets walking freely in an arena, the angle of each turn is related to the angle between the cricket's longitudinal axis and the loudspeaker just before the turn (Bailey and Thompson 1977; Latimer and Lewis 1986; Fig. 5.9A). When walking on a treadmill, crickets turn more vigorously (with greater angular velocity) the more lateral the loudspeaker (Schmitz, Scharstein, and Wendler 1982; Schildberger and Kleindienst 1989; Fig. 5.9B,C). Similarly, phonotactic steering movements during tethered flight are larger for more lateral loudspeaker positions (Pollack and Plourde 1982; Fig. 5.9D,E). This ability to turn in a graded fashion according to the target angle means that the neural correlate for sound direction must include information not only about which ear is more strongly stimulated but also how large the interaural intensity difference is.

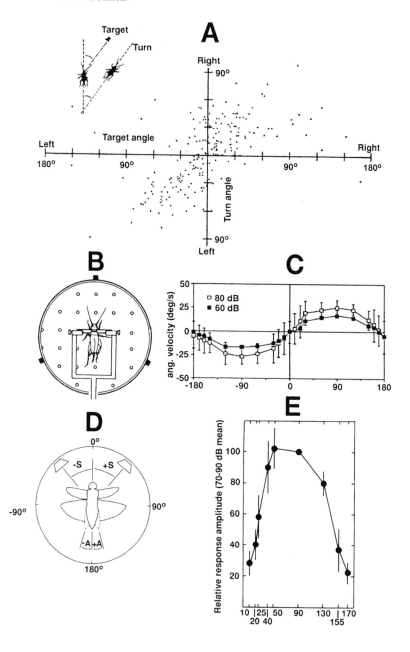

1.4. Summary

The experiments summarized earlier serve to identify those features of acoustic signals that are most important for behavior. These include sound frequency, the temporal pattern of the signal, and sound intensity at each ear. The remainder of this chapter examines how these features are analyzed by the nervous system.

2. The Auditory Periphery: Encoding of Signals by Auditory Receptors

Many hearing insects can discriminate between sound frequencies. The ability to discriminate derives from the fact that different auditory receptors are tuned to different sound frequencies. This was first discovered for locusts (Michelsen 1966) and has since been confirmed in crickets (e.g., Nocke 1972; Esch, Huber, and Wohlers 1980; Hutchings and Lewis, 1981; Oldfield, Kleindienst, and Huber 1986) and bushcrickets (e.g., Zhantiev 1971; Kalmring, Lewis, and Eichendorf 1978; Oldfield 1982; Römer 1983). Although single-unit receptor recordings have not yet been made from cicadas, behavioral evidence (summarized earlier, see Section 1.1) suggests that they too are capable of frequency discrimination.

As in the vertebrates, there is a close correlation between the position of a receptor within the ear and its physiological characteristics. For example, there are four anatomically distinct receptor groups in the locust ear, with

◄――――――――――――――――――――――――――――――――

FIGURE 5.9. Scaling of phonotactic turns in crickets. A: Relationship between the direction of a sound source and the direction of phonotactic turns in freely walking crickets. The larger the angle of the target, the larger the turn angle. B: Method for measuring the open-loop turning tendency in walking crickets. The cricket is fixed to a support and walks on the surface of an air-supported sphere, which consequently turns beneath it. The sphere is decorated with numerous reflecting dots that are monitored by a camera. The cricket's walking direction and velocity are inferred from the movements of the dots. C: Turning tendency, quantified as angular velocity of the sphere, for different loudspeaker azimuths. More lateral azimuths (±90 degrees) elicit more rapid turning. D: Method for measuring steering movements during tethered flight. A song model is switched between two speakers situated at equal angles from the midline, and steering movements of the abdomen are measured as the cricket "tracks" the switches in sound location. E: Normalized response amplitudes for different speaker azimuths. As in C, orientation movements are larger for more lateral azimuths. (A reprinted with permission from Bailey and Thomson 1977; copyright 1977 The Company of Biologists. B reprinted with permission from from Schildberger and Hörner 1988; C from Schildberger and Kleindienst 1989; D and E from Pollack and Plourde 1982; B,C,D–E copyright 1988, 1989, 1982, respectively, Springer-Verlag.)

receptors in the different groups associated with distinct cuticular specializations on the interior of the tympanic membrane (Gray 1960; Fig. 5.10A).
As a result of the mechanics of this system, different receptor groups are
stimulated most effectively by different sound frequencies (Michelsen
1971b; Stephen and Bennet-Clark 1982; Breckow and Sippel 1985). The
ears of bushcrickets and crickets also show clear tonotopic organizations,
with receptors tuned to low frequencies situated proximally and those
tuned to high frequencies found distally (Oldfield 1982; Oldfield,
Kleindienst, and Huber 1986; Stumpner 1996). In these insects, however,
the physical basis of tonotopy is not yet understood (Oldfield 1985). In
mantids, tonotopy has been taken to an extreme, with different frequency
ranges divided among completely separate, segmentally homologous ears;
low-frequency sensitivity is provided by a mesothoracic ear whereas sensitivity to ultrasound derives from the metathorax (Yager 1996). Frequency
coding has been studied most thoroughly in orthopteran insects, and these
form the focus of the following sections.

2.1. Frequency Coding in Grasshoppers

The grasshopper's ears are situated bilaterally on the first abdominal segment. The 60 to 80 receptor neurons in each ear were classed into four
groups on the basis of their anatomy within the hearing organ (Gray 1960;
Fig. 5.10A), and this scheme has generally been validated by physiological
studies (Michelsen 1971a; Römer 1976; Halex, Kaiser, and Kalmring 1988).
Both in *Locusta migratoria* and in *Schistocerca gregaria*, two groups of cells

FIGURE 5.10. Tonotopic organization of orthopteran ears. A: Ovals show the locations of the cell bodies of the four anatomically defined receptor groups of the
grasshopper ear; arrows indicate dendrite trajectories. B: Typical tuning curves of
receptors of each of the four physiologically defined groups. Comparison of mechanical properties at the attachment points of the dendrites with physiologically
measured tuning curves suggests that type 1, 2, 3, and 4 receptors correspond
respectively to groups a, c, b, and d (Breckow and Sippel 1985). C: The proximal
tibia of a tettigoniid, showing the subgenual organ (sgo), the intermediate organ
(io), and the crista acustica (ca). D: Tuning curves of individual receptors, the
positions of which in the crista acustica are shown below; the more distal a receptor,
the higher the frequencies to which it is sensitive. E: The receptor array in a cricket
ear, showing the proximal and distal cell groups; only every second receptor is
shown. F: Tuning curves of individual receptors, the positions of which are shown
below. (A reprinted with permission from Halex, Kaiser, and Kalmring 1988; B
reprinted with permission from Römer 1976; C reprinted with permission from
Lakes and Schikorski 1990; D reprinted with permission from Oldfield 1982; E
reprinted with permission from Young and Ball 1974; F reprinted with permission
from Oldfield, Kleindienst, and Huber 1986. A–F copyright 1988, 1976, 1990, 1982,
1974, and 1986, respectively, Springer-Verlag.)

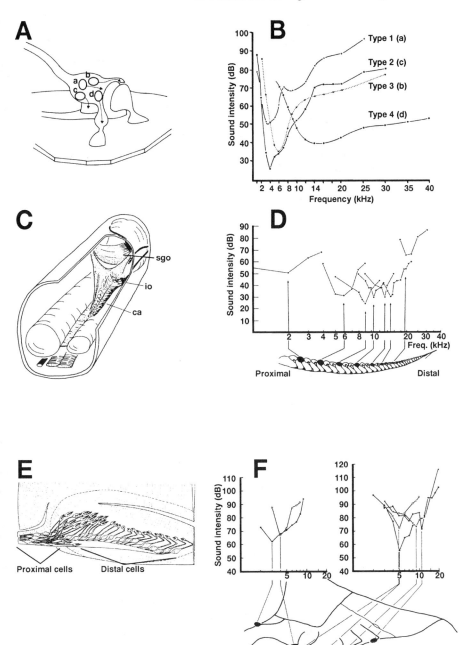

are tuned to low frequencies (3.5 to 4 kHz for *L. migratoria*, 2 to 3.5 kHz for *S. gregaria*), with one (designated type 1) with higher thresholds than the other (type 2). A third group (type 3) is tuned to somewhat higher frequencies (ca. 6 kHz for both species), and a fourth (type 4) is tuned to still higher frequencies (>10 kHz for both species) (Fig. 5.10B). However, the receptor population is clearly more diverse than this simple classification scheme might indicate. For example, receptors within a single "group" can have thresholds that span a range of 20 to 30 dB (Halex, Kaiser, and Kalmring 1988). Even the designation of a cell as "tuned" to a particular frequency range is not always as straightforward as it might seem because many receptors have multiple sensitivity peaks, with thresholds at the different "best" frequencies sometimes differing by only a few decibels (Michelsen 1971a).

The tympanal nerve enters the metathoracic ganglion, where receptors send processes into two distinct areas, the rostral and caudal neuropils. The relative density of branching into these two areas differs among receptor types, with type 1 branching abundantly in both areas, and types 2 to 4 favoring rostral the area (Halex, Kaiser, and Kalmring 1988). The positions of the branches within the rostral area also differ between receptor types, forming a tonotopic map within the neuropil (Halex, Kaiser, and Kalmring 1988; Römer, Marquart, and Hardt 1988; Fig. 5.11A). In addition to their metathoracic arbors, many receptors continue more anteriorly to the mesothoracic or even the prothoracic ganglion, where they branch in regions homologous to the rostral neuropil of the metathoracic ganglion (Halex, Kaiser, and Kalmring 1988).

2.2. Frequency Coding in Bushcrickets

The proximal tibia of the prothoracic legs of bushcrickets houses a population of scolopidial receptors that collectively respond to both airborne sound and substrate vibration. The most proximal portion of the so-called complex tibial organ (Lakes and Schikorski; Fig. 5.10C) is the subgenual organ, a sensory structure found in the legs of all insects. The approximately 40 receptor somata (Schumacher 1979) lie in the dorsal hemolymph channel of the tibia, and their dendrites insert into cap cells that form a fan-shaped membrane that attaches to the dorsal hypodermis of the leg. More distally is the intermediate organ, comprising 8 to 18 receptors (Schumacher 1979). Some of these are situated near those of the subgenual organ, while others form a continuum with those of the crista acustica (see later). Most distally, the crista acustica consists of a linear array of 15 to 49 somata along the dorsal surface of an anterior tracheal branch. Their dendrites insert into cap cells that form part of the tectoral membrane, which attaches at the dorsal surface of the leg (Schumacher 1979).

Single-unit recordings from the leg nerve at its entry to the prothoracic ganglion reveal some receptors that are sensitive to substrate vibration but

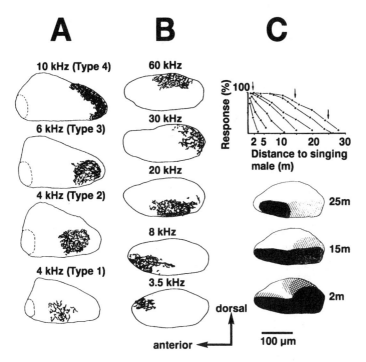

FIGURE 5.11. Central tonotopic organization of auditory receptors. A, B: The central terminals of individual receptors in a grasshopper (*Locusta migratoria*) and a bushcricket (*Tettigonia viridissima*), respectively. The drawings are parasagittal sections, with the auditory neuropil shown in outline. The best frequencies of the receptors are indicated. (Reprinted with permission from Römer, Marquart, and Hardt 1988; copyright 1988, John Wiley & Sons.) C: Odotopic organization within the central nervous system of the bushcricket *Mygalopsis marki*. The graph shows the responses of several receptors to sounds originating from different distances; some receptors respond only to nearby males, whereas others can respond at considerable distance. The sketches (parasagittal sections) show which regions of the auditory neuropil would be activated at different distances from a singing male. (Reprinted with permission from Römer 1987; copyright 1987, Springer-Verlag.)

not to airborne sound, some sensitive to sound but not vibration, and some sensitive to both sound and vibration (Kalmring, Lewis, and Eichendorf 1978). In most physiological studies, recordings were made either in the leg nerve or in the central nervous system, and the peripheral locations of the recorded neurons were not known. Therefore, it is not yet possible to assign the three response types described earlier to different regions of the tibial organ. However, recordings made in the tibial organ itself have demonstrated that cells of the crista acustica are highly sensitive to higher sound frequencies (Oldfield 1982). This conclusion has recently been confirmed by impaling axon terminals in the central nervous system and staining the cells intracellularly with neurobiotin, which travels throughout the neuron and reveals the location of its cell body in the ear (Stumpner 1996).

Within the crista acustica, proximal receptors are tuned to low frequencies and distal receptors to high frequencies (Oldfield 1982; Stumpner 1996; Fig. 5.10D). The shift in best frequency can be rather regular from receptor to receptor. For example, in the low-frequency portion of the crista of *Polichne* sp., best frequency changes by about 1 kHz per receptor. However, the same species has four receptors tuned to 20 kHz. In *Mygalopsis marki*, the gradient in best frequencies is irregular; some neighboring receptors have best frequencies that differ by as much as 7 kHz, whereas others share the same best frequency (Oldfield 1984). The anisotropy of the mapping of sound frequencies onto the receptor array may reflect the relative importance of different frequencies in the acoustic behavior of the insects, and is reminiscent of the "acoustic fovea" seen on the basilar membranes of CF/FM bats (Schuller and Pollak 1979).

As in grasshoppers, the peripheral tonotopic organization is maintained in the central nervous system (Oldfield 1983; Römer 1983; Stumpner 1996). Low frequencies are represented anteriorly within the auditory neuropil and as frequency increases its representation shifts posteriorly, wrapping dorsally around the posterior edge of the neuropil for the highest frequencies (Römer, Marquart, and Hardt 1987; Fig. 5.11B). Receptors that are tuned to frequencies that carry particular behavioral importance have more extensive terminal arbors, indicating that they may have more potent and/ or widespread effects on central neurons (Oldfield 1983; Römer 1983).

There is also relationship between a receptor's best frequency and the value of its minimum threshold. In *M. marki*, for example, the most sensitive receptors are those that are tuned to 10 to 20 kHz; minimum thresholds for other receptors can be as much as 40 dB higher than those of these highly sensitive receptors. This difference in sensitivity may allow the insects to gauge the distance to the sound source. The song of this species has a rather broad spectrum. Because high frequencies attenuate more rapidly with distance than low frequencies (Michelsen and Larsen 1983), the spectral characteristics of the signal at the receiver change with distance from the source Consequently, differently tuned receptors will reach threshold at different distances from the source. When this is taken into consideration, it becomes apparent that the *tono*topic receptor projection can also be described as *odo*topic; different regions within the neuropil receive input from receptors according to the distance from a singing male (Römer 1987; Fig. 5.11C). Behavioral studies have shown that perceived sound intensity is an important cue determining intermale distance in the field (Thiele and Bailey 1980; Römer and Bailey 1986). The highly ordered organization of the auditory neuropil may provide the anatomical substrate from which information about distance can be extracted.

Although the foregoing discussion considers only responses to sound, Kalmring, Lewis, and Eichendorf (1978) showed that single receptor cells may respond both to sound and to vibration of the leg, often with different best frequencies for the different stimuli. The peripheral locations of these neurons are not known with certainty, but Kalmring et al. suggested that they originate in the intermediate organ. More recently, Shaw (1994) showed that the subgenual organ of the cockroach, which has classically been considered a detector of substrate vibration, also responds robustly to airborne sound. Here, however, best frequencies were similar for acoustic and vibration responses.

2.3. Frequency Coding in Crickets

The cricket's ear, like that in bushcrickets, is situated in the prothoracic tibia, just distal to the subgenual organ. The 60 to 70 auditory receptor neurons fall into two main morphological groups: the proximal sensilla, whose dendrites are anchored via relatively short attachment cells to a tentlike membrane that itself attaches to the dorsal surface of the tibia, and the distal sensilla, whose dendrites are attached via long attachment cells to the dorsal hypodermis of the leg (Michel 1974; Young and Ball 1974; Fig. 5.10E).

As in the other orthopteran groups, the cricket ear is tonotopically organized (Fig. 5.10F). Oldfield, Kleindienst, and Huber (1986) recorded from the cell bodies of receptors of *G. bimaculatus* in the tibia and found that, as in bushcrickets, low frequencies are represented proximally and high frequencies distally. These experiments and others indicate a strong clustering of best frequencies around the calling-song frequency (ca. 5 kHz), and, in some studies, its higher harmonics as well (Zaretsky and Eible 1978; Esch, Huber, and Wohlers 1980; Hutchings and Lewis 1981; Imaizumi and Pollack 1996). Thus, particular sound frequencies are overrepresented at the receptor level.

In *T. oceanicus*, two frequency ranges, one centered on 4 to 5 kHz and the other ultrasonic, are particularly important for behavior. These correspond to the best frequencies for positive and negative phonotaxis, respectively (see Section 1.1). The neuronal strategies for detecting these two types of sounds are very different: the low-frequency range is encoded by a large number of receptors, each of which has only small effects on neurons in the central nervous system, whereas the ultrasonic range is encoded by only a few receptors, which have powerful central effects (Pollack 1994). These different ways of mapping the periphery onto the central nervous system are evident in the physiology of an identified interneuron, ON1, which is tuned to both ranges (Atkins and Pollack, 1986). The waveforms of synaptic potentials induced by 5-kHz stimuli are smooth in shape and increase in amplitude with increasing stimulus intensity, as might be expected if this input were provided by a large number of receptors, each of which has only a modest postsynaptic effect. In contrast, ultrasonic stimuli elicit trains of large, discrete synaptic potentials, suggesting that only a few receptors provide this input, each having powerful postsynaptic effects (Pollack 1994; Fig. 5.12A–E). The unequal peripheral representation of these two

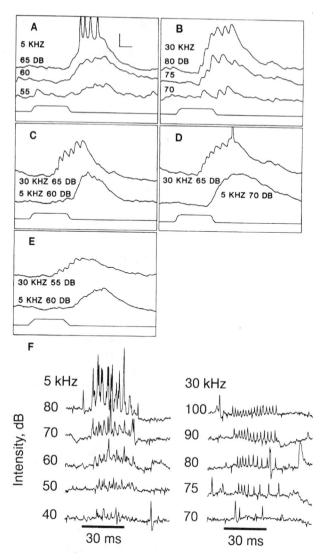

FIGURE 5.12. Representation of cricketlike and batlike sounds in *Teleogryllus oceanicus*. A–E: Intracellular recordings from an interneuron, ON1, that is tuning both to 4.5 to 5 kHz and to ultrasound (the neuron is hyperpolarized to suppress action potentials). Excitatory postsynaptic potentials (EPSPs) in response to 5 kHz are smooth in shape and graded in amplitude, whereas ultrasound elicits trains of discrete EPSPs that summate temporally. A and B from the same preparation, C–E from three additional preparations. F: The entire receptor population was recorded with a hook electrode on the tympanal nerve. The 5-kHz stimuli elicit compound responses that increase in amplitude with intensity as additional receptors are recruited; 30 kHz elicits activity from only a single receptor, which increases in firing rate with increases in intensity. This is an extreme example; usually several receptors (ca. 4) respond to 30 kHz. (A–E reprinted with permission from Pollack 1994; copyright 1994 Springer-Verlag. F: Pollack, unpublished observations.)

162

frequency ranges is also apparent in recordings from the entire tympanal nerve. Stimulation with 5 kHz evokes compound action potentials that increase markedly in size with stimulus intensity, indicating recruitment of additional receptors, but stimulation with ultrasound results in activity in far fewer receptors (Fig. 5.12F; Pollack and Faulkes 1998).

Information about the organization within the central nervous system of receptors tuned to different frequencies is scant. There is, so far, little indication of clear tonotopy such as exists in the other orthopteran groups (Esch, Huber, and Wohlers 1980).

2.4. General Comments on Receptor Physiology

Frequency tuning aside, auditory receptors do not appear to substantially filter or transform auditory signals. Although some exceptions have been noted (Nocke 1972; Hutchings and Lewis 1981), receptors generally respond to sound pulses with more-or-less tonic discharges that cease promptly with stimulus offset. They encode the temporal structures of intraspecific signals with high fidelity but do not show selectivity for these patterns. As is usual for sensory systems, stimulus intensity is encoded as firing rate. For a typical receptor, spike rate increases nearly linearly with sound intensity until about 20 to 30 dB above threshold, at which point the response saturates. This dynamic range is too small to account for the operating ranges of auditory interneurons or of behavioral responses (which, e.g., is ca. 50 dB for cricket phonotaxis). Intensity coding over this larger range is possible because of range fractionation. Different receptors have different thresholds, and thus as intensity increases new receptors are recruited so that the total afferent activity continues to increase over an intensity range that exceeds that of any single receptor.

3. Central Processing of Acoustic Signals

As the experiments outlined in Section 1 demonstrate, sound stimuli must be analyzed and evaluated in order to evoke behavioral responses. This processing is done chiefly within the central nervous system. Again, orthopteran insects have garnered the most attention and dominate this section. The discussion of central processing focuses on three main aspects: analysis of signal spectrum, processing of directional information, and recognition of temporal pattern.

3.1. Analysis of Sound Frequency

The first step in the neural analysis of sound frequency is the division of a complex signal into its different frequency components. This process begins at the receptor level (Section 2.1) and continues in the central nervous system, where the frequency characteristics of interneurons are determined by their connections with receptors and with each other.

3.1.1. Specificity of Receptor-to-Interneuron Wiring

The tonotopic organization of receptor terminals in the central nervous system provides the anatomical substrate for interactions between receptors and interneurons. Direct connections can only occur where receptor terminals overlap with interneuron dendrites. Because interneurons have characteristically different dendritic morphology, they sample different subsets of the receptor array and thus acquire different frequency sensitivities. For example, in the bushcricket *T. virridissima* a local interneuron (the omega neuron) has dendrites that branch throughout most of the auditory neuropil, avoiding only the anterior region where low-frequency receptors terminate. The tuning of the neuron correlates well with this dendritic anatomy; it is relatively insensitive to low frequencies, but in the remainder of its range its sensitivity mirrors that of the entire receptor population. By contrast, a descending interneuron in another bushcricket, *M. markii*, has dendrites that are restricted to the anterior end of the neuropil and is sensitive only to low-frequency sounds. Similarly, in *L. migratoria*, dendrites of the SN5 neuron invade only the posterior region of neuropil, which is the target of high-frequency receptors; accordingly, SN5 is tuned to high frequencies (Römer, Marquart, and Hardt 1988; Fig. 5.13).

In crickets, correlations between structure and frequency sensitivity are less clear. The ascending interneurons AN1 and AN2 are, respectively, most sensitive to low and to high frequencies. AN1 is sharply tuned to the main frequency component of the conspecific calling song (usually ca. 5 kHz), whereas AN2 is most sensitive to frequencies above 10 kHz but also has a secondary region of sensitivity at the calling-song frequency (Boyan and Williams 1982; Wohlers and Huber 1982; Hennig 1988; Hardt and Watson 1994). Not surprisingly, AN1 and AN2 receive monosynaptic input from low- and high-frequency receptors, respectively (Hennig 1988). It is not yet clear whether AN2's sensitivity to low frequencies is due to input from low-frequency receptors as well (if so, then this might be from a high-threshold subset of these, because AN2 is about 20 dB less sensitive to low frequencies than AN1), or whether it can be accounted for by the fact that high-frequency receptors can also show secondary sensitivity to low frequencies (Esch, Huber, and Wohlers 1980; Imaizumi and Pollack 1996).

The main dendrites of AN1 and AN2 overlap within the ventromedial auditory neuropil where, one would imagine, they would have access to the same population of receptors (Wohlers and Huber 1985; Hardt and Watson 1994). Thus, specific wiring between receptors and these neurons is unlikely to be explained solely by their anatomical relationships. A similar argument

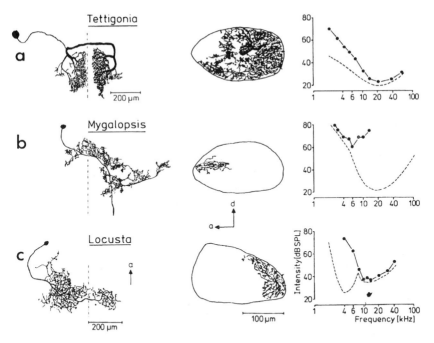

FIGURE 5.13. Receptor-to-interneuron connectivity. Three interneurons are shown in plane view at left and in parasagittal sections through the auditory neuropil (outlined) in the middle. Their tuning curves are shown at the right (solid lines), along with the overall sensitivity of the auditory system in their respective species (broken lines). The tuning of the interneurons matches that of the receptors that project to the regions occupied by the interneurons' dendrites. (Reprinted with permission from Römer, Marquart, and Hardt 1988; copyright 1988, John Wiley & Sons.)

applies as well to three other interneurons that have dendrites in the same region of neuropil: DN1, a descending neuron that is tuned to low frequencies (Wohlers and Huber 1982, 1985; Atkins and Pollack 1987a,b); and two more broadly tuned local interneurons, ON1 and ON2 (Wohlers and Huber 1982, 1985; Atkins and Pollack 1986). Thus, in crickets it seems unlikely that specific connectivity can be ascribed solely to the anatomical relationships between receptors and interneurons. Other factors, such as recognition of molecular labels on cell surfaces (Goodman 1996), are probably also important (and, of course, these other factors may also play a role in grasshoppers and bushcrickets).

The frequency sensitivity that is brought about by receptor-to-interneuron connectivity is often enhanced by inhibitory interactions between interneurons that are tuned to different frequencies. Several examples of this are now presented, organized according to their presumed functions.

3.1.1.1. Sharpening of Frequency Tuning

The song of the bushcricket *Caedicia simplex* has most of its energy near 16 kHz, and the ear includes receptors that are tuned to this frequency. Several interneurons in the ventral cord are also tuned to 16 kHz, but their threshold curves are considerably sharper than those of 16-kHz receptors. The enhanced selectivity is seen on the low-frequency side of the tuning curve, where roll-offs in sensitivity of interneurons (>50 dB/octave) greatly exceed those of receptors (Fig. 5.14). Two-tone experiments, in which a test tone of variable frequency was superimposed on a standard 16-kHz tone, revealed that the sharpening in tuning is due to inhibition of the interneuron (or of its high-frequency-tuned inputs) by lower frequencies (Oldfield and Hill 1983). Although the locus of the inhibition has not been identified in this example, the receptors themselves can be ruled out; responses of 16-kHz receptors are not suppressed by superposition of lower frequencies (Oldfield and Hill 1983). Similar sharpening of tuning has been ascribed to postsynaptic inhibition in other bushcrickets (Shul 1997, Stumpner 1997) as well as in crickets (Boyd et al. 1984; Schildberger 1984; Horseman and Huber 1994a). In all these cases, the increased selectivity of interneurons that results from inhibition presumably serves to enhance specificity for conspecific signals.

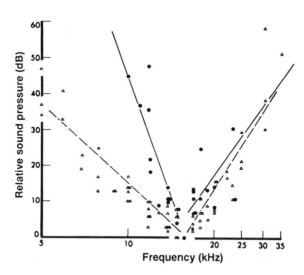

FIGURE 5.14. Increased sharpness of frequency tuning in interneurons. The filled circles and solid line represent the thresholds of four selected interneurons as a function of frequency; the open triangles and dotted line represent similar data from six receptors. All of the neurons are tuned to 16 kHz, but the interneurons are tuned more sharply. (Reprinted with permission from Oldfield and Hill 1983; copyright 1983 Springer-Verlag.)

3.1.1.2. Frequency-Specific Inhibition and Discrimination of Song Types

In *Gryllus* species, the calling and aggression songs are dominated by the fundamental frequency, ca. 4.5 kHz, with the third harmonic approximately 20 to 30 dB less intense (Nocke 1972). The courtship song, by contrast, consists chiefly of the third harmonic, which is required for the song's effectiveness (Libersat, Murray, and Hoy 1994). Information about these songs is carried to the brain by two different interneurons: AN1 is tuned to the frequency of calling and aggression songs, and AN2 is tuned to the higher frequencies of courtship song (Wohlers and Huber 1978, 1982; Boyan and Williams 1982; Schildberger 1984). Within the brain, different groups of neurons are tuned to one or the other of these two frequency bands (Boyan 1980; Schildberger 1984). One of the brain neurons, PABN2, is tuned to 12 to 15 kHz and its response is suppressed by 4 to 5 kHz sound (Boyan 1981). Boyan (1981) suggested that the function of this inhibition might be to ensure that neurons that are members of the courtship-song recognition network are not "confused" by stimulation, at close range, by the high-frequency components of calling or aggression songs.

AN2 is itself inhibited by low frequencies (Casaday and Hoy 1977; Wohlers and Huber 1978; Popov and Markovich 1982; Moiseff and Hoy 1983; Nolen and Hoy 1986b, 1987). If PABN2 is driven by AN2 (which, based on their anatomical relationships, seems plausible), then "inhibition" of PABN2 by low frequencies may simply represent inhibition of its main excitatory input. The same argument may also account for the inhibition, by low-frequency sounds, of high-frequency-tuned neurons recorded in the promesothoracic connective, many of which may be descending brain neurons (Pollack 1984).

3.1.1.3. Prevention of Crosstalk Between Positive and Negative Phonotaxis in Crickets

The behavioral threshold for negative phonotaxis is lowest for high sound frequencies (Moiseff and Hoy 1983; Nolen and Hoy 1986a), but the response can be evoked by intense low-frequency sound (Nolen and Hoy 1986a). In addition, the low-frequency songs of crickets contain higher harmonics, which, though at least 30 dB less intense than the fundamental frequency, can still exceed thresholds for negative phonotaxis at close range. Thus, there would appear to be a risk that negative phonotaxis might be elicited by conspecific signals.

Negative phonotaxis is triggered by high-frequency firing of AN2 (Nolen and Hoy 1984; see Section 3.2.2). The inhibition of AN2 by low-frequency sounds limits its firing rate and may serve to prevent inappropriate triggering of negative phonotaxis by cricket song.

In the preceding section AN2 was implicated in two very different behaviors: courtship and bat evasion. Whereas the evidence for its role in negative phonotaxis is decisive (see Section 3.2.2), the argument for its role in detecting courtship song rests on the correlation between the song's frequency content and AN2's selectivity for high frequencies. Although AN2 may indeed play some role in courtship, there is evidence that it cannot be exclusively responsible for song detection. The courtship song of *G. bimaculatus* has a dominant frequency of ca. 13.5 kHz and, as summarized earlier (Section 1.2.1), courtship success can be restored to muted males by broadcasting a courtship song model with this frequency. Libersat, Murray, and Hoy (1994) found that a song model with a frequency of 30 kHz was ineffective in restoring courtship success, even though it strongly excited AN2. Thus, whereas AN2 is the main player in commanding negative phonotaxis, it appears to play, at best, a supporting role in courtship song detection.

3.1.1.4. Inhibition and Intensity Maxima

In grasshoppers, most mesothoracic auditory interneurons receive mixed excitatory and inhibitory inputs; high frequencies tend to be excitatory and low frequencies have more varied effects, producing either pure excitation, pure inhibition, or, most commonly, a mixture of the two (Römer and Marquart 1984; Stumpner and Ronacher 1991a; Fig. 5.15A). As a consequence of the integration of these excitatory and inhibitory inputs, many interneurons exhibit maximal responses at particular intensities when stimulated with broadband sounds that mimic the songs of these insects

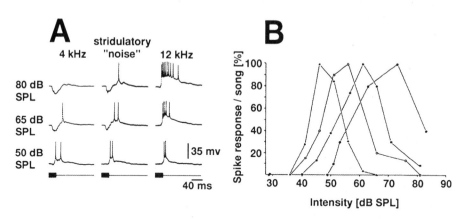

FIGURE 5.15. Intensity maxima in grasshopper interneurons. A: Responses of the ascending neuron AN2 to pure-tone low and high frequencies and to a noisy signal similar to the communication sounds of this species (*Locusta migratoria*). This neuron, like many others, receives excitatory input at high frequencies and mixed excitation/inhibition at low frequencies. As a result of these diverse inputs, responses to noise stimuli are highly complex and often show a maximum at a particular intensity. B: Intensity-response curves for four unidentified ascending interneurons recorded in the same animal. (Reprinted with permission from Römer and Seikowski 1985; copyright 1985 Springer-Verlag.)

(Römer and Seikowski 1985; Stumpner and Ronacher 1991a; Fig. 5.15B). The behavioral correlate, if any, of this complex intensity coding is not known for grasshoppers. However, intensity-tuned interneurons are also found in the bushcricket, *M. markii*, (Römer 1987), in which males are known to adjust their distance from their neighbors according to the intensity of their songs (Thiele and Bailey 1980).

3.2. Extraction of Information about Sound Direction

The behavior experiments described in Section 1.3 demonstrate that sound direction is determined by comparing auditory input at the two ears. In principle, two types of interaural cues are available: interaural time difference (because sound from one side will arrive first at the closer ear) and interaural intensity difference. Only the latter has been considered seriously as a cue for sound direction in insects. The possibility that insects might detect interaural time differences has been discounted because insects are small and time differences between the two ears would be on the order of a few microseconds at best. However, time differences of this magnitude, or even smaller, can be resolved by owls, bats, and electric fish (Carr 1993), Moreover, insects use a number of mechanical "tricks" to generate or amplify interaural differences despite their small body size (see Michelsen, Chapter 2 and Robert and Hoy, Chapter 6). Thus, the possibility that insects may make use of very small interaural time differences cannot be ruled out. Nevertheless, until now the assumption has been that the only pertinent cue for sound direction is interaural intensity difference.

3.2.1. Detection and Amplification of Interaural Intensity Differences

Interaural differences in effective sound intensity can be quite large for lateral sound sources (e.g., up to 35 dB for crickets: Boyd and Lewis 1983; cf. Michelsen, Chapter 2). However, insects are able to determine the laterality of sound sources that are much closer to the midline (5° to 10° in crickets: Oldfield 1980; Pollack and Plourde 1982; Latimer and Lewis 1986), which generate interaural intensity differences of only a few decibels. That such small differences are sufficient to determine the direction of phonotaxis was shown by Helversen and Rheinlaender (1988), who presented song models simultaneously from the left and right sides of male grasshoppers, *C. biguttulus*, and studied the effect of interaural intensity differences on the direction of phonotactic turns. The sound sources in this experiment were free-field loudspeakers, and thus each source could stimulate both ears. However, by calculating the contribution of each source at each ear (taking into account the measured directionality of this insect's auditory system), Helversen and Rheinlaender were able to determine the effective binaural sound intensities for various combinations of loudspeaker intensities. They demonstrated that *C. biguttulus* males turned in the "correct" direction (i.e., toward the more intense source) 75% of the time when binaural intensities differed by only 0.6 dB; performance was errorless at intensity differences of 1 to 2 dB.

A common mechanism for amplifying subtle interaural intensity differences is contralateral inhibition. Among mammals, for example, contralateral inhibition is evident in the responses of neurons in the superior olivary complex, many of which are excited by one ear and inhibited by the other (Irvine 1992). The existence of contralateral inhibitory circuits in insects was first inferred from comparisons of responses to monaural and binaural stimuli (Suga and Katsuki 1961). More recently, intracellular recordings have confirmed that many interneurons integrate excitatory input from one ear with inhibitory input from the other. For example, several interneurons in the metathoracic ganglion of the locust receive mixed excitatory and inhibitory inputs, with their relative weights varying with sound direction (Römer, Rheinlaender, and Dronse 1981; Fig. 5.16).

The most thoroughly investigated example of contralateral inhibition is that mediated by the ON1 (Omega Neuron 1) neuron of crickets. This was among the first morphologically identified auditory interneurons of insects, and its striking bilateral branching pattern (Fig. 5.17A) immediately suggested that it might play a role in processing of information about sound direction (Casaday and Hoy 1977; Popov, Markovich, and Andjan 1978).

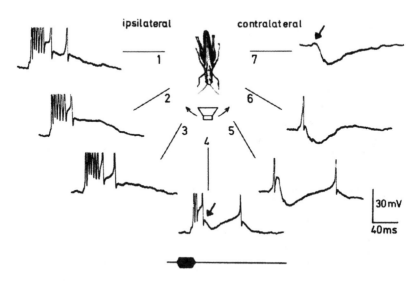

FIGURE 5.16. Directional sensitivity in an unidentified interneuron in the metathoracic ganglion of *Locusta migratoria*. The neuron is strongly excited by ipsilateral sound, inhibited by contralateral sound, and shows intermediate responses for intermediate azimuths. (Reprinted with permission from Römer, Rheinlaender, and Dronse 1981; copyright 1981 Springer-Verlag.)

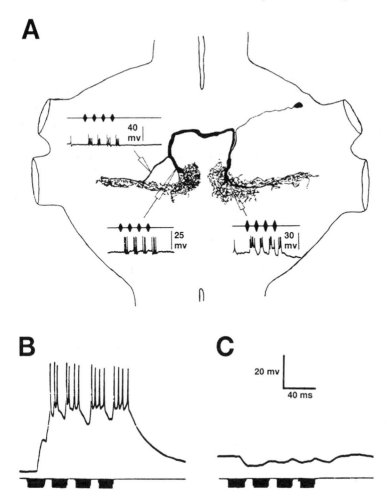

FIGURE 5.17. The ON1 neuron of crickets. A: Sketch of the neuron in the protho-
racic ganglion with insets showing the waveforms recorded at different sites. B, C:
The neuron is excited by soma-ipsilateral stimuli and inhibited by contralateral
stimuli. (A reprinted with permission from Wohlers and Huber 1978; copyright 1978
Springer-Verlag. B reprinted with permission from Selverston, Kleindienst, and
Huber 1985; copyright 1985 Society for Neuroscience.)

ON1 is excited by soma-ipsilateral sound due to direct input from auditory
receptors (Hennig, personal communication; Pollack, unpublished observations) and is inhibited by contralateral sound, because of activity in the
contralateral ON1 (Selverston, Kleindienst, and Huber 1985; Fig. 5.17B,C).
    In mature crickets ON1's processes are restricted to the prothoracic
ganglion (although it frequently has an ascending axon in larvae and young
adults; Atkins and Pollack 1986). However, the brain is also required for

phonotaxis, because it is there that song-recognition circuits analyze acoustic stimuli and accept or reject them (Pollack and Hoy 1981b; Schildberger and Hörner 1988). It may also be in the brain that the direction of a phonotactic response is decided by comparing inputs ascending bilaterally from the prothoracic ganglion (Pollack 1986; Stabel, Wendler, and Scharstein 1989; Doherty 1991). Much of the auditory input to the brain is carried by two identified ascending neurons, AN1 and AN2, both of which have been shown to affect phonotactic course direction (see Section 3.2.2). Both of these neurons are inhibited by contralateral sound. ON1 is known to be the source of this inhibiton for AN2 (Selverston, Kleindienst, and Huber 1985), and it is the best candidate for the inhibitory input to AN1 (Horseman and Huber 1994a). Thus, even though ON1 is an intraganglionic neuron, it can influence ascending neurons that are demonstrably involved in phonotaxis.

By using dichotic stimuli, Horseman and Huber (1994b) were able to quantify the inhibitory effect of contralateral sound on all three of these interneurons. They combined these data with information about the directional sensitivity of the periphery to construct a quantitative model of the enhancement of directionality resulting from contralateral inhibition (Fig. 5.18). The predicted enhancement of the left-right difference in spike count is a factor of 3 for AN1, and a factor of 1.4 for AN2. Moreover, contralateral inhibition is predicted to increase the steepness of the relationship between binaural spike-count difference and sound azimuth. This would allow corrective turns to be triggered by smaller course deviations.

The inhibition mediated by ON1 occurs very early within the auditory pathway; indeed, ON1, AN1, and AN2 are all first-order auditory interneurons (Hennig 1988; Pollack, unpublished observations). Contralateral inhi-

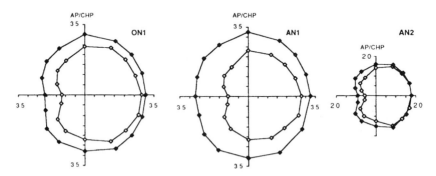

FIGURE 5.18. Predicted effects of contralateral inhibition on the directionality of ON1, AN1, and AN2. Graphs are polar plots of the number of action potentials elicited per stimulus as a function of sound direction. Open circles show directionality in the intact system, filled circles in the absence of contralateral inhibition. (Reprinted with permission from Horseman and Huber 1994b; copyright 1994 Springer-Verlag.)

bition is also evident further downstream. For example, an identified ultra-sound-sensitive brain neuron, LBN-ei, receives mixed excitatory and inhibitory input for ipsilateral stimuli but predominantly inhibitory inputs for contralateral stimuli (Brodfuehrer and Hoy 1990). Similarly, a prothoracic descending neuron, DN3, and a prothoracic T-shaped neuron (i.e., with axons that both ascend and descend), TN3, are excited by sound from one side but mainly inhibited from the other side (Atkins and Pollack 1987a).

3.2.2. Cricket Auditory Interneurons and Phonotaxis

The binaural difference detected at the receptor level is thus passed on, with amplification, to interneurons. The importance of bilateral imbalances in interneuron activity in determining sound direction has been studied in crickets by observing the effects on phonotaxis of manipulating the activity of single cells. This has been done in two ways: (1) Individual neurons were filled with fluorescent dye and killed selectively by illumination with the appropriate wavelength. The effects of this treatment were determined by comparing phonotactic performance in freely walking animals before and after this procedure (Atkins et al. 1984, 1992). (2) Individual neurons were implaled with microelectrodes in restrained animals as they performed phonotaxis, either while in tethered flight or while walking on a treadmill. Neuronal activity was manipulated by injection of current, and the effects on ongoing behavior were noted (Nolen and Hoy 1984; Schildberger and Hörner 1988).

3.2.2.1. AN2 and Negative Phonotaxis

Nolen and Hoy (1984) studied the neural basis for negative phonotaxis in *T. oceanicus* by monitoring the activity of flight steering muscles in preparations that permitted simultaneous intracellular studies of individual neurons. When such a preparation is stimulated with ultrasound, AN2 responds strongly to the sound stimulus and, shortly thereafter, the steering muscles are activated (Fig. 5.19A). If AN2 is prevented from responding normally by injecting hyperpolarizing current through an intracellular microelectrode, then the motor response is suppressed. Conversely, the motor response can be elicited in the absence of acoustic stimulation by electrically stimulating AN2 (Nolen and Hoy 1984; Fig. 5.19B,C).

Thus, activity of AN2 is both necessary and sufficient for negative phonotaxis, with the following two qualifications: (1) The animal must be actively locomoting. In Nolen and Hoy's experiment, this meant that the cricket's "flight motor" had to be "running"; stimulation of a quiescent preparation failed to elicit phonotactic motor responses. Brodfuehrer and Hoy (1989) showed that AN2's response to ultrasound is unaffected by flight but that the responses of ultrasound-driven neurons that descend from the brain to the thoracic ganglia, and may be involved in triggering motor responses, are enhanced. This suggests that the brain may be the site

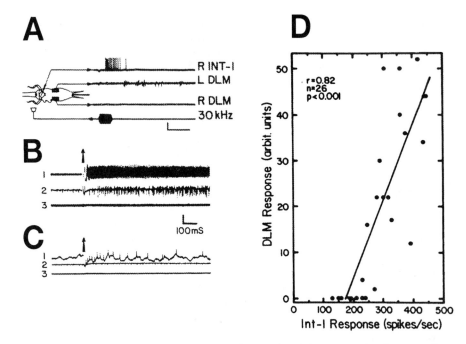

FIGURE 5.19. AN2 and negative phonotaxis. A: Stimulating a restrained preparation with 30 kHz from the right elicits a burst of spikes in the right AN2 (labeled here as INT-1, the designation, in *T. oceanicus*, for AN2), followed by activity in the left abdominal dorsal longitudinal muscles (DLM), which flex the abdomen to the left. This preparation is thus exhibiting negative phonotactic steering. B: The right AN2 (trace 1) is caused to fire at a high rate by release from hyperpolarization; this elicits a steering response (activation of left DLM, trace 2) despite the absence of sound stimulation. C: AN2 is hyperpolarized as 30-kHz sound is played, decreasing its firing rate in response to the sound stimulus; under these conditions no motor response ensues. D: Motor responses only occur if AN2 (INT-1) fires at high rates, $\geq$180 spikes per second. (Reprinted with permission from Nolen and Hoy 1984; copyright 1984 American Association for the Advancement of Science.)

at which motor activity exerts its gating effect on the behavioral response. Crickets can also perform negative phonotaxis to ultrasound while walking (Pollack, Huber, and Weber 1984), indicating that activity of the flight motor per se is not a strict requirement for negative phonotaxis. (2) AN2 must fire at a high rate, $\geq$180 spikes/s. When AN2 is active at lower rates, motor responses do not occur (Fig. 5.19D).

3.2.2.2. Identified Interneurons and Positive Phonotaxis

ON1, as a putative enhancer of directional information, is expected to influence phonotactic direction. As predicted, killing this neuron on one side (Atkins et al. 1984) or hyperpolarizing it to decrease its spiking rate

(Schildberger and Hörner 1988) causes systematic errors in walking direction during positive phonotaxis. The animal veers toward the side on which ON1 is undisturbed, that is, if the ON1 that receives input from the left side is killed or hyperpolarized, then the errors are to the right (Fig. 5.20A). This is the expected error direction given that ON1 inhibits contralateral cells; decreasing or eliminating the activity of the left ON1 would remove inhibition of right ascending neurons, generating a binaural difference in activity of ascending neurons that is appropriate for a sound source located further to the right than the actual source.

Manipulation of the ascending neurons themselves also affects positive phonotactic direction. Hyperpolarization of AN1 causes crickets to attempt to turn toward the undisturbed side, no matter on which side sound is broadcast. Hyperpolarization of AN2 has more modest effects, producing smaller errors toward the undisturbed side (Schildberger and Hörner 1988; Fig. 5.20B). Similarly, killing the neurons L1 and L3 by photoinactivation leads to errors or circling towards the intact side (Atkins et al. 1992; L1 and L3 are the designations of two ascending neurons in *Acheta domesticus*: L1 may be homologous to AN1 of other cricket species; L3's counterpart in other species is less clear).

3.2.3. Interaural Latency Differences as a Cue for Sound Direction

In the foregoing discussion, the code for intensity (and thus for sound direction) was presumed to be the magnitude of the neural response, that is, spike count. However, another response parameter that varies with stimulus intensity is latency, which becomes shorter with increasing intensity. Response latency can vary by nearly 10 ms over the 20 to 30 dB dynamic range of a receptor neuron. Variations are somewhat less pronounced for the more modest differences in effective intensity that are associated with changes in sound direction during phonotaxis but are still large enough to result in interaural latency differences on the order of 6 ms; thus the physical time-of-arrival difference of a few microseconds is amplified, through the intensity dependence of latency, a thousandfold (Mörchen, Rheinlaender, and Schwartzkopff 1978).

Both physiological and behavioral experiments show that the central nervous system is able to make use of interaural latency differences of this magnitude. Rheinlaender and Mörchen (1979) used direct mechanical stimulation of the tympanic membrane of the locust, rather than airborne sound, in order to present dichotic stimuli in which interaural timing relationships could be controlled with precision. When equally intense stimuli were delivered to the two ears (which presumably resulted in equal spike counts in tympanal receptors), the activity of an identified interneuron could be driven from complete inhibition to maximal excitation by shifting interaural timing over a range of only 4 ms (Fig. 5.21A). Helversen and Rheinlaender (1988) varied the relative timing of stimuli that were

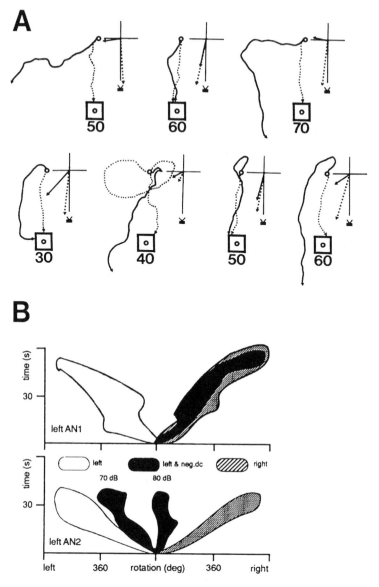

FIGURE 5.20. Bilateral imbalances in interneuron activity affect phonotactic direction. A: Killing ON1 affects phonotactic paths of *Acheta domesticus*. The upper and lower sets of diagrams show the phonotactic paths of two different females before (dotted lines) and after (solid lines) the left ON1 was killed by photoinactivation. The squares represent the loudspeaker (numbers below them indicate the pulse periods used in the song models); insets show mean vectors of the paths. B: Phonotactic paths reconstructed from measurements of turning tendencies in restrained crickets walking on a spherical treadmill (see Fig. 5.9B for technique). Open and hatched areas show the range of path directions to sounds broadcast from the left and right, respectively; dark areas show the effect of hyperpolarizing the left AN1 (top diagram) or the left AN2 (bottom diagram). (A reprinted with permission from Atkins et al. 1984; copyright 1984 Springer-Verlag. B reprinted with permission from Schildberger 1994; copyright 1994 Gustav Fischer Verlag.)

delivered dichotically to freely behaving male grasshoppers and found that the animals turned toward the leading stimulus with 75% accuracy, with an interaural delay of only 400 µs and with nearly errorless performance at a time difference of 1.5 ms (Fig. 5.21B).

3.2.4. Analyzing One Signal Among Many

A problem that is closely related to determining sound direction is that of analyzing the signals emanating from a single source in an acoustically distracting environment. Humans are familiar with the "cocktail party problem," that is, the need to listen to a single conversation despite the simultaneous presence of distracting sounds (Cherry 1953). Insects often encounter similar situations, because a single listener may find itself within earshot of several different singing individuals. Behavior experiments show that crickets are able to choose a particular song model when another, less attractive but still acceptable, model is played simultaneously, provided the two are sufficiently separated in space, for example, with one song played from the left side and the other from the right (Pollack 1986). This implies that rather than mutually obliterating one another, the temporal patterns of the simultaneous signals are still individually discernable.

The basis for this may be a sort of "selective attention" that is shown by the ON1 neurons. These cells receive biphasic input; their excitatory response is followed by a long-lasting hyperpolarization, the depth and duration of which are intensity dependent (Pollack 1988; Fig. 5.22A). Only the most intense sound pulses are able to override the hyperpolarization and elicit spiking responses from the neuron; less intense inputs are filtered out. Under some circumstances, for example, when different, equally intense signals are broadcast from the two sides of the animal, each ON1 responds selectively to the ipsilateral signal. This is because, due to the directional characteristics of the auditory system, each signal is of higher effective intensity for the ipsilateral ON1 than for its contralateral partner. Thus, two different temporal patterns can be represented simultaneously and independently on the two sides of the nervous system (Pollack 1986, 1988; Fig. 5.22B). Under other circumstances, for example, with the listener close to one singer while several others also sing in the background, the listener's nervous system will selectively encode the signal from the nearby source, filtering out more distant songs (Römer 1993; Fig. 5.22C).

The ability to listen selectively to the most intense signal is caused, at least in part, by a calcium-activated current, perhaps a calcium-activated potassium current. Optical monitoring with the calcium-sensitive dye fura 2 shows that intracellular $[Ca^{++}]$ increases and decreases with a time course similar to the buildup and decay of the poststimulus hyperpolarization. Moreover, hyperpolarization can be induced in the absence of acoustical stimulation by the photolytic release of calcium from intracellularly injected DM-Nitrophen (Sobel and Tank 1994). In *T. oceanicus*, in which

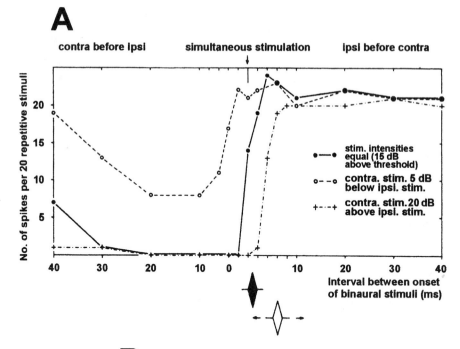

A

contra before ipsi   simultaneous stimulation   ipsi before contra

No. of spikes per 20 repetitive stimuli

Interval between onset
of binaural stimuli (ms)

- stim. intensities
  equal (15 dB
  above threshold)
- contra. stim. 5 dB
  below ipsi. stim.
- contra. stim. 20 dB
  above ipsi. stim.

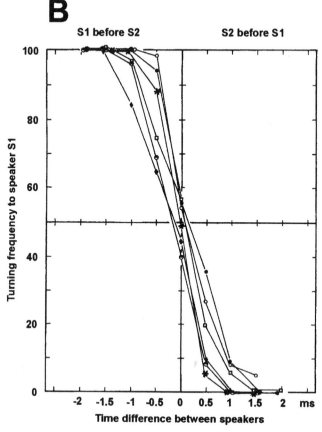

B

S1 before S2   S2 before S1

Turning frequency to speaker S1

Time difference between speakers

ON1 is tuned both to cricketlike frequencies and to ultrasound, poststimu-lus hyperpolarization is frequency specific; it is much more pronounced for cricketlike sounds than for ultrasound (Pollack 1988). Thus, it cannot be accounted for solely by calcium entry through voltage-senstive channels opened by the neuron's spiking activity. Rather, at least a portion of the effect appears to be due to frequency-specific synaptic inputs to the cell.

3.3. Analysis of Temporal Pattern

As the behavior experiments summarized in Section 1 demonstrate, tempo-ral pattern is a crucial cue for the recognition of acoustic signals. The neural basis for pattern selectivity has been studied most extensively in crickets and grasshoppers. Surprisingly, these two families of Orthoptera use quite different strategies for solving this problem. In crickets, recognition seems to be accomplished largely, perhaps exclusively, within the brain, whereas in grasshoppers substantial processing occurs within the thoracic ganglia.

3.3.1. Temporal Pattern Recognition in the Cricket Brain

Studies of temporal pattern recognition have focused on the calling song. Neither auditory receptors nor first-order interneurons in the prothoracic ganglion are strongly selective for this signal; rather, these neurons copy a wide variety of temporal patterns. Selectivity first appears in the brain.

Acoustic information is delivered from the prothoracic ganglion (the termination site of auditory receptors) to the brain mainly by the two identified ascending neurons AN1 and AN2. Both neurons project to the protocerebrum, with their main terminal arbor in an anterior-dorsal region. Here, their terminals overlap with processes of a diverse population of local brain neurons, the BNC1 cells (Brain Neuron Class 1; defined as brain neurons with processes that overlap the projections of the ascending neu-rons). Some of the BNC1 cells project to another brain region, near the protodeutocerebral border, where they overlap with processes of the BNC2 cells (defined as neurons that do not overlap with the projections of

---

FIGURE 5.21. Sensitivity to interaural time differences. A: An ascending interneuron in *Locusta migratoria* switches from complete absence of a response to maximal response when interaural timing is varied by only 4 ms (solid curve). Dotted curves show that time differences can be offset by interaural intensity differences; that is, that time-intensity tradeoffs occur. B: Behavioral sensitivity of *Chorthippus biggutulus* males to differences in the timing of sounds broadcast from the two sides. The animals turn toward the leading sound, even with time differences of less than 1 ms. (A reprinted with permission from Rheinlaender and Mörchen 1979; copyright 1979 Macmillan Magazines Ltd. B reprinted with permission from Helversen and Rheinlaender 1988; copyright 1988 Springer-Verlag.)

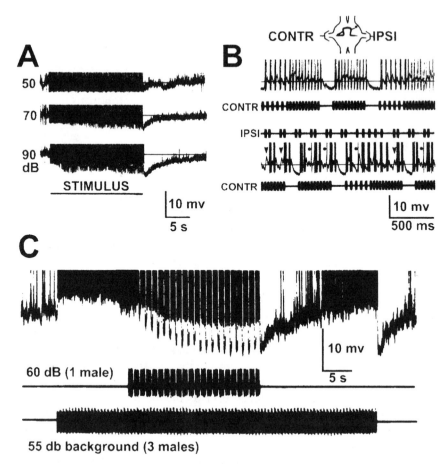

FIGURE 5.22. Selective attention in the ON1 neuron. A: In addition to being excited by song stimuli, the ON1 neuron of *Teleogryllus oceanicus* shows a slowly building, long-lasting hyperpolarization, which is intensity dependent. B: The neuron can respond to a sufficiently intense contralateral song model (top traces), but when this is paired with an ipsilateral stimulus of equal nominal intensity (but greater effective intensity, due to the directionality of the auditory system), only the latter stimulus can overcome the inhibition; the contralateral stimulus is filtered out of the neuron's spiking response. Horizontal lines through the recordings indicate prestimulus membrane potential. C: Recordings from the ON neuron of a bushcricket, *Tettigonia viridissima*. This neuron is similar to, and perhaps homologous with, the ON1 neuron of crickets. When stimulated with the combined songs of three males, the neuron fires tonically at a high rate, but when a single male's song is broadcast at slightly higher intensity, the additional hyperpolarization filters out the background sounds, allowing the neuron to encode selectively the more intense song. (A and B reprinted with permission from Pollack 1988; copyright 1988 Society for Neuroscience. C reprinted with permission from Römer 1993; copyright 1993 The Royal Society.)

AN1 and AN2). Both the anatomical relationships and the relative response latencies of these neuron classes (BNC2 > BNC1 > AN) suggest that the flow of information is from the ANs (which are directly post-synaptic to receptors; see Section 3.1.1) to BNC1s to BNC2s (Schildberger 1984).

Figure 5.23A illustrates responses of some of these neurons to three stimulus patterns. All three patterns have pulse-to-interval ratios of 1:1, and so total sound energy is identical across patterns. Only the central pattern, however, with a pulse period of 34 ms, is behaviorally effective (Thorsen, Weber, and Huber 1982). Both AN1 and BNC1a (one member of the class of BNC1 neurons) respond equally well to all three patterns. BNC1d, however, acts as a "low-pass" filter for pulse rate; it responds to the two slower rates but not to the fastest one. High-pass neurons also occur (within the BNC2 class). The most striking selectivity is shown by BNC2a, which responds strongly only to the middle temporal pattern; it exhibits "band-pass" behavior. Filter properties were evident in only 3 of the 18 individual BNC1 neurons that were studied but were more frequent in BNC2 neurons. Nine of 15 BNC2 neurons showed temporal selectivity, with 7 of these being band-pass neurons. The cutoff frequencies and sharpness of these neuronal low-, high-, and band-pass filters closely match the selectivity for temporal pattern that emerges in behavior experiments (Fig. 5.23B), providing strong, though as yet only correlational, evidence that these neurons may be causally responsible for behavioral selectivity (Schildberger 1984).

Another striking feature of this putative recognition circuit is the progressive loss of information about the fine structure of the stimulus at higher levels in the circuit. As Figure 5.23A shows, the individual sound pulses and intervals are reflected clearly in AN1's spike train for a wide range of temporal patterns. The BNC1 neurons only show good copying of the stimulus for slow pulse rates, and in BNC2 neurons the relationship between the individual sound pulses of the stimulus and the structure of the spike train is further obscured, despite the fact that the magnitude of the neuronal response is highly sensitive to the stimulus temporal pattern. Information about sound intensity is also lost. The responses of BNC2 neurons are nearly independent of intensity once the stimulus is more than 10 dB above threshold (Schildberger 1984).

Interestingly, neuronal responses to different stimulus temporal patterns that are phenomenologically identical to those in cricket brains have been recorded in the brains of frogs and toads, which also communicate acoustically and use temporal pattern as an important cue for signal identification. In anurans, as in crickets, low-pass, high-pass, and band-pass temporal filters are found, and, as in crickets, there is little relationship between the structure of the spike trains of these cells and the fine structure of the stimulus (Rose and Capranica 1983; Feng, Hall, and Gooler 1990). In both animals it has been suggested that band-pass filters might be constructed by

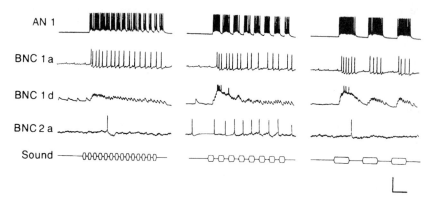

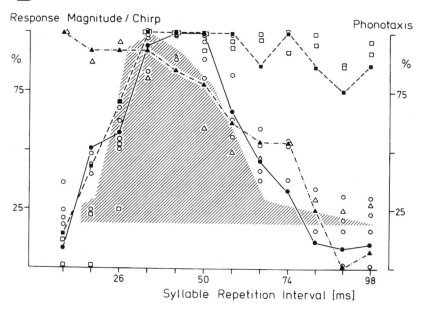

FIGURE 5.23. Analysis of temporal pattern in the cricket's brain. A: Recordings from an ascending neuron (AN1), two different BNC1 neurons, and a BNC2 neuron in response to stimuli with different syllable rates. Only the middle rate is behaviorally effective. B: Correspondence between behaviorally measured selectivity for syllable rate (shaded) and the filter properties of low-pass, high-pass, and band-pass neurons in the brain (lines). (Reprinted with permission from Schildberger 1984; copyright 1984 Springer-Verlag.)

"ANDing" the outputs of low- and high-pass filters; that is, a BNC2 neuron with band-pass characteristics might receive input from both low- and high-pass BNC1 or BNC2 cells and only respond when both inputs were active simultaneously. The cellular mechanisms underlying the selectivity of the low- and high-pass filter neurons are not known in either crickets or anurans, although several well-known processes, such as temporal summation, synaptic depression, and facilitation, are obvious candidates.

3.3.2. Pattern Recognition in Grasshoppers

The recognition network in grasshoppers appears to be more distributed than that in crickets, with substantial analysis of temporal pattern occurring in the metathoracic ganglion. Unlike crickets, in which behaviorally relevant information is believed to be delivered to the brain mainly by two neurons (AN1 and AN2), at least 15 ascending neurons have been identified in grasshoppers. At least one of these can copy a wide range of temporal patterns, but the majority fail to encode the behaviorally important parameters of syllable duration and the interval between syllables (Römer and Seikowski 1985; Stumpner and Ronacher 1991b). Some ascending neurons respond phasically and thus mark the onsets of syllables but not their durations; others encode song structure only irregularly. Moreover, responses are often restricted to intensity ranges that are far narrower than the operating range of behavior. Because various features of song stimuli seem to be distributed among a number of parallel ascending channels, it has been suggested that the role of recognition circuits in the brain might be to integrate information across these different channels and, additionally, to combine temporal pattern-related information with binaural cues for sound direction, carried along still other ascending channels (Stumpner and Ronacher 1991b).

Within the brain, neurons exist that are tuned to the pulse rate of intraspecific signals, though these have not yet been identified anatomically (Römer and Seikowski 1985). Unlike crickets, in which low-pass, high-pass, and band-pass neurons occur, only band-pass neurons have been reported thus far in grasshoppers. It thus seems possible that the circuit designs underlying pulse-rate sensitivity might differ between these two groups. Römer and Seikowski (1985) suggested that the tuning of locust brain neurons to pulse rate might result from their receiving both direct and delayed versions of the stimulus, with simultaneous arrival of the two required to activate the brain neurons (at the "correct" pulse rate, the ith pulse of the delayed input would coincide temporally with the $(i + 1)^{th}$ pulse of the direct input). They proposed that the direct input might be provided by neurons that ascend directly from the thorax to the brain, and a delayed version by parallel channels that traverse one or more additional synapses in the subesophageal ganglion (Boyan and Altman 1985). This interesting notion, for which there is as yet no direct evidence, is reminiscent of the use of delay lines to measure interaural time differences in owls, pulse-echo delays in echolocating bats, and phase differences in electrical signals at different points on the body surface in electric fish (see Carr 1993 for review).

Another aspect of pattern analysis in grasshoppers, the rejection by females of songs that contain gaps within the syllables (Fig. 5.24A; see Section 1.2.2), has a clear neural correlate. An identified ascending neuron, AN4, responds well only to syllables that lack gaps (Fig. 5.24B). The similarity in the shapes of the curves describing behavioral and neural responses suggests that activity of AN4 may be required for the female to respond (Ronacher and Stumpner 1988).

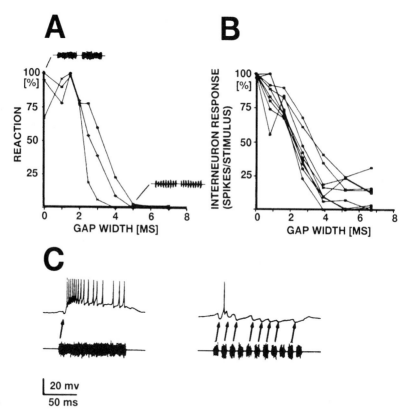

FIGURE 5.24. Rejection of gaps in *Chorthippus biggutulus*. A: Response rates of three different females to model songs with different gap widths; insets illustrate syllables with no gaps and with 5-ms gaps. B: Response of the neuron AN4 as a function of gap width; each line represents a different individual. C: Mechanism for gap rejection by AN4. Sound onset evokes an IPSP in AN4. The repeated onsets in syllables with gaps evoke a train of IPSPs that summate and suppress the neuron's response. (Reprinted with permission from Ronacher and Stumpner 1988; copyright 1988 Springer-Verlag.)

Gaps prevent AN4 from responding because the sound onset following each gap elicits an IPSP that hyperpolarizes the cell. A similar IPSP occurs at the onset of gapless syllables, but it is nullified by the strong excitation that follows it. The repeated IPSPs in syllables with gaps, however, summate and effectively suppress the neuron's response (Ronacher and Stumpner 1988; Fig. 5.24C).

3.4. Interactions Between Recognition and Localization

In the preceding discussion, two basic tasks that the auditory system must accomplish, song recognition and sound localization, were treated separately. However, both tasks must be accomplished together during behavior, and, therefore, neural analysis of these two features of sound must somehow be integrated. The overall neuronal "strategies" for this have been examined in crickets and grasshoppers and appear to differ drastically between these two groups.

3.4.1. Parallel Processing in Grasshoppers

In grasshoppers, recognition and localization are performed in parallel, by at least partly separate sets of neurons. The first indication of this came from experiments that demonstrated decisions about a signal's temporal pattern were made after pooling the input from both ears (Helverson 1984; Helversen and Helversen 1995). *C. biguttulus* are highly selective for the relationship between syllable duration and syllable interval within the song (see Section 1.2.2). If a song with an acceptable temporal pattern is altered such that every other syllable is deleted, it becomes ineffective because the syllable interval is now too long. But if such a song is played from one side, with the "missing" syllables played in an interleaved fashion from the other side, the song regains effectiveness. This observation suggests that the patterns at the two ears, each of which is ineffective by itself, are combined internally to recreate an effective temporal pattern.

By contrast, if two acceptable songs are played from the two sides, again in an interleaved fashion, behavioral responses are lost. In this case the combination of two patterns creates an ineffective signal because the syllable interval after combination is too short. (It is, of course, conceivable that the combination of the signals from the two sides is done at the level of the tympanal membrane rather than internally, i.e., that each ear is stimulated by both stimuli, albeit at different effective intensities. This possibility was ruled out by a number of control experiments. See Helverson 1984; Helversen and Helversen 1995 for details.) The pooling of input from both ears that these experiments imply would necessarily obliterate binaural cues for sound localization. Thus, the circuitry for localization must be at least partially separate from that for recognition. This notion is supported by neurophysiological recordings, which show that neurons that are best suited to encode information about sound direction are poor at encoding temporal features of the song (Stumpner and Ronacher 1991b). Lesion experiments, in which various parts of the nervous system were surgically isolated (Ronacher, Helversen, and Helversen 1986), suggest that the division of information into separate recognition and localization channels occurs quite early along the central neuronal pathway, within the metathoracic ganglion (see Helversen and Helversen 1990 for review).

3.4.2. Serial Processing in Crickets

In crickets, recognition and localization appear to be serial processes, with the "localizer" acting on the output of the "recognizer." This type of organization was first suggested in order to account for the observation that crickets (unlike grasshoppers) can discriminate reliably between two signals (both acceptable, but differing in attractiveness) that are played simultaneously from the two sides (Pollack 1986). This would be possible if the direction of phonotactic turns were determined by comparing the outputs of two recognition circuits, one located on each side of the brain. This notion was reinforced when it was discovered that left and right neural homologues were able to listen selectively to the patterns played from their respective sides (Pollack 1988; see Section 3.2.4), that is, that the cricket's nervous system is able to avoid the pooling of temporal patterns from the two sides that occurs in grasshoppers. Consistent with the idea of bilaterally paired recognition circuits, the brain neurons that show selectivity for stimulus temporal pattern have their processes restricted to one half of the brain, and are driven by input from the ipsilateral ear (Schildberger 1984). Perhaps the most convincing evidence for serial processing is the following experiment of Stabl, Wendler, and Scharstein (1989). They stimulated crickets with calling song from a loudspeaker placed above the animal (which performed phonotaxis on a treadmill). This arrangement provides an attractive temporal pattern but produces no bilateral cues that would indicate that the sound is to one side or the other. When the song from above was paired with a constant tone played from a loudspeaker situated in the cricket's horizontal plane to one side, which provides strong directional cues, crickets did not orient towards this new source (as the rule "turn towards the most strongly stimulated ear" might suggest); rather, they turned away from it. This apparent paradox was resolved by recording from the two ascending interneurons, AN1 and AN2. Although total activity in the neuronal pair was greater on the side of the loudspeaker broadcasting the constant tone, the high level of activity on this side induced by the continuous tone obscured the song's temporal pattern, which was better represented by the contralateral AN1 and AN2. This experiment suggests that the correct phrasing of the rule underlying positive phonotaxis in crickets is "turn toward the side on which the song's temporal pattern is represented most strongly."

3.4.3. Why Are Recognition and Localization Organized so Differently in Crickets and Grasshoppers?

Helversen and Helversen (1995) suggest that the different schemes used by grasshoppers and crickets for recognizing and localizing songs may reflect the evolution of their respective communication systems. In crickets, hearing and stridulation probably share a long, tightly coupled evolutionary history (Sharov 1971; Hoy 1992; Otte 1992). In other words, cricket auditory systems probably evolved in the context of intraspecific communication, and can thus be expected to be highly specialized for this task. One such specialization may be the requirement that signals must first pass a recognition filter before being localized, which allows crickets to locate conspecific males in noisy environments and to choose between simultaneously singing males. Hearing and stridulation are less closely associated in grasshoppers. In this group, hearing appeared before stridulation, which later evolved several times independently (Dirsh 1975). The neuronal organization underlying auditory behavior in grasshoppers may thus reflect its more primitive functions, for example, predator detection.

Modern crickets use hearing not only for communication but also for detecting predators, specifically, echolocating bats. The notion that crickets' auditory systems became highly specialized for intraspecific communication early in their evolutionary history, some 200 to 300 million years before bats evolved (Hoy 1992), may help to explain two puzzling aspects of the neural basis for bat evasion. First, only a few receptors are sensitive to ultrasound compared with many that respond to cricket songs (see Fig. 5.12). One might suppose that a larger number of ultrasound-sensitive receptors would lead to greater sensitivity for bat cries or allow more precise estimation of direction or distance. Second, the interneuron that triggers the bat evasion response, AN2 (see Fig. 5.15), projects from the prothoracic ganglion to the brain, and intact connections with the brain are necessary for this behavior to occur (Pollack and Hoy 1981b), despite the fact the motor aspects of the response occur mainly in more posterior segments. The additional conduction distance and circuitry within the brain (Brodfuhrer and Hoy 1990) must surely increase the latency of this escape response. These apparent anomalies in circuitry are less jarring if one considers that the ability to detect and avoid bats was probably incorporated during evolution into a system that was already highly specialized to detect, analyze, and localize conspecific signals.

4. Summary

Insects, like all animals that must extract information from sound signals, face several general problems: discrimination of signal from noise, recognition of and discrimination among different signals, and determination of the direction of the source. Observations of behavioral responses to synthetic signals reveal that insects recognize signals by their frequency content and temporal pattern and locate them in space on the basis of interaural intensity differences. Sound frequency is first analyzed by the array of frequency-tuned receptors in the ear, whose output is delivered to the central nervous system in a tonotopically organized manner. There, different interneurons acquire different frequency sensitivity both through specific connections with receptors and through inhibitory interactions between interneurons. Differences in sound intensity at the two ears, which are initially imposed by the acoustics of the auditory system, are amplified by contralateral inhibition. That temporal features of stimuli are analyzed is evident from behavioral observations and from the temporal-pattern selectivity of some interneurons, but the mechanisms for this are understood only in broad outline.

Because the Orthoptera are the acoustical virtuosos of the insect world, they have accounted for a great deal of the experimental work on the neural processing of acoustic signals and thus for the bulk of this chapter. Within the Orthoptera, similarities in neural processing are evident across groups. The central organization of the projections of auditory receptors are strikingly similar between grasshoppers and bushcrickets (see Fig. 5.11), despite the different structures and locations of their ears (see Fig. 5.10). Some auditory interneurons of bushcrickets, for example, the omega neuron, are similar both in morphology (compare Figs. 5.13 and 5.17) and in physiology (see Fig. 5.22) to their cricket counterparts.

There are also differences in neural processing among Orthoptera. In crickets, for example, analysis of stimulus temporal pattern is done mainly in the brain, whereas in grasshoppers significant analysis of temporal features occurs in the metathoracic ganglion. Even more striking, the neuronal organization underlying recognition and localization of signals is grossly different in these groups, with the two processes operating in parallel, essentially independent of one another in acridids but serially, with localization dependent on recognition, in crickets. These differences in organization may reflect the different evolutionary histories of hearing in these two groups.

References

Atkins G, Pollack GS (1986) Age dependent occurrence of an ascending axon on the omega neuron of the cricket, *Teleogryllus oceanicus*. J Comp Neurol 243:527–534.

Atkins G, Pollack GS (1987a) Response properties of prothoracic, interganglionic, sound-activated interneurons in the cricket *Teleogyyllus oceanicus*. J Comp Physiol A 161:681–693.

Atkins G, Pollack GS (1987b) Correlations between structure, topographic arrangement, and spectral sensitivity of sound-sensitive interneurons in crickets. J Comp Neurol 266:398–412.

Atkins G, Ligman S, Burghardt F, Stout J (1984) Changes in phonotaxis by the female cricket after killing identified acoustic interneurons. J Comp Physiol A 154:795–804.

Atkins G, Henley J, Handysides R, Stout J (1992) Evaluation of the behavioral roles of ascending auditory interneurons in calling song phonotaxis by the female cricket (*Acheta domesticus*). J Comp Physiol A 170:363–372.

Bailey WJ, Robinson D (1971) Song as a possible isolating mechanism in the genus *Homorocoryphus* (Tettigonioidea, Orthoptera). Anim Behav 19:390–397.

Bailey WJ, Thomson P (1977) Acoustic orientation in the cricket *Teleogryllus oceanicus* (Le Guillou). J Exp Biol 67:61–75.

Balakrishnan R, Pollack GS (1996) Recognition of courtship song in the field cricket, *Teleogryllus oceanicus*. Anim Behav 51:353–366.

Balakrishnan R, Pollack GS (1997) The role of antennal sensory cues in female responses to courting males in the cricket *Teleogryllus oceanicus*. J Exp Biol 200:511–522.

Boyan GS (1980) Auditory neurones in the brain of the cricket *Gryllus bimaculatus* (DeGeer). J Comp Physiol A 140:81–93.

Boyan GS (1981) Two-tone suppression of an identified auditory neurone in the brain of the cricket *Gryllus bimaculatus* (De Geer). J Comp Physiol A 144:117–125.

Boyan GS, Altman JS (1985) The suboesophageal ganglion: a "missing link" in the audiory pathway of the locust. J Comp Physiol A 156:413–428.

Boyan GS, Williams JLD (1982) Auditory neurones in the brain of the cricket *Gryllus bimaculatus* (DeGeer): ascending interneurones. J Insect Physiol 28:493–501.

Boyd P, Lewis B (1983) Peripheral auditory directionality in the cricket (*Gryllus campestris* L., *Teleogryllus oceanicus* Le Guillou). J Comp Physiol A 153:523–532.

Boyd P, Kühne R, Silver S, Lewis B (1984) Two-tone suppression and song coding by ascending neurones in the cricket *Gryllus campestris* L. J Comp Physiol A 154:523–532.

Breckow J, Sippel M (1985) Mechanics of the transduction of sound in the tympanal organ of adults and larvae of locusts. J Comp Physiol A 157:619–629.

Brodfuehrer PD, Hoy RR (1989) Integration of ultrasound and flight inputs on descending neurons in the cricket brain. J Exp Biol 145:157–171.

Brodfuehrer PD, Hoy RR (1990) Ultrasound sensitive neurons in the cricket brain. J Comp Physiol A 166:651–662.

Cade W (1979) The evolution of alternative male reproductive strategies in field crickets. In: Blum MS, Blum NA (eds) Sexual Selection and Reproductive Competition in Insects. New York: Academic Press, pp. 343–379.

Campbell DJ, Shipp E (1979) Regulation of spatial pattern in populations of the field cricket *Teleogryllus commodus* (Walker). Z Tierpsychol 51:260–268.

Carr CE (1993) Processing of temporal information in the brain. Ann Rev Neurosci 16:223–243.

Casaday GB, Hoy RR (1977) Auditory interneurons in the cricket *Teleogryllus oceanicus*: physiological and anatomical properties. J Comp Physiol 121:1–13.

Cherry EC (1953) Some experiments on the recognition of speech, with one and with two ears. J Acoust Soc Am 25:975–979.

Dirsh VM (1975) A preliminary revision of the families and subfamilies of Acridoidea (Orthoptera, Insecta). Bull Br Mus (Nat Hist) Entomol 10:349–419.

Dobler S, Heller K-G, Helversen O von (1994) Song pattern recognition and an auditory time window in the female bushcricket *Ancistrura nigrovittata* (Orthoptera: Phaneropteridae). J Comp Physiol A 175:67–74.

Doherty JA (1985a) Trade-off phenomena in calling song recognition and phonotaxis in the cricket, *Gryllus bimaculatus* (Orthoptera, Gryllidae). J Comp Physiol A 156:787–801.

Doherty JA (1985b) Temperature coupling and "trade-off" phenomena in the acoustic communication system of the cricket, *Gryllus bimaculatus* De Geer (Gryllidae). J Exp Biol 114:17–35.

Doherty JA (1991) Song recognition and localization in the phonotaxis behavior of the field cricket, *Gryllus bimaculatus* (Orthoptera: Gryllidae). J Comp Physiol A 168:213–222.

Doherty JA, Callos JD (1991) Acoustic communication in the trilling field cricket, *Gryllus rubens* (Orthoptera: Gryllidae). J Insect Behav 4:67–82.

Doherty JA, Hoy RR (1985) The auditory behavior of crickets: some views of genetic coupling, song recognition, and predator detection. Q Rev Biol 60:457–472.

Doolan JM, Pollack GS (1985) Phonotactic specificity of the cricket *Teleogryllus oceanicus*: intensity-dependent selectivity for temporal parameters of the stimulus. J Comp Physiol A 157:223–233.

Doolan JM, Young D (1989) Relative importance of song parameters during flight phonotaxis and courtship in the bladder cicada *Cystosoma saundersii*. J Exp Biol 141:113–131.

Elsner N, Popov AV (1978) Neuroethology of acoustic communication. Adv Insect Physiol 13:229–355.

Esch H, Huber F, Wohlers DW (1980) Primary auditory neurons in crickets: physiology and central projections. J Comp Physiol 137:27–38.

Ewing AW (1983) Functional aspects of *Drosophila* courtship. Biol Rev 58:275–292.

Feng AS, Hall JC, Gooler DM (1990) Neural basis of sound pattern recognition in anurans. Prog Neurobiol 34:313–329.

Fullard JH, Yack JE (1993) The evolutionary biology of insect hearing. Trends Ecol Evol 8:248–252.

Goodman CS (1996) Mechanisms and molecules that control growth cone guidance. Ann Rev Neurosci 19:341–377.

Gray EG (1960) The fine structure of the insect ear. Philos Trans R Soc Lond B Biol Sci 243:75–94.

Halex H, Kaiser W, Kalmring K (1988) Projection areas and branching patterns of the tympanal receptor cells in migratory locusts, *Locusta migratoria* and *Schistocerca gregaria*. Cell Tissue Res 253:517–528.

Hardt M, Watson AHD (1994) Distribution of synapses on two ascending interneurones carrying frequency-specific information in the auditory system of the cricket: evidence for GABAergic inputs. J Comp Neurol 345:481–495.

Heller K-G, Helvesen D von (1986) Acoustic communication in phaneropterid bushcrickets: species-specific delay of female stridulatory response and matching male sensory time window. Behav Ecol Sociobiol 18:189–198.

Helversen D von (1972) Gesang des Männchens und Lautschema des Weibchens bei der Feldheuschrecke *Chorthippus biguttulus* (Orthotpera, Acrididae). J Comp Physiol 81:381–422.

Helversen D von (1984) Parallel processing in auditory pattern recognition and directional analysis by the grasshopper *Chorhtippus biguttulus* L. (Acrididae). J Comp Physiol A 154:837–846.

Helversen D von (1993) "Absolute steepness" of ramps as an essential cue for auditory pattern recognition by a grasshopper (Orthoptera; Acrididae; *Chorthippus biguttulus* L.). J Comp Physiol A 172:633–639.

Helversen D von, Helversen O von (1983) Species recognition and acoustic localization in acridid grasshoppers: a behavioral approach. In: Huber F, Markl H (eds) Neuroethology and Behavioral Physiology. Berlin: Springer-Verlag, pp. 95–107.

Helversen D von, Helversen O von (1990) Pattern recognition and directional analysis: routes and stations of information flow in the CNS of a grasshopper. In: Cribakin FG, Wiese K, Popoav AV (eds) Sensory Systems and Communication in Arthropods. Basel: Birhäuser Verlag, pp. 209–216.

Helversen D von, Helversen O von (1994) Forces driving coevolution of song and song recognition in grasshoppers. In: Schildberger K, Elsner N (eds) Neural Basis of Behavioural Adaptations. Stuttgart: Gustav Fischer Verlag, pp. 253–284.

Helversen D von, Helversen O von (1995) Acoustic pattern recognition and orientation in orthopteran insects: parallel or serial processing? J Comp Physiol A 177:767–774.

Helversen D von, Helversen O von (1997) Recognition of sex in the acoustic communication of the grasshopper *Chorthippus biguttulus* (Orthoptera, Acrididae). J Comp Physiol A 180:373–386.

Helversen D von, Rheinlaender J (1988) Interaural intensity and time discrimination in an unrestrained grasshopper: a tentative behavioural approach. J Comp Physiol A 162:333–340.

Hennig RM (1988) Ascending auditory interneurons in the cricket *Teleogryllus commodus* (Walker): comparative physiology and direct connections with afferents. J Comp Physiol A 163:135–143.

Hennig RM, Weber T (1997) Filtering of temporal parameters of the calling song by cricket females of two closely related species: a behavioral analysis. J Comp Physiol A 180:621–630.

Hill KG (1974) Carrier frequency as a factor in phonotactic behaviour of female crickets (*Teleogryllus commodus*). J Comp Physiol 93:7–18.

Horseman G, Huber F (1994a). Sound localisation in crickets. I. Contralateral inhibition of an ascending auditory interneuron (AN1) in the cricket *Gryllus bimaculatus*. J Comp Physiol A 174:389–398.

Horseman G, Huber F (1994b) Sound localisation in crickets. II. Modeling the role of a simple neural network in the prothoracic ganglion. J Comp Physiol A 175:399–413.

Hoy RR (1992) The evolution of hearing in insects as an adaptation to predation from bats. In: Webster DB, Fay RR, Popper AN (eds) The Evolutionary Biology of Hearing. New York: Springer-Verlag, pp. 115–129.

Hutchings M, Lewis B (1981) Response properties of primary auditory fibers in the cricket *Teleogryllus oceanicus* (Le Guillou). J Comp Physiol 143:129–134.

Imaizumi K, Pollack GS (1996) Anatomy and physiology of auditory receptors of the cricket *Teleogryllus oceanicus*. Soc Neurosci Abstr 22:1082.

Irvine DRF (1992) Physiology of the auditory brainstem. In: Popper AN, Fay RR (eds) The Mammalian Auditory Pathway: Physiology. New York: Springer-Verlag, pp. 153–231.

Kalmring K, Lewis B, Eichendorf A (1978) The physiological characteristics of the primary sensory neurons of the complex tibial organ of *Decticus verrucivorus* L. (Orthoptera, Tettigonioidae). J Comp Physiol 127:109–121.

Lakes R, Schikorski T (1990) Neuroanatomy of tettigoniids. In: Bailey WJ, Rentz DCF (eds) The Tettigoniidae: Biology, Systematics and Evolution. Berlin: Springer-Verlag, pp. 166–190.

Latimer W, Lewis DB (1986) Song harmonic content as a parameter determining acoustic orientation behaviour in the cricket *Teleogryllus oceanicus* (Le Guillou). J Comp Physiol A 158:583–591.

Libersat F, Hoy RR (1991) Ultrasonic startle behavior in bushcrickets (Orthoptera; Tettigoniidae). J Comp Physiol A 169:507–514.

Libersat F, Murray JA, Hoy RR (1994) Frequency as a releaser in the courtship song of two crickets, *Gryllus bimaculatus* (de Geer) and *Teleogryllus oceanicus*: a neuroethological analysis. J Comp Physiol A 174:485–494.

Michel K (1974) Das Tympanalorgan von *Gryllus bimaculatus* DeGeer (Saltatoria, Gryllidae). Z Morphol Tiere 77:285–315.

Michelsen A (1966) Pitch discrimination in the locust ear: observations on single sense cells. J Insect Physiol 12:1119–1131.

Michelsen A (1971a) The physiology of the locust ear. I. Frequency sensitivity of single cells in the isolated ear. Z Vergl Physiologie 71:49–62.

Michelsen A (1971b) The physiology of the locust ear. II. Frequency discrimination based upon resonances in the tympanum. Z Vergl Physiologie 71:63–101.

Michelsen A, Larsen ON (1983) Strategies for acoustic communication in complex environments. In: Huber F, Markl H (eds) Neuroethology and Behavioral Physiology. Berlin: Springer-Verlag, pp. 321–331.

Miller LA, Olsen J (1979) Avoidance behavior in green lacewings. I. Behavior of free flying green lacewings to hunting bats and ultrasound. J Comp Physiol 131:113–120.

Moiseff A, Hoy RR (1983) Sensitivity to ultrasound in an identified auditory inter-neuron in the cricket: a possible neural link to phonotactic behavior. J Comp Physiol 152:155–167.

Moiseff A, Pollack GS, Hoy RR (1978) Steering responses of flying crickets to sound and ultrasound: mate attraction and predator avoidance. Proc Natl Acad Sci USA 75:4052–4056.

Mörchen A, Rheinlaender J, Schwartzkopff J (1978) Latency shift in insect auditory nerve fibers. Naturwissenschaften 65:656–657.

Nocke H (1972) Physiological aspects of sound communication in crickets (*Gryllus campestris* L.). J Comp Physiol 80:141–162.

Nolen TG, Hoy RR (1984) Initiation of behavior by single neurons: the role of behavioral context. Science 226:992–994.

Nolen TG, Hoy RR (1986a) Phonotaxis in flying crickets. I. Attraction to the calling song and avoidance of bat-like ultrasound are discrete behaviors. J Comp Physiol A 159:423–439.

Nolen TG, Hoy RR (1986b) Phonotaxis in flying crickets. II. Physiological mechanisms of two-tone suppression of the high frequency avoidance steering behavior by the calling song. J Comp Physiol A 159:441–456.

Nolen TG, Hoy RR (1987) Postsynaptic inhibition mediates high-frequency selectivity in the cricket *Teleogryllus oceanicus*: implications for flight phonotaxis behavior. J Neurosci 7:2081–2096.

Oldfield BP (1980) Accuracy of orientation in female crickets. *Teleogryllus oceanicus* (Gryllidae): dependence on song spectrum. J Comp Physiol 141:93–99.

Oldfield BP (1982) Tonotopic organisation of auditory receptors in Tettigoniidae (Orthoptera: Ensifera). J Comp Physiol 147:461–469.

Oldfield BP (1983) Central projections of primary auditory fibres in Tettigoniidae (Orthoptera: Ensifera). J Comp Physiol A 151:389–395.

Oldfield BP (1984) Physiology of auditory receptors in two species of Tettigoniidae (Orthoptera: Ensifera): alternative tonotopic organisations of the auditory organ. J Comp Physiol A 155:689–696.

Oldfield BP (1985) The tuning of auditory receptors in bushcrickets. Hearing Res 17:27–35.

Oldfield BP, Hill KG (1983) The physiology of ascending auditory interneurons in the tettigoniid *Caedicia simplex* (Orthoptera: Ensifera): response properties and a model of integration in the afferent auditory pathway. J Comp Physiol A 152:495–508.

Oldfield BP, Kleindienst HU, Huber F (1986) Physiology and tonotopic organization of auditory receptors in the cricket *Gryllus bimaculatus* DeGeer. J Comp Physiol A 159:457–464.

Otte D (1992) Evolution of cricket songs. J Orthopt Res 1:24–46.

Pollack GS (1982) Sexual differences in cricket calling song recognition. J Comp Physiol 146:217–221.

Pollack GS (1984) Ultrasound-sensitive neurons descending in the thoracic nervous system of the cricket *Teleogryllus oceanicus*. Can J Zool 62:555–562.

Pollack GS (1986) Discrimination of calling song models by the cricket, *Teleogryllus oceanicus*: the influence of sound direction on neural encoding of the stimulus temporal pattern and on phonotactic behavior. J Comp Physiol A 158:549–561.

Pollack GS (1988) Selective attention in an insect auditory neuron. J Neurosci 8:2635–2639.

Pollack GS (1994) Synaptic inputs to the omega neuron of the cricket *Teleogryllus oceanicus*: differences in EPSP waveforms evoked by low and high sound frequencies. J Comp Physiol A 174:83–89.

Pollack GS, El-Feghaly E (1993) Calling song recognition in the cricket *Teleogryllus oceanicus*: comparison of the effects of stimulus intensity and sound spectrum on selectivity for temporal pattern. J Comp Physiol A 171:759–765.

Pollack GS, Faulkes Z (1998) Representation of behaviorally relevant sound frequencies by auditory receptors in the cricket *Teleogryllus oceanicus*. J Exp Biol 201:155–163.

Pollack GS, Hoy RR (1979) Temporal pattern as a cue for species-specific calling song recognition in crickets. Science 204:429–432.

Pollack GS, Hoy RR (1981a) Phonotaxis to individual rhythmic components of a complex cricket calling song. J Comp Physiol A 144:367–373.

Pollack GS, Hoy RR (1981b) Phonotaxis in flying crickets: neural correlates. J Insect Physiol 27:41–45.

Pollack GS, Plourde N (1982) Directionality of acoustic orientation in flying crickets. J Comp Physiol 146:207–215.

Pollack GS, Huber F, Weber T (1984) Frequency and temporal pattern-dependent phonotaxis of crickets (*Teleogryllus oceanicus*) during tethered flight and compensated walking. J Comp Physiol A 154:13–26.

Popov AV, Markovich AM (1982) Auditory interneurons in the prothoracic ganglion of the cricket, *Gryllus bimaculatus*. II. A high-frequency ascending neuron (HF1AN). J Comp Physiol A 146:351–359.

Popov AV, Shuvalov VF (1977) Phonotactic behaviour of crickets. J Comp Physiol 119:111–126.

Popov AV, Shuvalov VF, Svetogorskaya ID, Markovich AM (1974) Acoustic behavior and auditory system in insects. In: Schwartzkopff J (ed) Mechanoreception. Abh Rheinisch-Westfäl Akad Wiss 53:281–306.

Popov AV, Markovich AM, Andjan AS (1978) Auditory interneurons in the prothoracic ganglion of the cricket *Gryllus bimaculatus* deGeer. I. The large segmental auditory neuron (LSAN). J Comp Physiol 126:183–192.

Regen J (1913) Über die Anlocking des Weibchens von *Gryllus campestris* durch telephonische übertragung der Stridulation des Männchens. Pflügers Arch Eur J Physiol 155:193–200.

Rheinlaender J, Mörchen A (1979) "Time-intensity trading" in locust auditory interneurones. Nature 281:672–674.

Robert D (1989) The auditory behaviour of flying locusts. J Exp Biol 147:279–301.

Robinson DJ (1980) Acoustic communication between the sexes of the bush cricket, *Leptophyes punctatissima*. Physiol Entomol 5:183–189.

Robinson D, Rheinlaender J, Hartley JC (1986) Temporal parameters of male-female sound communication in *Leptophyes punctatissima*. Physiol Entomol 11:317–323.

Roeder KD (1967) Turning tendency of moths exposed to ultrasound while in stationary flight. J Insect Physiol 13:873–888.

Römer H (1976) Die Informationsverarbeitung tympanaler Rezeptorelemente von *Locusta migratoria* (Acrididae, Orthoptera). J Comp Physiol 109:101–122.

Römer H (1983) Tonotopic organization of the auditory neuropile in the bushcricket *Tettigoni viridissima*. Nature 306:60–62.

Römer H (1987) Representation of auditory distance within a central neuropil of the bushcricket *Mygalopsis marki*. J Comp Physiol A 161:33–42.

Römer H (1993) Environmental and biological constraints for the evolution of long-range signalling and hearing in acoustic insects. Philos Trans R Soc Lond B 340:179–185.

Römer H, Bailey WJ (1986) Insect hearing in the field. II. Male spacing behaviour and correlated acoustic cues in the bushcricket *Mygalopsis marki*. J Comp Physiol A 159:627–638.

Römer H, Marquart V (1984) Morphology and physiology of auditory interneurons in the metathoracic ganglion of the locust. J Comp Physiol A 155:249–262.

Römer H, Rheinlaender J, Dronse R (1981) Intracellular studies on auditory processing in the metathoracic ganglion of the locust. J Comp Physiol 144:305–312.

Römer H, Seikowski U (1985) Responses to model songs of auditory neurons in the thoracic ganglia and brain of the locust. J Comp Physiol A 156:845–860.

Römer H, Marquart V, Hardt M (1988) Organization of a sensory neuropile in the auditory pathway of two groups of orthoptera. J Comp Neurol 275:201–215.

Ronacher B, Stumpner A (1988) Filtering of behaviourally relevant temporal parameters of a grasshopper's song by an auditory interneuron. J Comp Physiol A 163:517–523.

Ronacher B, Helversen D von, Helversen O von (1986) Routes and stations in the processing of auditory directional information in the CNS of a grasshopper, as revealed by surgical experiments. J Comp Physiol A 158:363–374.

Rose G, Capranica RR (1983) Temporal selectivity in the central auditory system of the leopard frog. Science 219:1087–1089.

Schildberger K (1984) Temporal selectivity of identified auditory neurons in the cricket brain. J Comp Physiol A 155:171–185.

Schildberger K (1994) The auditory pathway of crickets: adaptation for intraspecific acoustic communication. In: Schildberger K, Elsner N (eds) Neural Basis of Behavioural Adaptations. Stuttgart: Gustav Fischer Verlag, pp. 210–225.

Schildberger K, Hörner M (1988) The function of auditory neurons in cricket phonotaxis. I. Influence of hyperpolarization of identified neurons on sound localization. J Comp Physiol A 163:621–631.

Schildberger K, Kleindienst HU (1989) Sound localization in intact and one-eared crickets: comparison of neuronal properties with open-loop and closed-loop behaviour. J Comp Physiol A 165:615–626.

Schmitz B, Scharstein H, Wendler G (1982) Phonotaxis in *Gryllus campestris* L. (Orthoptera, Gryllidae). I. Mechanism of acoustic orientation in intact female crickets. J Comp Physiol A 148:431–444.

Schul J (1997) Neuronal basis of phonotactic behaviour in *Tettigonia viridissima*: processing of behaviourally relevant signals by auditory afferents and thoracic interneurons. J Comp Physiol A 180:573–583.

Schuller G, Pollak G (1979) Disproportionate frequency representation in the inferior colliculus of Doppler-compensating greater horseshoe bats: evidence for an acoustic fovea. J Comp Physiol 132:47–54.

Schumacher R (1979) Zur funcktionellen Morphologie des auditiven Systems der Laubheuschrecken (Orthoptera: Tettigonioidea). Entomol Gen 5:321–356.

Selverston AI, Kleindienst HU, Huber F (1985) Synaptic connectivity between cricket auditory interneurons as studied by selective photoinactivation. J Neurosci 5:1283–1292.

Sharov AG (1971) Phylogeny of the Orthopteroidae. Jerusalem: Israel Program for Scientific Translations.

Shaw SR (1994) Detection of airborne sound by a cockroach "vibration detector": a possible missing link in insect auditory evolution. J Exp Biol 193:13–47.

Sobel EC, Tank DW (1994) In vivo Ca^{2+} dynamics in a cricket auditory neuron: an example of chemical computation. Science 263:823–826.

Spangler HG (1988a) Hearing in tiger beetles (Cicindelidae). Physiol Entomol 13:447–452.

Spangler HG (1988b) Moth hearing, defense, and communication. Ann Rev Entomol 33:59–81.

Spooner JD (1968) Pair-forming acoustic systems of phaneropterine katydids (Orthoptera, Tettigoniidae). Anim Behav 16:197–212.

Stabel J, Wendler G, Scharstein H (1989) Cricket phonotaxis: localization depends on recognition of the calling song pattern. J Comp Physiol A 165:165–177.

Stephen RO, Bennet-Clark HC (1982) The anatomical and mechanical basis of stimulation and frequency analysis in the locust ear. J Exp Biol 99:279–314.

Stout JF, DeHaan CH, McGhee RW (1983) Attractiveness of the male *Acheta domesticus* calling song to females. I. Dependence on each of the calling song features. J Comp Physiol 108:1–9.

Stumpner A (1996) Tonotopic organization of the hearing organ in a bushcricket—physiological characterization and complete staining of auditory receptor cells. Naturwissenschaften 83:81–84.

Stumpner A (1997) An auditory interneurone tuned to the male song frequency in the duetting bushcricket *Ancistrura nigrovittata* (Orthoptera, Phaneropteridae). J Exp Biol 200:1089–1101.

Stumpner A, Ronacher B (1991a) Auditory interneurones in the metathoracic ganglion of the grasshopper *Chorthippus biguttulus*. I. Morphological and physiological characterization. J Exp Biol 158:391–410.

Stumpner A, Ronacher B (1991b) Auditory interneurones in the metathoracic ganglion of the grasshopper *Chorthippus biguttulus*. II. Processing of temporal patterns of the song of the male. J Exp Biol 158:411–430.

Suga N, Katsuki Y (1961) Central mechanism of hearing in insects. J Exp Biol 38:545–558.

Thiele D, Bailey WJ (1980) The function of sound in male spacing behaviour in bushcrickets (Tettigoniidae, Orthoptera). Aust J Ecol 5:275–286.

Thorson J, Weber T, Huber F (1982) Auditory behavior of the cricket. II. Simplicity of calling-song recognition in *Gryllus*, and anomalous phonotaxis at abnormal carrier frequencies. J Comp Physiol A 146:361–378.

Walker TJ (1957) Specificity in the response of female tree crickets (Orthoptera: Gryllidae: Oecanthinae) to calling songs of the males. Ann Entomol Soc Am 50:626–636.

Wohlers DW, Huber F (1978) Intracellular recording and staining of cricket auditory interneurons (*Gryllus campestris* L., *Gryllus bimaculatus* DeGeer). J Comp Physiol 127:11–28.

Wohlers DW, Huber F (1982) Processing of sound signals by six types of neurons in the prothoracic ganglion of the cricket, *Gryllus campestris* L. J Comp Physiol 146:161–173.

Wohlers DW, Huber F (1985) Topographical organization of the auditory pathway within the prothoracic ganglion of the cricket, *Gryllus campestris* L. Cell Tissue Res 239:555–565.

Wyttenbach RA, May ML, Hoy RR (1996) Categorical perception of sound frequency by crickets. Science 273:1542–1544.

Yager DD (1996) Serially homologous ears perform frequency range fractionation in the praying mantis, *Creobroter* (Mantodea, Hymenopodidae). J Comp Physiol A 178:463–475.

Yager DD, May ML, Fenton MB (1990) Ultrasound-triggered, flight-gated evasive maneuvers in the praying mantis, *Paresphendale agrionina* (Gerst.). I. Free-flight. J Exp Biol 152:17–39.

Young D (1980) The calling song of the bladder cicada, *Cystosoma saundersii:* a computer analysis. J Exp Biol 88:407–411.

Young D, Ball EE (1974) Structure and development of the auditory system in the prothoracic leg of the cricket *Teleogryllus commodus* (Walker). Z Zellforsch 147:293–312.

Zaretsky MD, Eible E (1978) Carrier frequency-sensitive primary neurons and their anatomical projection to the central nervous system. J Insect Physiol 24:87–95.

Zhantiev RD (1971) Frequency characteristics of tympanal organs in grasshoppers (Orthoptera, Tettigoniidae). Zool Zh 50:507–514.

6
The Evolutionary Innovation of Tympanal Hearing in Diptera

DANIEL ROBERT and RONALD R. HOY

1. Introduction

Earlier chapters in this volume emphasize the many variations in form and function that play out in the auditory system of insects. As pointed out in Chapter 1, the diverse developmental origins of insectan hearing organs lead to an assortment of tympanal ears. In comparison with the singular origin and location of the vertebrate ears, insectan hearing organs present a bewildering diversity of form and location. These differences stand in contrast to the similarity of the behavioral tasks that are subserved by the auditory system in both vertebrates and insects. It is fascinating to the comparative physiologist to observe the dynamic interplay between the convergence of behavioral function and the divergence of form in the auditory systems of vertebrates and insects. The morphological diversity of insectan auditory systems is very well suited to the comparative study of auditory function, through its biomechanics, transduction mechanisms, or neural processing capabilities. Our understanding of the evolutionary adaptive process certainly benefits from study of the variation and diversity of forms.

The biological problems investigated here are at the crossroads of physiology and behavior. They must also be considered in the larger context of evolutionary tempo, patterns, and processes. Remarkably, the sense of hearing in insects evolved repeatedly, and thus independently, in at least seven different orders. From these multiple origins results a diversity that provides an excellent opportunity to investigate an important evolutionary question: How (and possibly why) should a new sensory modality arise in a group of animals lacking that sense in its phylogenetic history? More specifically, how do new hearing organs, with their associated mechanical and neural systems, come into being? What is the nature of such evolutionary innovation? Is it possible to identify some of the constraints (e.g., biomechanical, physiological, ecological) that guide the selective process of evolution? Can we understand the adaptive function of the diverse and particular auditory structures observed in insects in order to relate them to particular evolutionary histories? The study of the structure and function of the sense of hearing in flies illustrates the value of the comparative approach, which, it is our hope, allows us to better understand the evolution and adaptive function of innovative sensory systems.

In this chapter the sense of hearing in flies by means of tympanal ears is presented as an adaptation to the sensory ecological context of prey detection and localization. So far, investigations have concentrated on flies of the Tachinidae family. As the most documented case, flies in the ormiine tribe are used as the basis for illustrating and understanding the emergence of tympanal hearing in a major order of insects — the Diptera. Tympanal hearing in Diptera is the result of a convergent evolutionary process that allowed the ormiine flies to parasitize singing Orthoptera. Endowed with the sense of hearing, the ormiine flies could take advantage, for their own reproductive purposes, of the channel of acoustic communication so important for the reproductive behavior of their host species, field crickets or bushcrickets (Hoy, Chapter 1). Acoustic parasitism, while being a rare function for hearing in insects, allowed the ormiine flies access to an ecological niche otherwise relatively free of competition.

Some of the biophysical constraints that presumably guided the evolutionary development of the sense of audition are presented and discussed. Particular attention is given to directional hearing in small animals with regard to the following constraints: physical acoustics, the biophysics of hearing, the behavioral biology of parasitoids, and the acoustic ecology of their hosts. Finally, a phylogenetic scenario of the evolutionary innovation of tympanal hearing in Diptera is proposed.

2. Acoustic Parasitism: An Unusual Behavioral Function of Hearing in Insects

Parasitism based on acoustic host detection was first demonstrated by Cade (1975). In these experiments, the parasitoid tachinid fly *Ormia ochracea* (previously named *Euphasiopteryx ochracea*) could be attracted to the calling song of its host, the field cricket, *Gryllus integer*. Sound alone was shown to be sufficient for successful detection and approach of a calling cricket. Experimentally, the flies were shown to approach a loudspeaker that was broadcasting the calling song of their hosts (Cade 1975; Walker 1986).

Cade's study established the presence of another, then rarely observed, function of hearing in insects: the detection of prey. Earlier studies based on taxonomic evidence had reported the parasitic association of tachinid fly species with orthopteran hosts, in particular field crickets (Sabrosky 1953; Léonide 1969). These studies established that the life cycle of ormiine flies is closely associated with night-active singing orthopterans, and Cade's behavioral experiments pointed to the importance of sound in host detection (Cade 1975). Subsequent studies considered ormiine flies, through their parasitic lifestyle, as potential control agents for some introduced orthopteran species such as mole crickets (Walker 1986; Fowler 1987). In particular, phonotactic trapping experiments conducted in the field showed the flies to be preferentially attracted to the song of their specific cricket host (Walker 1986, 1993). These and other studies amply illustrate the role of sound detection in the sensory ecology of ormiine flies (see also Mangold 1978; Fowler 1988).

The sensory basis for this particular mode of host detection and localization, a key element of any parasitoid's life, was given very little attention until recently (Lakes-Harlan and Heller 1992; Robert, Amoroso, and Hoy 1992). There has been a renewal of general entomological interest in the behavior, evolution, and sensory ecology of parasitoids (see Godfray 1994). In particular, acoustic parasitism in Diptera (ormiine flies) and its possible influence on the acoustic behavior and behavioral ecology of the host species has been the subject of several recent studies (Zuk et al. 1993; Adamo et al. 1995b; Allen 1995; Wagner 1996).

Although rare, acoustic parasitism is not restricted to flies of the family Tachinidae. An early study reported the attraction of a sarcophagid parasitoid fly to the mating song of its host, a cicada (Soper, Shewell, and Tyrrell 1976). A brief account of current knowledge on the sense of hearing in Sarcophagids (another very large family of parasitoids), and its remarkable convergence to the Tachinids, is presented in this chapter.

The reproductive behavior of the fly *Ormia ochracea* is typical of many insect parasitoids. The gravid female must find her host species (the field cricket *Gryllus rubens* in Florida or *Gryllus integer* in Texas) as a food source for her offspring. The female fly deposits a few first instar larvae on or around the host, leaving the small maggots to enter the cricket by themselves. After about 1 week of endoparasitic nutrition and larval development, the third instar larvae emerge from the still-living host and form pupana outside (Adamo et al. 1995a).

Although the phonotactic behavior of the fly has not been examined in detail yet, our preliminary observations strongly suggest that, either in flight or on the ground, these flies have an acute auditory sense of direction. Whether in flight or walking on leaf litter, gravid females briskly align their body toward the source of the sound and locomote directly toward it (Klein and Robert, Hage and Robert, unpublished observations). When a sound source simulating the host song is switched on at some azimuthal angle, the female fly orients her body in the desired direction almost instantanously and runs rapidly toward the sound source (Hsu and Hoy, unpublished observations; H. Spangler, personal communication). These laboratory tests, along with numerous field observations of female flies homing in and landing on acoustic traps (Walker 1993), suggest that ormiine flies are endowed with keen directional hearing. The directional sensitivity of the ormiine ears and the unique way by which it is achieved are presented here in some detail.

3. Comparative Auditory Anatomy of Diptera

Numbering perhaps more than several hundred thousands of species, Diptera is one of the most speciose and diverse groups of insects (Shewell 1987; Wood 1987). Higher flies have been the subject of historically significant studies of the sensory physiology of taste and olfaction (Dethier 1976) and of vision (Reichardt and Wenking 1969; Strausfeld 1989). By comparison, the auditory capacity of higher flies and its sensory basis have only occasionally been studied, leaving us with much uncharted territory. An example of such an elusive acoustic communication system is the short-range "love song" used for courtship in Drosophilidae (Bennet-Clark and Ewing 1967; Hoy, Hoikkala, and Kaneshiro 1988). Hearing, that is, the reception of airborne vibrations, takes two distinct forms in Diptera. The next sections briefly presents the fundamental differences in the mechanisms by which different Diptera, such as fruitflies and tympanate parasitoids, can detect airborne sounds.

3.1. Nontympanal Sense of Hearing

The "ears" of fruitflies (Drosophilidae) and mosquitoes, the other group of Diptera known to use sound for communication (Bennet-Clark 1984; Michelsen and Larsen 1985), are structurally and functionally different from the tympanal ears of ormiine tachinids. The sensory structures mediating near-field sound reception are sensory hairs borne on the insect's body or head (McIver 1985). The arista of the antennae of fruitflies, or Johnston's organs of mosquitoes, are used to detect the movement of air particles in the close vicinity of a sound source, and therefore are described as "near-field detectors" (Römer and Tautz 1992). These detectors are insensitive to the variations of air pressure propagated in the "farfield" of the sound field. In effect, the detection range of near-field detectors in fruitflies is limited to a few body lengths, whereas tympanal far-field detectors are sensitive over much greater distances of up to thousands of body lengths.

3.2. Comparative Auditory Anatomy of Tympanate Flies

The striking fact about ormiine flies is that, in contrast to many tens of thousands of species of other higher flies, they possess tympanal hearing organs. To better understand the function of such organs in host phonotaxis, it is important to become familiar with the unusual auditory anatomy of ormiine flies.

The ears of ormiine flies are positioned on the ventral prothorax, above and anterior to the first pair of legs and just behind the head (Fig. 6.1A) (Lakes-Harlan and Heller 1992; Robert, Amoroso, and Hoy 1992). The general design features of the ormiine hearing organs is consistent with the design of other insectan tympanal organs (Robert, Read, and Hoy 1994; Hoy and Robert 1996). A tympanal ear in insects essentially consists of three basic, major morphological components:

1. A localized thinning of the cuticle to provide a tympanal membrane, which is formed by the apposition of a modified, thin exocuticle and a tracheal air sac
2. An air chamber backing the tympanal membrane, which is formed by the associated tracheal air sac
3. A mechanoreceptive sensory organ of the scolopidial type, which is in direct contact with either the tympanal membrane or the modified tracheal system abutting it.

The hearing organs of *O. ochracea* are no exception to this general architecture. Like any other insect, however, *Ormia* presents morphological variations that reflect, in structure and function, its adaptation to a particular sensory ecology and species-specific anatomical constraints. The external and internal anatomy of the tympanal ears of *O. ochracea* have been described in greater detail elsewhere (Robert, Read, and Hoy 1994; Edgecomb et al. 1995; Robert et al. 1996a), but it will be presented here in comparison with the prosternal anatomy of an atympanate species, the closely related parasitoid tachinid fly *Myiopharus doryphorae* (Fig. 6.2A,B). It is from such comparisons that the key anatomical innovations leading to the presence of the sense of hearing in Ormiine Tachinid flies can best be visualized (Edgecomb et al. 1995; Robert et al. 1996a).

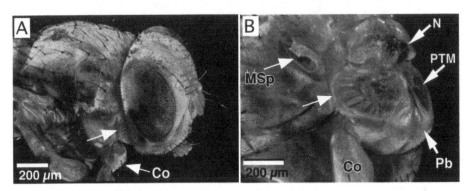

FIGURE 6.1. External auditory anatomy of *Ormia ochracea*. In ormiine flies, the tympanal ears are located between the first pair of legs and the base of the neck. The light scanning micrographs show a semilateral view of an intact fly (A), and (B) of a fly with the head removed to see the prostenal hearing organs. The black arrows indicate the lateralmost border of the fly's right tympanal membrane. Co, prothoracic coxa; N, neck; PTM, prosternal tympanal membranes; Pb, probasisternum; MSp, mesothoracic spiracle.

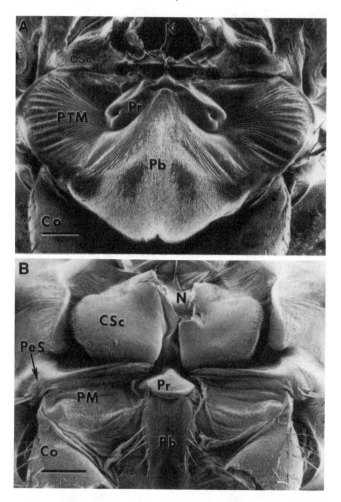

FIGURE 6.2. Scanning electron micrographs of two closely related species of tachinid flies illustrating, in frontal view, the anatomical differences of the prosternal region. A: The tympanate fly *Ormia ochracea*. B. The atympanate fly *Myiopharus doryphorae*. PeS, proepisternal setae; PTM, prosternal tympanal membrane; PM, prosternal membrane; Pb, probasisternum; Pr, presternum; Co, prothoracic coxa; N, neck; CSc, cervical sclerite. Notably, in the tympanate species (A) the Pb and the Pr are conspicuously larger, and the PTMs present a larger surface area and are thrown with radial corrugations. The small cuticular depressions at the distal ends of the Pr are the points of insertion of the sensory organs with the tympanal system. Scale 200 μm.

A series of eight morphological specializations that are particular to ormiines have been identified:

1. Enlargement of the prosternal membranes (PM), giving rise to thin prosternal tympanal membranes (PTM; see Fig. 6.1A), which present a relatively large surface area (compare Figs. 6.2A and 6.2B).
2. Inflation of the ventral probasisternum (Pb) to provide structural support for the tympanal membranes.
3. Bilateral extensions of an unpaired sclerite associated with the tympanal membranes, the presternum (Pr), to which attach the sensory organs at the tympanal pits (TP). The biomechanical function of the presternum is a key to the process of directional hearing and is presented in Section 7.
4. Enlargement of the prosternal air sac forming the acoustic chamber backing the tympanal membranes.
5. Location of the two scolopidial sensory organs in the unpartitioned prosternal air sac. A scolopidium is the elementary multicellular arrangement, including a mechanoreceptive neuron that is commonly used in insects for vibration reception.
6. Cuticular apodemes, establishing a stiff mechanical link between these sensory organs and the presternum.
7. Reduction in size of the prosternal cervical sclerites (CSc).
8. Structural reorganization of the internal endoskeleton of the prosternal region.

Two striking features of the ormiine auditory anatomy are its small size and the immediate adjacency of its two tympanal ears (Fig. 6.2A). The two tympanal membranes span only 1.68 mm (SD 0.19 mm, $n = 16$) across the prosternal region. Also, the tympanal pits (the points of insertion of the auditory sensory organs to the PTMs) are separated by only 520 μm (SD 20 μm, $n = 16$). Although very small in absolute numbers (the total volume taken by both hearing organs is on the order of 1 mm^3), these adjacent tympanal ears still occupy about 80% of the breadth of the prosternal anatomy. When compared with the size of the ormiine fly, these hearing organs appear to be the dominant feature in the prosternal region. This is in contrast to the prosternal anatomy of the atympanate fly, in which the probasisternum, the presternum, and the presternal membranes are relatively small and remain recessed between the prothoracic coxae (Co; Fig. 6.2B). It seems then that, given their anatomical position, the ormiine tympanal organs could hardly grow any larger.

A comparison of the internal prosternal anatomies of tympanate and atympanate tachinids is presented in Figure 6.3. In *Ormia*, the sensory organs (bulba acoustica) are very close to each other and, remarkably, are both located in the unpaired prosternal air sac. As in other tympanate insects, the sensory organs of *Ormia* are scolopophorous, consisting of a bundle, or an array, of mechanoreceptive scolopales (Robert, Read, and Hoy 1994). In the atympanate tachinid *M. doryphorae*, the air sac behind

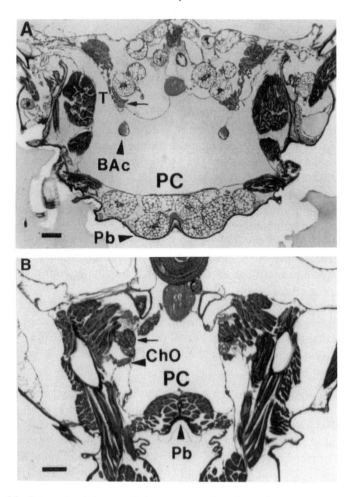

FIGURE 6.3. Internal anatomy of the prosternal region of a tympanate and an atympanate tachinid. The histological cross sections (up is dorsal) are stained with toluidine blue. A: The tympanate fly *Ormia ochracea*. B: The atympanate fly *Myiopharus doryphorae*. BAc, bulba acustica; ChO, chordotonal organ; PC, prosternal chamber; Pb, probasisternum; T, trachea. In the tympanate fly, the PC is larger, is undivided, and contains both auditory sensory organs, the bulba acustica. The Pb of the tympanate fly occupies a much great portion of the ventral prothorax. Scale 100 μm.

the prosternal membranes is reduced in volume, leaving the chordotonal sensory organs (ChO) located laterally from the air sac, embedded in muscle mass, fat bodies, and hemolymph (Fig. 6.3B). The reduced size and concave shape of the probasisternum in the atympanate tachinid can also be seen from Figure 6.3B. The response properties, and therefore the exact function, of these chordotonal organs in *M. doryphorae* remain unknown.

Although proprioception would seem a reasonable putative original function for the prosternal chordotonal organ, electrophysiological and biomechanical experiments are still needed to test the organ's response to sound, vibration, respiration, or head, leg, or wing movements.

3.3. Tympanal Hearing in Sarcophagid Flies

Earlier behavioral evidence reported the acoustic parasitic lifestyle of some species of another big family of parasitoid flies, the Sarcophagidae (Soper, Shewell, and Tyrrell 1976). More recently, playback experiments in the field, similar to those previously performed by Walker (1986), but broadcasting cicada calling songs, successfully attracted sarcophagid female flies (Farris and Hoy, unpublished observations). Our observations of several specimens of the (taxonomically still uncertain) genus *Emblemasoma* established the presence a well-developed prosternal inflation, with thin membranes and a related chordotonal organ reminiscent of the auditory arrangement of ormiines (Robert and Hoy, unpublished observations). The external prosternal anatomy of the Sarcophagid parasitoid *Emblemasoma* sp. is shown in Figure 6.4. Its anatomical similarity to that of *Ormia* is striking.

In *Emblemasoma*, the probasisternum (Pb) is also inflated, protruding anteriorily between the prothoracic coxae (Co). A thin membrane spans

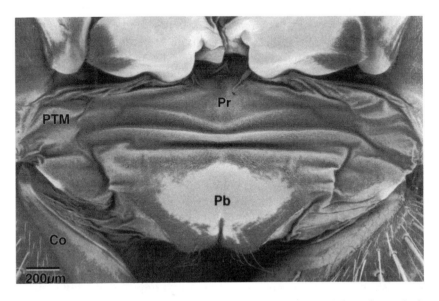

FIGURE 6.4. Scanning electron micrograph of the prosternal region of the sarcophagid fly *Emblemasoma* sp. PTM, prosternal tympanal membrane; Pb, probasisternum; Pr, presternum; Co, prothoracic coxa.

the whole width of the prosternal area (PTM). Unlike *Ormia*, these tympanal membranes do not present fine radial corrugations converging to the lateral process of the presternum (as seen in Fig. 6.2A) but form a few deep folds that run horizontally from one side of the prosternum to the other (see Fig. 6.4). Another noticeable difference is the less clearly differentiated medial unpaired sclerite, the presternum (Pr). The gross internal auditory anatomy of *Emblemasoma* is similar to that of *Ormia*; the air sac is unpaired and contains two scolopidial sensory organs that directly connect to the median corrugation of the PTM through a cuticular apodeme. On the basis of such striking similarity, along with preliminary biomechanical evidence, it seems quite reasonable to conclude that the observed prosternal structures are functional tympanal hearing organs.

4. The Endoparasitic Condition and Its Morphometric Consequences

As a general rule, parasitoids are limited in their size because, by definition, their larval instars rely on another insect as a food source. Because they are obligatory endoparasites during their larval stages, ormiine flies can only grow to a fraction of the size of their host. It has been shown experimentally that the size of the emerging flies decreases as the number of endoparasitic larvae increases (Adamo et al. 1995b). It was also shown in this study that the number of endoparasitic larvae per host is not under the control of the larviparous flies. This is unlike the capacity of some hymenopteran parasitoids to monitor the number of offspring invested in a single host and thus to assess the presence of parasitic competition for that host (Godfray 1994). The apparent inability of *O. ochracea* to assess parasitized hosts (by host probing or marking) thus implies the risk of multiparasitism. By extension, given the finite resource represented by the cricket host and the uncertainty of parasitic load, ormiine flies are restricted to relatively small body size, smaller than a single fly larva in a single host could theoretically attain (Adamo et al. 1995b). Consequently, this size constraint acts as a limitation on the size of the hearing organs and, as seen in the next section, has important implications for the biophysics of directional hearing.

5. The Problem of Small Size for Directional Hearing

Animals use two basic acoustic cues, interaural time and amplitude differences, for the directional binaural detection of a sound source. From the physical laws of sound propagation in air and the diffraction of sound around solid bodies, it is apparent that very small animals such as insects face serious limitations in their ability to extract detectable interaural time and amplitude acoustic cues from the incident sound field.

It is important to mention, in passing, that time and amplitude acoustic cues are not necessarily required to localize an incident sound. Monaural sound localization by means of spectral cues is possible in humans under certain circumstances (Middlebrooks and Green 1991; but see also Wightman and Kistler 1997). The processing of spectral information for directional hearing in insects has not been given much attention and can therefore not be excluded from consideration.

As the speed of sound is 344 m/s in air, an interaural distance of 1 cm will only generate an interaural time difference of the incident sound wave (ITD) of about 30 μs. Such, and even shorter, time delays admittedly pose a severe challenge to the nervous system in its typical temporal range of operation on the order of a millisecond. The other basic cue, the interaural intensity difference, has operational size limits that are related to the wavelength of the incident sound and the size, shape, and density of the body or head carrying the ears. It is generally accepted from acoustical diffraction theory that a ratio of 1:10 between size and wavelength does not give rise to measurable diffractive effects (Morse and Ingard 1968). For example, a spherical body of 7 mm in diameter does not significantly diffract sounds below a frequency of 5 kHz (with a wavelength of 68 mm), and therefore very little or even no difference in interaural sound pressure occurs. We use 5 kHz as an example because it is the dominant frequency of field crickets' calling song and therefore is salient to its acoustic parasitoid as well (Robert, Amoroso, and Hoy 1992).

In theory, given the small body size of the ormiine flies (a few millimeters in breadth) and the very short interaural distance (1.68 mm total width for both tympanal membranes taken together), the generation of reliable (and physiologically relevant) interaural acoustic cues seems highly problematic. Practically, how does the fly acoustically localize her cricket host singing at 5 kHz? In the fly's case, the body size to wavelength ratio is about 1:130, thus precluding the effects of diffraction as the source of interaural intensity difference. An angle of incidence of 90° is the best case for sound localization, and sound will travel from one side to the other of the most lateral margins of the tympanal membranes (1.68 mm apart) in about 4.9 μs. In addition, the calculated interaural time difference at the location of the sensory organs is about 1.5 μs, given their separation of 520 μm.

Actual measurements made with custom-made probe microphones confirm these theoretical arguments. At 5 kHz, the interaural intensity difference is too small to be measured, (≤ 1 dB) either across the fly's body or at the tympanal membranes. The maximal interaural time delay measured by two phase-calibrated probe microphones positioned directly in front of the tympanal membranes is 1.45 μs (SD 0.49, $n = 10$). Given such tight biophysical constraints, how is sound localization achieved by O. ochracea?

6. Biomechanics of Directional Hearing in *O. ochracea*

6.1. Process of Hearing

For any animal, the process of hearing involves a three-stage conversion of sound energy into neural energy. The obligatory first step in this chain of events is the conversion of sound energy into mechanical energy. At this stage, the tympanal membranes are set into vibration, with some degree of efficiency, by the forces resulting from the variations in the incident acoustic pressure. These vibrations are then transmitted to the mechanoelectrical transducers of the auditory sensory organs. The description of the mechanical response of the tympanal membranes in a sound field is therefore a first, essential step toward understanding the mechanical response of the auditory system as a whole.

The mechanical response of the tympanal ears of the fly *O. ochracea* to an incident sound field was measured by laser Doppler vibrometry. This noninvasive and sensitive remote-sensing technique permits the detection of vibration velocities as low as 0.5 µm/s over a frequency range of 0.1 Hz to 500 kHz (Miles, Robert, and Hoy 1995; Robert, Miles, and Hoy 1996b).

6.2. Mechanical Interaural Time and Amplitude Differences (mITD and mIAD)

The mechanical response of the tympanal pits (TP in Fig. 6.6A) to a stimulus simulating the cricket calling song is shown in Figure 6.5. Anatomically, the tympanal pits (TP) are the small depressions located laterally on the presternum (see Fig. 6.6A). They also mark the site of mechanical attachment of the sensory organs to the tympanal membranes. In this experiment the sound stimulus has a carrier frequency of 5 kHz (10-ms tone duration, 1-ms rise/fall time) and is delivered at an angle of incidence of 45° off the longitudinal axis of the fly (zero degree elevation). Given the short intertympanal distances involved, the interaural time delay of the incident acoustic pressures is measured to be less than 2 µs (Fig. 6.5A). As expected, the two acoustic pressures incident at each ear vary sinusoidally with equal amplitudes and, seen at that time scale, almost in synchrony (Fig. 6.5A).

The vibratory, mechanical response of the tympanal pits is, however, quite different. The ipsilateral TP (the one near the sound source) oscillates in the sound field and leads in phase the oscillations of the contralateral TP. In this example, the mITD between the ipsilateral and contralateral tympanal pits is 58 µs. At the carrier frequency of the stimulus, this delay corresponds to about a 100.8° phase shift, or over a quarter of the stimulus period. For eight animals, the mean mechanical intertympanal time delay is 48.3 µs (SD 11.2 µs; range 35.3 to 68.0 µs). It is also remarkable that the amplitude of oscillation of the contralateral TP is reduced compared with the ipsilateral TP.

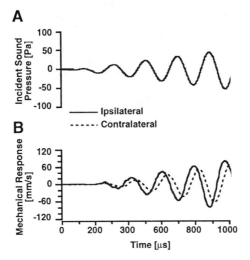

FIGURE 6.5. Mechanical response of the tympanal pits measured by laser vibrometry in response to an incident tone pulse (5 kHz; 10 ms length; 1 ms rise/fall; 45° incidence). A: Incident sound pressure at the tympanal pits. The measured interaural time delay is 1.7 μs and is not resolvable at this time scale. B: Mechanical response of the tympanal pits. The oscillations of the ipsilateral tympanal pit (solid line) lead those of the contralateral tympanal pit (dotted line). In this example the mechanical interaural time delay is 57 μs and is much larger than the 1.7-μs delay in the interaural incident sound pressure.

The tympanal mechanical response was also investigated by random noise analysis. In this analysis, the stimulus was a band-limited random noise burst (1 to 25 kHz bandwidth; 10-ms duration; 104 dB re 20 μPa) delivered at different angles of incidence. The tympanal vibrations elicited by the broadband stimulus are shown in Figure 6.6. The mechanical response of the ipsilateral and contralateral tympanal membranes (as measured by positioning the beam of the laser vibrometer on the locations indicated in Fig. 6.6A) shows a pronounced asymmetry above 4.5 kHz (Fig. 6.6B). The ipsilateral side vibrates in the sound field with significantly higher amplitude than the contralateral side. The average mIAD between 4.5 and 25 kHz is 13.6 dB (SD 4.0 dB). Control measurements were also taken at the midline of the probasisternum, a sclerite made of thicker cuticle that supports the tympanal membranes (central dot). The relatively low level of vibration of the probasisternum (−15 to −30 dB relative to the PTMs; Fig. 6.6B, center) shows that the vibration amplitudes measured for the tympana are not caused by the overall vibrations of the preparation and thus are not stimulation artifacts. Hence, the acoustic stimulus applied here specifically generates interaural differences in tympanal vibrations that are much larger than the interaural intensity difference in the sound pressure

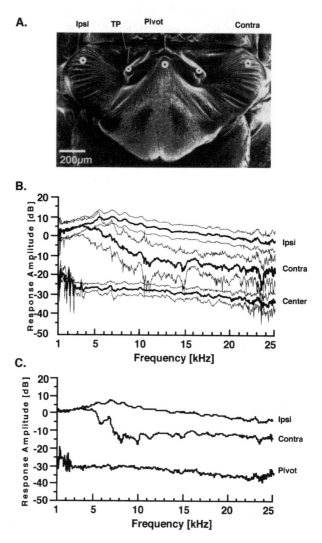

FIGURE 6.6. Mechanical response of the tympanal membranes to a band-limited noise stimulus (1 to 25 kHz, 45° incidence). A: Scanning electron micrograph (SEM; as in Fig. 6.2A) of the prosternal hearing organs of *O. ochracea*. The five dots placed on the SEM indicate the position of the laser beam during the vibrometric measurements. The two lateralmost dots indicate the ipsilateral and contralateral measurements made on the PTMs. Two more measurements were made on the tympanal pits (TP), and a control measurement was performed on a thick process of the probasisternum (central dot). B: Displacement amplitude spectra of the tympanal membranes at the locations shown in A. Amplitude transfer functions are given in dB re 31.6 nm/Pa, with their associated standard deviation (±1 SD, thinner lines). The ipsilateral PTM (ipsi, average of $n = 8$ animals) responds with greater amplitude than the contralateral PTM ($n = 8$ animals), while the probasisternum remains relatively immobile. C: Displacement amplitude spectra of the tympanal pits (ipsi, contra: $n = 5$ animals) and of the pivot of the intertympanal bridge ($n = 3$ animals). Above 4.5 kHz, the two hearing organs do not move with the same amplitude in response to an incident sound noise burst. The mean difference for all frequencies above 4.5 kHz is 12.4 dB (SD 3.4 dB).

(<1 dB). These observations lead to the conclusion that the observed mean mIAD (13.6 dB) is not due to the difference in the acoustic pressures acting on the tympanal membranes but must be the result of some acoustico-mechanical process at the tympanal level.

Other measurements made under the same experimental conditions indicate that the mechanical response of the ipsilateral and contralateral pits is very similar to that of the tympanal membranes. The pit ipsilateral to the sound source moves significantly more than the contralateral pit (Fig. 6.6C). Furthermore, the level of vibration of the central point of the presternum (the pivot, Fig. 6.6A) is very low compared with both tympanal pits.

6.3. The Ears' Sensitivity to the Direction of Sound

In order to test the directional abilities of this tympanal system, mITDs and mIADs were measured for different angles of incidence of the sound stimulus. At 5 kHz, the frequency of the calling song of the cricket host, a critical frequency for the directional detection of sound, the mechanical response of the tympanal system is a function of sound azimuthal angle (Fig. 6.7A,B). Significant mITDs and mIADs are observed for angles of incidence 45° or more off the longitudinal axis of the animal. Both time and amplitude differences decrease with smaller angles of incidence to about zero at zero degree incidence (Fig. 6.7A,B). From these measurements it can be concluded that the mechanical response reflects changes in sound direction and that this tympanal system seems, in principle, well suited for the coding of sound direction, at least at the mechanical level.

6.4. Tympanal Dynamics

The data presented earlier show that, despite their extreme proximity, the tympanal membranes (and the tympanal pits of the intertympanal bridge) move asynchronously and with different amplitudes in an incident sound field. Describing the dynamics of tympanal vibrations is necessary to better understand the mechanism by which such interaural differences are produced. For this purpose, laser measurements were taken across the breadth of the tympanal membranes and the intertympanal bridge. The mechanical displacements of 15 locations across the ears were monitored with respect to their amplitude and phase (Fig. 6.8A). Real and complex transfer functions were computed for each point with reference to the acoustic pressure at the ears. From the amplitude and phase data of each measurement point, it is then possible to know the instantaneous amount of deflection of each of the 15 points measured across the ears at any chosen time.

A "freeze frame" representation of the tympanal deflections is shown in Figure 6.8B. This deflection shape is in response to a sound stimulus of 5 kHz incident at 45°. In this example, time T_0 has been chosen with reference to the time of minimal sound pressure at the fly's right ipsilateral

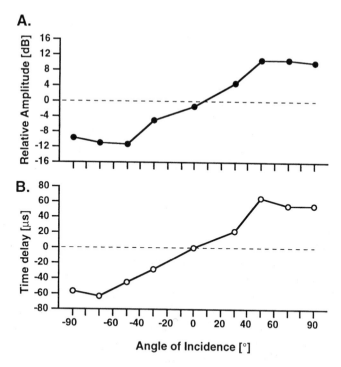

FIGURE 6.7. Directional sensitivity of the mechanical response to a 5-kHz incident tone stimulus. A: Difference in the amplitude of the mechanical response (mIAD) between the tympanal pits for different angles of sound incidence. B: Mechanical interaural time delay (mITD) for the tympanal pits. Positive values indicate greater amplitudes (in A) or a lead (in B) for the animal's right hearing organ. Averages from five animals.

tympanal membrane (left side of the diagram). At this time, the right tympanal membrane is deflected outwards, while the left PTM experiences an inward deflection. The points circled 1 and 2 indicate the positions of the ipsilateral and contralateral tympanal pits, respectively, while point 3 shows the pivot of the intertympanal bridge. Thus, in response to a 5-kHz incident sound pressure, an outward deflection of one tympanal pit is accompanied by an inward deflection of the other tympanal pit (Fig. 6.7B), while the central pivot point of the intertympanal pit remains immobile. The relative immobility of the pivot point confirms the observation of Figure 6.6C. This effect can also be observed by manually gently deflecting one tympanal pit with a shaft of human hair. As one tympanal pit is pushed inward, the other tympanal pit is deflected outwards, somewhat like a seesaw.

The dynamic response of the tympanal system can be visualized graphically by stacking a series of time intervals, T_0 to T_{40}, equally spaced every 5 μs along the 200-μs period of the 5-kHz stimulus (Fig. 6.9). These deflection shapes show that outward displacements of one side (light-gray peak)

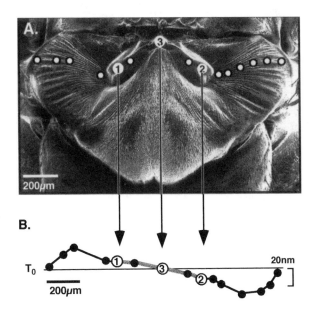

FIGURE 6.8. Mechanical displacement of the tympanal ears to a 5-kHz incident sound. A: SEM illustrating the 15 points of laser measurement taken across the ears. Dots 1, 2, and 3 indicate the ipsilateral pit, the contralateral pits, and the pivot of the intertympanal bridge, respectively. B: Deflection shape at time T_0, when sound pressure is minimum at the ipsilateral side. In this "frozen time frame," the outward displacement (above baseline) of the ipsilateral prosternal tympanal membrane (PTM) and tympanal pit (TP) is accompanied by an inward displacement (below baseline) of the contraleteral PTM and TP. The segments joining dots 1 and 2 represent the intertympanal bridge. The vertical scale indicates linear mechanical displacement (20 nm).

correspond to inward displacements of the other side (dark-gray valley). The tympana and the intertympanal bridge rock back and forth about the almost immobile pivot point. The timing of these oscillations can be better seen on the contour plot of the graphic's floor. Maximal outward deflections (delay between light-gray peaks marked *-*) occur about 60 μs apart, corroborating the delay observed with the 5-kHz pure tone stimulus (Fig. 6.5B). The phase delay between the maximal deflection of the two tympanal pits is thus ca. 100° at 5 kHz (keeping in mind that the phase delay of incident sound is about 3°).

6.5. Neurophysiological Relevence of the Mechanical Interaural Differences

An interaural time delay on the order of 60 μs does not seem to be a sufficient localization cue when it comes to the presumed time coding

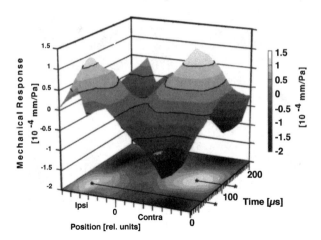

FIGURE 6.9. Deflection shapes of the tympanal organs. The position axis indicates the 15 ipsilateral and contralateral measurement locations across the ears (as shown in Fig. 6.8). The time axis spans one period of the 5-kHz incident stimulus (200 μs). The vertical axis gives the mechanical displacement (re. 10^{-4} mm/Pa). The contour plot of the graphic's floor points to the alternance of displacement maxima (light gray) and minima (dark gray). The time delay between the deflections of maximal amplitudes of the ipsilateral and the contralateral ears is highlighted by the oblique black lines (----) on the graphic's floor.

capabilities of the insectan nervous system (Mörchen, Rheinlaender, and Schwartzkopff 1978). The activity of the auditory primary afferents was monitored by extracellular recordings made from the auditory nerve (the frontal nerve of the prothoracic ganglion; Robert, Read, and Hoy 1994). First, probe microphones were used to monitor the acoustic interaural time delays under the experimental conditions of the electrophysiological setup. As expected from previous measurements, the acoustical ITD measured at the tympana were about 2 μs (data not shown). However, when the sound source was switched from 90° to the left to 90° to the right of the fly, the same auditory afferent fired with an absolute time difference of 320 μs (Fig. 6.10). When measured for six different animals, this neural ITD averaged 313 μs (SD 137 μs). This neural ITD, prior to any central neural processing, is considered sufficient to provide the first interneuronal processing station in the prothoracic ganglion with directional information (Mörchen, Rheinlaender, and Schwartzkopff 1978; Oshinski and Hoy 1995). These data show that the mITDs and mIADs reported earlier are reflected and amplified in the temporal activity of the primary afferents. This also shows that the peripheral mechanical processing reported here is not simply a bizarre biomechanical event unrelated to physiological reality. This tympanal mechanical response has functional significance for directional hearing.

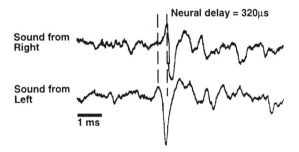

Figure 6.10. Directional sensitivity of the primary sensory afferents. The activity of the fly's left auditory nerve was monitored by extracellular neurophysiological recording. Because sound is delivered from the left of the fly, the left afferents lower trace) are active earlier than when sound originates from the fly's right side (upper trace). The neural time delay is 320 μs. In these conditions, the interaural time delay in the sound pressure measured by the probe microphones was 1.5 μs.

These anatomical, biomechanical, and electrophysiological results phenomenologically constitute the substrate for directional hearing in the presence of minimal acoustic cues. The mechanical process of intertympanal coupling by which such interaural differences are generated has been described and modeled elsewhere (Miles, Robert, and Hoy 1995; Robert, Miles, and Hoy 1996b). Because intertympanal coupling is only briefly presented in the next section, we encourage the reader to refer to the original publications for more complete explanations.

7. The Process of Intertympanal Coupling

The analysis of the deflection shapes of the tympanal membranes, and more particularly of the intertympanal bridge, indicates that the two ears, joining at the midline of the animal, do not move in an independent way. Given the measured interaural acoustic cues, two independent tympana would move almost in phase (within about 3°, e.g., 2 μs at 5 kHz) and with the same amplitude. However, as shown in Figures 6.8 and 6.9, the intertympanal bridge undergoes an asymmetrical displacement about its center, much like a flexible seesaw rocking back and forth about its pivot point.

The intertympanal bridge that links the tympanal membranes is a key feature of the auditory mechanics of the ormiine ears. The ability of the intertympanal bridge to rock back and forth in a flexible manner is what can drive both membranes with such phase and amplitude differences. It is true that a nonflexible intertympanal bridge (a rigid lever) oscillating at 5 kHz (200-μs period) would produce a phase delay between its extremities (the tympanal pits) of half a period (180° phase shift, i.e., 100 μs). Also, the extremities of a such a hypothetical rigid intertympanal bridge would oscillate about the pivot point with equal amplitudes. In contrast, data taken at 5 kHz show (a) significant amplitude differences and (b) a time delay between the tympanal pits of 50 to 60 µs rather than 100 µs. The relative "floppiness" of the intertympanal bridge explains why the phase delay between the tympanal pits is 90° to 100° rather than 180° (as would be the case for a rigid lever).

The ability of the incident sound pressures to drive this tympanal system depends on the relative phase of the pressures acting on the tympana (Miles, Robert, and Hoy 1995). In the same way, a seesaw is put out of static equilibrium by two weights of equal mass applied at different times on each of its ends; these forces (pressures) result in the rocking motion. As one arm of the bridge is deflected downward, the other arm will move upward due to the stiffness of the bridge (see Figs. 6.8 and 6.9).

A more complete analysis of this system shows that the behavior of this mechanical system with two degrees of freedom can be explained by the interaction of two basic modes of vibration (one rotational, one translational). The relative contributions of these modes at different driving frequencies depend on the difference or sum of the forces acting on the system and on its resonant properties (see Miles, Robert, and Hoy 1995; Robert, Miles, and Hoy 1996b).

This morphological and biomechanical evidence led to the development of a simple mechanical analogue (Fig. 6.11A), the mathematical formulation of which is presented in Miles, Robert, and Hoy (1995). In this model, the flexible intertympanal bridge is represented by two rigid bars connected medially by a torsional spring and a dash pot (Fig. 6.11A). Both extremities of the bridge are connected to a spring and a dash pot that represent the stiffness and damping characteristics of the auditory sensory organs attached to the tympanal pits.

The deflection shapes computed from laser vibrometry measurements illustrate the unusual rocking and translational motions of the intertympanal bridge. The resulting ipsilateral and contralateral motion amplitudes can be visualized in the sketch of Figure 6.11B. This represents the idealized effect of the flexibility of the intertympanal bridge during the deflection time sequence from 1 to 5 (Fig 6.11B). Unfortunately, this representation does not illustrate the phase delay introduced by this floppy connection. To visualize this delay, one could think of the respective points in this time sequence (1 on the left is associated with 1 on the right) as being delayed by about 50 µs ($1_{\text{left}} = 1_{\text{right}} + 50\,\mu\text{s}$).

However, the present evidence does not formally exclude that an alternative mechanism could account for, or contribute to, the observed tympanal dynamics. Indeed, it is still possible that the air space behind both tympanal membranes could have a sufficient stiffness to act as a resonator and thus provide the basis for a directional response, as is the case for the acoustically coupled ears of pressure-difference receivers (Michelsen, Chapter 2). In a pressure-difference system, the volume of the air chamber backing to

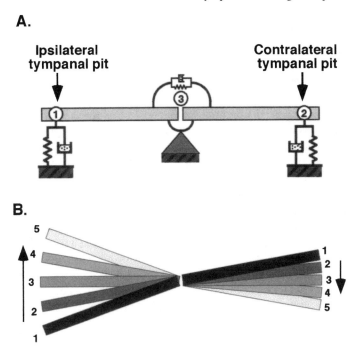

FIGURE 6.11. Mechanical analogue of the tympanal system. A: Abstractly, the intertympanal bridge is drawn as two beams connected by a torsional spring, accounting for the finite flexibility observed in the mechanical response. Points 1, 2, and 3 as in Figure 6.8. B: Schematized response of the intertympanal bridge. Because the ipsilateral side moves up from position 1 to 5, the contralateral side, due to the flexibility of the intertympanal bridge, deflects in the opposite direction but with lesser amplitude. Because this representation does not take into account the time delay introduced by the flexible pivot, a more accurate image may be reached by mentally delaying the contralateral downwards 1 to 5 sequences by about 50 μs.

tympana is crucial for the directionality of the mechanical response of the tympana. Recent experiments of direct mechanical stimulation were performed to test whether intertympanal coupling (as described earlier) was influenced by the air volume. Results clearly show that intertympanal coupling is not influenced by the volume of the air space backing the tympana (Robert et al., unpublished observations). These experiments establish that a pressure-difference receiver system, as proposed for cicadas (Fletcher 1992), is not at work in *O. ochracea*.

In conclusion, the asymmetrical mechanical response of the peripheral auditory apparatus of the fly *O. ochracea* results from the coupling of the two tympanal membranes by the intertympanal bridge. As an evolutionary innovation, this mechanism of peripheral processing accounts for the fly's directional sense of hearing.

8. The Evolutionary Origin of Hearing in Ormiine Flies

8.1. Phylogenetic Origins

The phylogenetic comparative approach can help clarify the origin of some specialized characters such as the hearing organs described earlier. Our comparisons are based on tympanate and atympanate flies, which were deliberately selected for their decreasing taxonomic proximity to ormiine tachinids. This analysis remains somewhat incomplete because it ignores dozens of fly families, each containing several thousands (or tens of thousands) of species. Such shortcomings do not, however, jeopardize the validity of this analysis if one assumes that fly species of nonparasitic families are quite unlikely, considering their life history and sensory ecology, to be under the evolutionary pressure of detecting far-field sounds. Hence, a more focused and complete survey of the presence of tympanal hearing in Diptera would necessitate the examination of numerous families of parasitoid Diptera and their relation to singing insects, especially those for whom long-distance acoustic communication is important. The next sections present some facts and ideas on the possible origin of tympanal hearing in ormiine flies and its taxonomic position within higher Diptera.

As shown earlier, the closely related non-ormiine tachinid fly *M. Doryphorae* does not possess the prosternal anatomical modifications that are typical of tympanal ears. Is it also the case for other higher Diptera unrelated to tachinids? The detailed anatomical examination of different families of higher Diptera (Fig. 6.12) revealed a very conservative prosternal anatomy among atympanate flies (Edgecomb et al. 1995). Atympanate specimens issued from different dipteran families

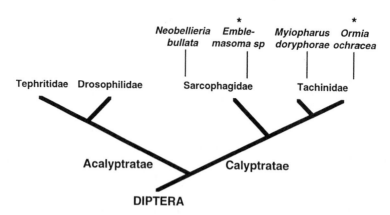

Figure 6.12. Cladogram of the Schizophora, showing the phyletic relationships (from McAlpine 1989) of a number of species selected for the morphological analysis of prosternal structures (Adapted from Edgecomb et al. 1995). The asterisks indicate the presence of tympanal hearing.

(Tephritidae, Drosophilidae, Sarcophagidae, and Tachinidae) all show strikingly similar prosternal anatomies that are very likely to be homologous. It is tempting to surmise that it is from this ancestral condition (pleisiomorphism) that tympanal hearing organs (the derived character) evolved in ormiine tachinids (Edgecomb et al. 1995).

Other comparative anatomical work has shown the similarity between prosternal morphologies among closely related ormiines, as well as how such morphologies differ from an atympanate tachinid sister species (Robert et al. 1996). A broader survey was thus necessary to estimate how widespread hearing organs were within the family Tachinidae. Because it is possible to establish the presence of prosternal tympanal ears on museum specimens, a survey was undertaken at the Natural History Museum in London. This work established that all flies within the tribe Ormiini examined (34 species from 6 different genera of Ormiini) present the prosternal anatomical modifications related to hearing organs (Robert, unpublished observations).

Remarkably, no specimens issued from seven tribes closely related to the Ormiini show prosternal specializations. More precisely, no tympanal hearing organs were observed on 120 specimens representing 15 genera within these 7 tribes. This is all the more remarkable because some of these species are reported to parasitize orthopteran species. Thus, a parasitic lifestyle, even related to an orthopteran singing host, does not necessarily imply the presence of acoustic parasitism. However, the examination of 140 specimens belonging to 6 different genera established the unequivocal and ubiquitous presence of prosternal tympanal ears in the tribe Ormiini. These observations imply that among tachinids, only members of the Ormiini are endowed with tympanal hearing organs; the sister tribes are clearly lacking this anatomical specialization. At this point, however, it would be unreasonable to formally exclude the presence of hearing organs in any other tachinid; in effect, the taxonomic ground covered by this first survey is still quite small compared with the extent of the family Tachinidae.

Tentatively, it seems that acoustic parasitism in tachinids (and hence the sense of hearing) is confined to a single tribe, suggesting a monophyletic origin for this character. Tympanal hearing in ormiine tachinids thus is a very local phenomenon phylogenetically. It can therefore be regarded as a key evolutionary innovation that opened up an adaptive radiation for some tachinids, namely, the acoustic parasitism of numerous species of field crickets, mole crickets, and katydids (Léonide 1969; Cade 1975; Walker 1986).

8.2. Precursors

The phylogenetic comparative approach says little about the process by which hearing evolved in Diptera. The inspection of the tachinid sister species *M. doryphorae* provided some insights on the morphological precursors of the ormiine tympanal hearing organs (Fig. 6.2A,B) (Edgecomb et al. 1995; Robert et al. 1996). The direct comparison between *O. ochracea* and *M. doryphorae* strongly suggests that hearing organs could have evolved through the concerted modification of several anatomical components. The hypothetical scheme of the evolution of hearing in ormiines would then encompass the lateral extension of the presternum into an intertympanal bridge (compare Figs. 6.2A and 6.2B). This component, as shown earlier, is functionally crucial for the implementation of directional hearing. Another element is the inflation of the probasisternum, sending lateral branches to isolate the tympanal membranes from the adjacent coxal membranes. To convert acoustic energy into mechanical energy (the function of tympanal membranes), the prosternal membranes become larger, much thinner ($\leqslant 1\,\mu m$), and radically corrugated. Also, in the atympanate flies the mechanoreceptive (chordotonal) organs are actually attached to the presternum in a way that is strongly reminiscent of the tympanate condition.

It then appears that several essential building bricks are already present as precursors in atympanate Diptera (some of them are even parasitoids of orthopteran species). Thus, the different functional anatomical elements identified in the ormine ears can be reasonably associated with homologous prosternal structures already present in distantly, as well as closely, related atympanate flies (Edgecomb et al. 1995).

8.3. A Case for Evolutionary Convergence

The evolutionary context within which the ormiine ears find their origin is simple. *Ormia ochracea* is a parasitoid on field crickets (*Gryllus rubens* and *G. integer*) and finds its host by hearing and homing in on the cricket's calling song. The female parasitoid fly and the female cricket must solve the same problem to reproduce: to detect and localize a male cricket by his calling song. In caricature, for a fly to act like a cricket, it must hear like one. Earlier, we presented evidence that ormiine flies possess tympanal hearing organs that are built similarly, with regard to their general layout, to other insectan tympanal ears (i.e., orthopteran insects like crickets; Michel 1974). As a general principle, the conversion of sound energy into mechanical energy seems to require the presence of structures that can easily vibrate in sound pressure (but see Römer and the case of the bladder grasshoppers *Bullacris*, Chapter 3). To this end, a thin membrane backed by an air cavity seems to represent a very common (and phylogenetically widespread) solution to allow the reception of airborne vibrations. The presence of analogous design in insectan tympanal ears can be seen as another example of convergent evolution. As stated earlier, insects probably evolved tympanal hearing organs some 15 times independently (Hoy and Robert 1996). It is reasonable to think that a good or efficient design for the conversion of sound energy would have been selected through the adaptive process of evolution. It is then comforting to observe that the problem of acoustic wave detection has repeatedly converged on adaptive solutions (and variations thereupon) that employ ears of the tympanal type. As shown earlier, remarkably, the convergence toward tympanal design has also occurred within Diptera, as witnessed by the independent evolution of prosternal hearing organs in Sarcophagidae (see Figs. 6.4 and 6.12).

Of course, the assertion of convergence on tympanal design has limited reach because it necessarily relies on currently limited knowledge of the design principles of hearing organs. It is, however, thoroughly possible that other types of far-field acoustic sensors (e.g., nontympanal) exist in insects and represent another evolutionary solution to the problem of sound-pressure detection.

The description of a convergence in the evolutionary process arises from the recognition of similarities in the structure and function between two independently evolved characters. Such convergence occurred in the present case between two unrelated groups of insects (crickets and flies), for which the parasitoid–host relationship presumably constituted the driving force for the evolution of tympanal hearing organs in Diptera. While the convergent similarities in the general functional design of the orthopteran and dipteran hearing organs are undisputable, their differences are of even greater interest. As seen earlier, the ways by which crickets and flies obtain directional information from the sound field are fundamentally different. The process of intertympanal coupling represents an innovative variation on the theme of tympanal hearing. Thus, evolutionary convergence is tightly linked to another adaptive process, evolutionary innovation.

8.4. A Case for Evolutionary Innovation

The detection of sound waves in the farfield is a common problem that promoted common solutions across many orders of insects. Such a ubiquitous solution is the evolutionary design of tympanal hearing organs. As presented earlier (Section 5), the endoparasitic condition of *O. ochracea* severely limits the size of these tympanal hearing organs. Such a constraint certainly played a role the emergence of intertympanal coupling, another distinct way of processing interaural sound cues. The comparative study of the detailed functional variations of tympanal organs, in conjunction with their specific biomechanical and ecophysiological constraints, is likely to reveal innovative processes of sound detection and possibly to provide an addition to our understanding of the basic mechanisms of hearing and their diversity.

Notably, the innovation of tympanal hearing occurs probably at least twice in higher Diptera (Edgecomb et al. 1995; Robert et al. 1996). So far, tympanal organs in tachinids (see Fig. 6.2) and sarcophagids (see Figs. 6.4 and 6.12) have been observed only in a few taxonomically localized genera (seven in the former, two in the latter). The vast majority (in the tens of thousands) of tachinids and sarcophagids are very likely to be atympanate, a fact that ensures the highly derived character of tympana in these families. Thus, the evolution of the sense of hearing is also convergent within the Diptera.

9. A New Mechanism for Directional Hearing

In vertebrates, tympanal hearing evolved presumably once (Northcutt 1985). This innovation was to place, once and for all, the ears on either side of the head of vertebrates. Allowing for some variation in form and sensitivity, the vertebrate ears are always found embedded in the skull, more or less laterally. This is in striking contrast to insects, in which hearing organs can be found virtually anywhere on the body (Hoy and Robert 1996; Hoy, Chapter 1). The diversity of insects, through their propensity to take very diverse shapes, sizes, and colors, warrants an equal diversity of the structure and function of their sense of hearing. This diversity provides prime material for the study of particular, adaptive solutions to the general problem of acoustic wave detection.

The vast majority of tympanal auditory organs involved in directional hearing belong to two functional categories: pressure receivers and pressure-difference receivers (Fig. 6.13A,B). In larger mammals, each ear is acoustically isolated (uncoupled) from the other ear. Because the incident sound pressure acts only on the external surface of the tympanum, such receivers operate as pressure receivers (Fig. 6.13A). The human auditory system can detect the direction of an incident sound because of its relatively large interaural distance, which generates a respectable time difference at each ear of ca. 500 μs as well as a significant amplitude difference (e.g., 16 dB at 5 kHz) due to sound diffraction by the head (Middlebrooks, Makous, and Green 1989; Schlegel 1994). In contrast, in small animals such as birds (Hill et al. 1980; Knudsen 1980; Calford and Piddington 1988), frogs (Henson 1974; Narins, Ehret, and Tautz 1988), insects (Michel 1974), and even some mammals such as moles (Coles et al. 1982), the ears may be linked by an internal air passage. In such systems, the incident sound can reach the tympana through two (and sometimes more) pathways: an external pathway around the head or body of the animal and an internal pathway acoustically linking the two bilateral auditory receivers. Thus, the vibration of each tympanum is determined by external and internal sound pressures that may differ in their amplitude and phase characteristics. These acoustically coupled ears are called *pressure-difference receivers* (Fig. 6.13B). By amplifying interaural time and amplitude differences, pressure-difference receivers, in effect, endow small animals with directional hearing (frogs: Eggermont 1988; Narins, Ehret, and Tautz 1988; birds: Hill et al. 1980, Knudsen 1980; crickets: Michelsen, Popov, and Lewis 1994; cicadas: Fletcher 1992; Fonseca 1993).

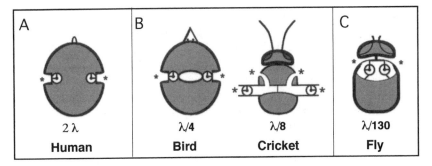

FIGURE 6.13. Mechanisms of directional hearing. A: Pressure receivers. B: Pressure-difference receivers. C: Mechanically coupled receivers. The ratio between the interaural distance (ID) of the animal and the wavelength (λ) at 5 kHz is given as a fraction of λ. The sound inputs to the respective auditory systems are indicated by the asterisks. In pressure receivers (A), variations in sound pressure are exerted only on the external side of the tympanal membrane. The two auditory organs are acoustically isolated by the head. In humans, the interaural distance (ID) is about twice the wavelength at 5 kHz (ca. 2λ). In smaller animals such as birds and crickets (ID-wavelength ratio: λ/4 to λ/8) (B), directional hearing is based on a mechanism in which the receivers are acoustically coupled so that sound is transmitted from one side of the animal to the other by one or more internal interaural sound channels. In the case of ormiine flies, the size mismatch between the ID and λ at 5 kHz is about 130. As presented in this chapter, directional hearing relies on a pair of mechanically coupled receivers (C). The extremely small interaural distance is one of the reasons that impose severe limitations on the directional detection of a sound source using the conventional mechanisms of A and B.

The biomechanical analysis presented earlier demonstrates that the motion of the tympana of *O. ochracea* in an incident sound field is strongly asymmetric. The reason for this unexpected behavior lies in the mechanical coupling between the tympanal membranes through a particular cuticular structure, the intertympanal bridge, which acts like a flexible lever. The use of such coupling in *O. ochracea* constitutes a novel mechanism for directional hearing (Fig. 6.13C) and represents a third, distinct kind of directional receiver for terrestrial animals.

10. Summary

The investigation of hearing in acoustic parasitoid flies reveals the "ingenuity" of nature in solving a vexing biophysical problem. In order to take on its role as an acoustic parasite, the ormiine fly had to "invent" hearing organs that are capable of localizing the acoustic emissions of its host, a field cricket. However, these hearing organs had to evolve within the morphological and developmental constraints of the dipteran *bauplan*. In effect, the fly did not evolve a cricket's hearing organ, but one that accomplishes the same result. The ormiine fly is considerably smaller than a field cricket, and moreover, its closest relatives, other parasitoids of the family Tachinidae, do not possess tympanal hearing organs, so there were no evolutionary precursors for such an ear. As described in the text, tympanal hearing in ormiine flies constitutes an evolutionary innovation of a high order. Tympanal hearing constitutes a novelty for the order Diptera. On one hand, this innovation is conservative because it incorporates the general characteristics of an archetypical insect tympanal hearing organ, and, on the other hand, it is unique because of the way it achieves directional sensitivity. The mechanism of mechanical coupling of the eardrums had not been heretofore reported in any other animal. How phylogenetically widespread the ormiine solution to directional hearing is remains a tantalizing question. Keeping in mind the overwhelming diversity of insects, acoustic prospecting in tropical areas rich in orthopteran species may yield many more ormiine species that are parasitoids upon them. It is then likely that the tribe underwent an evolutionary radiation that promoted the emergence of several species, specialized on particular hosts or cluster of hosts. These flies' specific adaptations to a particular, relatively uncommon in insects, sensory ecological niche may be reflected in their auditory anatomy and tympanal mechanics. The diversity of auditory morphology observed in museum specimens corroborates this idea and encourages the further documentation of the diversity of auditory structure and function in ormiine flies.

Although it is a just-emerging story, it is remarkable that some sarcophagid flies that are also acoustic parasitoids possess hearing organs that bear much closer resemblance to the ormiine ear than to the hearing organs of the sarcophagid's host, a singing cicada. Whereas sarcophagid flies are only distantly related to tachinid flies, both are nonetheless flies, and so it may not be so surprising that evolutionary convergence led to the development of similar hearing organs with the dipteran *bauplan*.

Finally, it seems likely that the principle of mechanical coupling is an adaptation to the fly's small size. There are other animals besides flies that are small in size and that have a severe mismatch between the wavelength of sound to be directionally perceived and the interaural separation. It might then be interesting to examine the hearing organs of some small birds and mammals to see if they possess insect-like, if not ormiine-like, ears that endow them with directional hearing and help them overcome the problem of size. In particular, the ears of fish, which live in a medium where the speed of sound is faster than in air, creates the same mismatch between wavelength and interaural separation. With this in mind, the investigation of how small fish accomplish directional hearing might well reveal other schemes of evolutionary convergence and thus enrich our general knowledge of the fascinating process of hearing.

References

Adamo SA, Robert D, Hoy RR (1995) Effects of a tachinid parasitoid, *Ormia ochracea*, on the behaviour and reproduction of its male and female field crickets hosts (*Gryllus* spp). J Insect Physiol 41:269–277.

Adamo SA, Robert D, Perez J, Hoy RR (1995) The response of an insect parasitoid, *Ormia ochracea* (Tachinidae), to the uncertainty of larval success during infection. Behav Ecol Sociobiol 36:111–118.

Allen GR (1995) The biology of the phonotactic parasitoid *Homotrixa* sp. (Diptera: Tachinidae), and its impact on the survival of male *Sciarasaga quadrata* (Orthoptera: Tettigoniidae) in the field. Ecol Entomol 20:103–110.

Bennet-Clark HC, (1984) Insect hearing: Acoustics and transduction. In: Lewis T (ed) Insect Communication. London: Academic Press, pp. 49–82.

Bennet-Clark HC, Ewing AW (1967). Stimuli provided by courtship of male *Drosophila melanogaster*. Nature 215:669–671.

Cade WH (1975) Acoustically orienting parasitoids: fly phonotaxis to cricket song. Science 190:1312–1213.

Calford MB, Piddington RW (1988) Avian interaural canal enhances interaural delay. J Comp Physiol A 162:503–510.

Coles RB, Gower DM, Boyd PJ, Lewis DB (1982) Acoustic transmission through the head of the common mole, *Talpa europaea*. J Exp Biol 101:337–341.

Dethier VG (1976) The Hungry Fly. Cambridge, MA: Harvard University Press.

Edgecomb RS, Robert D, Read M, Hoy RR (1995) The tympanal hearing organ of a fly: phylogenetic analysis of its morphological origins. Cell Tissue Res 282:251–268.

Eggermont JJ (1988) Mechanisms of sound localization in anurans. In: Fritzsch B, Ryan MJ, Wilczynski W, Hetherington TE, Walkowiak W (eds) The Evolution of the Amphibian Auditory System. New York: John Wiley and Sons, pp. 307–336.

Fletcher NH (1992) Acoustic Systems in Biology. New York: Oxford University Press.

Fonseca P (1993) Directional hearing of a cicada: biophysical aspects. J Comp Physiol A 172:767–774.

Fowler HG (1987) Field confirmation of the phonotaxis of *Euphasiopterix depleta* (Diptera: Tachinidae) to calling males of *Scapteriscus vicinus* (Orthoptera: Gryllotalpidae). Fla Entomol 70:409–410.

Fowler HG (1988) Traps for collecting live *Euphasiopterix depleta* (Diptera: Tachinidae) at a sound source. Fla Entomol 71:654–656.

Godfray HCJ (1994) Parasitoids. Behavioral and Evolutionary Ecology. Princeton, NJ: Princeton University Press.

Henson OW (1974) Comparative anatomy of the middle ear. In: Keidel WD, Neff WD (eds) Handbook of Sensory Physiology Vol. 1. Berlin: Springer-Verlag.

Hill KG, Lewis DB, Hutchings ME, Coles RB (1980) Directional hearing in the Japanese quail (*Coturnix coturnix japonica*) I. Acoustic properties of the auditory system. J Exp Biol 86:135–151.

Hoy RR, Robert D (1996) Tympanal hearing in insects. Ann Rev Entomol 41:433–450.

Hoy RR, Hoikkala A, Kaneshiro KY (1988) Hawaiian courtship songs: Evolutionary innovation in communication signals of *Drosophila*. Science 240:217–219.

Knudsen EI (1980) Sound localization in birds. In: Popper AN, Fay RR (eds) Comparative Studies of Hearing in Vertebrates. Berlin: Springer-Verlag, pp. 289–322.

Lakes-Harlan R, Heller K-G (1992) Ultrasound sensitive ears in a parasitoid fly. Naturwissenschaften 79:224–226.

Léonide J-C (1969) Les Ormiini — étude de *Plesiooestrus leonidei* Mesnil. In: Recherches sur la Biologie de Divers Diptères Endoparasites d'Orthoptères. Série A, Zoologie, Vol. 53. Mémoires du Muséum National d'Histoire Naturelle, pp. 129–138.

Mangold JR (1978) Attraction of *Euphasiopterix ochracea, Corothrella* sp. and gryllids to broadcast songs of the southern mole cricket. Fla Entomol 61:57–61.

McAlpine JF (1989) Phylogeny and classification of Muscomorpha. In: McAlpine JF, Wood DM (eds) Manual of neartic Diptera, Vol. 3. (Agric Can Monogr 32). Research Branch, Agriculture Canada, pp. 1397–1518.

McIver SB (1985) Mechanoreception. In: Kerkut G, Gilbert L (eds) Comprehensive Insect Physiology, Biochemistry, and Pharmacology, Vol. 6. New York: Pergamon, pp. 71–132.

Michel K (1974) Das tympanal Organ von *Gryllus bimaculatus* Degeer (Saltatoria: Gryllidae). Z Morph Tiere 77:285–315.

Michelsen A, Larsen ON (1985) Hearing and sound. In: Kerkut G, Gilbert L (eds) Comprehensive Insect Physiology, Biochemistry, and Pharmacology, Vol. 6. New York: Pergamon, pp. 495–556.

Michelsen A, Popov AV, Lewis B (1994) Physics of directional hearing in the cricket *Gryllus bimaculatus*. J Comp Physiol A 175:153–164.

Middlebrooks JC, Green DM (1991) Sound localization by human listeners. Ann Rev Psychol 42:135–159.

Middlebrooks JC, Makous JC, Green DM (1989) Directional sensitivity of sound pressure levels in the human ear canal. J Acoust Soc Am 86:89–108.

Miles RN, Robert D, Hoy RR (1995) Mechanically coupled ears for directional hearing in the parasitoid fly *Ormia ochracea*. J Acoust Soc Am 98:3059–3070.

Mörchen A, Rheinlaender J, Schwartzkopff J (1978) Latency shift in insect auditory fibers. Naturwissenschaften 65:657.

Morse PM, Ingard KU (1968) Theoretical Acoustics. New York: McGraw-Hill, pp. 418–422.

Narins PM, Ehret G, Tautz J (1988) Accessory pathway for sound transfer in a neotropical frog. Proc Natl Acad Sci USA 85:1508–1512.

Northcutt RG (1985) The brain and sense organs of the earliest vertebrates: reconstruction of a morphotype. In: Foreman RE, Gorbman A, Dadd JM, Olssen R (eds) Evolutionary Biology of Primitive Fishes. New York: Plenum.

Oshinsky ML, Hoy RR (1995) Response properties of auditory afferents in the fly *O. ochracea*. In: Burrows M, Matheson T, Newland PL, Schuppe H (eds) Nervous Systems and Behaviour. Stuttgart: Georg Thieme Verlag, p. 359.

Reichardt W, Wenking H (1969) Optical detection and fixation of objects by fixed flying flies. Naturwissenschaften 56:424–425.

Robert D, Amoroso J, Hoy RR (1992) The evolutionary convergence of hearing in a parasitoid fly and its cricket host. Science 258:1135–1137.

Robert D, Read MP, Hoy RR (1994) The tympanal hearing organ of the parasitoid fly *Ormia ochracea* (Diptera, Tachinidae, Ormiini). Cell Tissue Res 275:63–78.

Robert D, Edgecomb RS, Read MP, Hoy RR (1996) Tympanal hearing in tachinid flies (Diptera, Tachinidae, Ormiini): the comparative morphology of an innovation. Cell Tissue Res 284:435–448.

Robert D, Miles RN, Hoy RR (1996) Directional hearing by mechanical coupling in the parasitoid fly *Ormia ochracea* J Comp Physiol A 179:29–44.

Römer H, Tautz J (1992) Invertebrate auditory receptors. Adv Comp Environ Physiol 10:185–212.

Sabrosky CW (1953) Taxonomy and host relations of the tribe *Ormiini* in the western hemisphere. Proc Entomol Soc Wash 55:167–183.

Schlegel PA (1994) Azimuth estimates by human subjects under free-field and headphone conditions. Audiology 33:93–116.

Shewell GE (1987) Sarcophagidae. In: Manual of Nearctic Diptera. Research Branch Agriculture Canada. Monograph No. 28, Vol. 2, pp. 1159–1186.

Soper RS, Shewell GE, Tyrrell D (1976) *Colcondamyia auditrix* nov. sp. (Diptera: Sarcophagidae), a parasite which is attracted by the mating song of its host, *Okanagana rimosa* (Homoptera: Cicadidae). Can Entomol 108:61–68.

Strausfeld NJ (1989) Beneath the compound eye: neuroanatomical analysis and physiological correlates in the study of insect vision. In: Stavenga V, Hardie R (eds) Facets of Vision. Berlin: Springer-Verlag.

Wagner WE (1996) Convergent song preferences between female field crickets and acoustically orienting parasitoid flies. Behav Ecol 7:279–285.

Walker TJ (1986) Monitoring the flights of field crickets (*Gryllus* spp) and a tachinid fly (*Euphasiopterix ochracea*) in north Florida. Fla Entomol 69:678–685.

Walker TJ (1993) Phonotaxis in female *Ormia ochracea* (Diptera: Tachinidae), parasitoid of field crickets. J Insect Behav 6:389–410.

Wightman FL, Kistler DJ (1997) Monaural directional hearing revisited. J Acoust Soc Am 101:1050–1063.

Wood DM (1987) Tachinidae. In: Manual of Nearctic Diptera. Research Branch Agriculture Canada. Monograph No. 28, Vol. 2, pp. 1193–1270.

Zuk M, Simmons LW, Cupp L (1993) Calling characteristics of parasitized and unparasitized populations of the field cricket *Teleogryllus oceanicus*. Behav Ecol Sociobiol 33:339–343.

7
The Vibrational Sense of Spiders

FRIEDRICH G. BARTH

1. Introduction

Sensory physiology, like other disciplines of science, tends to be biased toward the *obviously* spectacular and, of course, toward the sensory world we humans are living in. In this sense, spiders and their vibration sense may seem to be a rather exotic topic to study. Animals, however, look at the world through windows that may differ drastically from our own. At first sight the spectacular may not be obvious to us at all. Hopefully, this chapter will convince the reader that studying such seemingly aberrant subjects as spiders is well worth the effort.

Spiders are a highly successful group of about 35,000 known species. They have existed for about 400 million years. If we trust the value of a comparative approach to hearing, spiders should not be neglected for the sheer reason of number and ecological significance. More importantly, however, the behavior of spiders is guided and controlled by vibrations to a greater degree than that of most other animal groups. Spiders, then, are a good choice of animals to study if we are interested in understanding the vibration sense and its behavioral significance. In addition, spiders teach us a lot about more general issues in sensory biology.

Spiders live on widely differing substrates and thus receive vibrations through different media. The most spectacular of these media may be the delicate webs of spider silk. Many spiders, however, live on solid substrates such as plants, and a few semiaquatic ones even receive vibrations through the water surface. If we strive to understand any vibration sense in the larger context of its behavioral significance, both the biologically relevant signals and the receiver's sensory and behavioral reaction to them have to be considered. In addition, an understanding of the physics of signal transmission through the various media and of the distortion and filtering of the signals on their way to the receiver is essential.

From a physics point of view, rhythmic or even irregular vibrations of the substrate are not unlike airborne sound waves. The close relationship between airborne sound and vibrations is underlined by the circumstance that even in what is commonly referred to as *farfield hearing*, airborne acoustic events are transmitted to the sensory cells by vibrating solid bodies, such as the ossicles in our middle ear and the basilar membrane in the inner ear. In both vibration and sound reception, pressure changes are exerted onto the receiving structures. Apart from vibration receptors, many arthropods, including spiders, have hairlike sensory receptors ("ears") that are specialized to respond to the *movement* of the surrounding medium, to both rhythmic oscillations and true air (or water) currents. The long evolutionary history of spiders has led to a perfection of both the vibration and air movement sense, which is impressive not only from a biological but also from a technical point of view. There are several reviews on the arthropod and spider vibration sense in the literature to which the reader is referred (Markl 1969, 1973, 1983; Tautz 1979, 1989; Barth 1982, 1985a, 1986; Kalmring and Elsner 1985).

2. Behavioral Significance

Among all stimuli, vibrations are of particular importance in the life of a spider (Barth 1982, 1985a, 1986), even though some spider families, such as many jumping spiders (Salticidae) and net-casting spiders (Dinopidae), are mainly visual with eyes of an extremely sophisticated organization (Blest 1985; Forster 1985; Land 1985). Examples to illustrate this are given later, together with a description of the vibrations on the various substrates where spiders are found.

One of the behaviors obviously guided by vibrations in most spiders is *prey capture* (Barth 1985a). Every nature lover has watched an orb weaver alarmed by the vibrations emitted by a prey insect entangled in the spider web and trying to free itself. The reaction of the spider is quick and its approach to its victim precise, both with regard to direction and distance. Similar reactions are elicited in spiders that do not build webs for prey capture but instead receive prey-generated vibrations when sitting on plants or the water surface.

Another behavior that much depends on vibratory signals in many if not most spiders is *courtship and precopulatory behavior* (Barth 1993). Often both the male and the female emit courtship vibrations and thus communicate reciprocally. Their signals, in particular those of the male, are species specific and in some cases at least were shown to be important filters that ensure reproductive isolation of sympatric species (Uetz and Stratton 1982; Stratton and Uetz 1983; Barth and Schmitt 1991). Substratum-coupled vibratory communication not only serves the ethological separation of species but also suppresses aggressiveness so that the male will not be mistaken as prey by the female. Vibratory communication may also be relevant in male competition (Schmitt, Schuster, and Barth 1992) and in sexual selection, but there is no experimental evidence for the latter as yet.

In addition to the above, spiders use vibrations actively for a kind of *vibratory echolocation* to detect and localize motionless prey and other particles in their web (Liesenfeld 1956; Klärner and Barth 1982). *Zygiella x-notata* (Araneidae), the sector spider, was reported to detect objects as light as 0.05 mg in its orb web by such active localization (Klärner and Barth 1982).

3. Vibrations on Different Substrates

It seems reasonable to assume that properties of an animal's habitat have shaped the evolution of its senses and their integration into behavior. As a consequence, properties of the sense organs are likely to reflect properties of the relevant stimuli. In the present context we therefore have to ask which vibrations spiders are exposed to and how the various substrates on which they live transmit vibrations from the sender to the receiver.

3.1. The Spider Web

The web of a spider is a remarkable product of animal behavior (Fig. 7.1). It serves the spider not only as a trap to intercept and retain insect prey but also as a self-made extension of its sensory space. Spiders receive and emit vibrations in their webs. They use it not only to receive information on entangled prey but also to avoid predators and as a communication channel for intraspecific communication. From an evolutionary point of view, a spider web is the result of a compromise: It has to serve architectural needs, that is, to guarantee sufficient stability as a shock-absorbing lightweight structure, and it has to effectively transmit vibrations. Our understanding of the interaction of these two trends is still incomplete.

We do have many largely descriptive stories on fascinating uses of spider webs in the literature (Barth 1982, 1985a; Shear 1986) but very few experimental reports on the vibrations in spider webs. The two main reasons for this situation are (a) that the physics of a spider web, even that of a rather regular structure such as the orb web, is very complicated indeed and (b) that the proper measurement of vibrations of the fine silken threads and of their transmission asks for contactless methods and considerable technical effort. The availability of laser vibrometers was an important breakthrough when used to study the orb web of *Nuctenea sclopetaria* (Masters and Markl 1981; Masters, Markl, and Moffat 1986). We recently used laser vibrometry in our laboratory to study the orb web of *Nephila clavipes*, the golden silk spider (Landolfa and Barth 1996) and now hope to extend the analysis to other types of spider webs. Even with the laser vibrometer (and modern photodiode technology), however, a significant problem remains: the sequential measurement at specific points in the web cannot fully satisfy the need to understand what the whole web is doing at the same and subsequent instants of time. A truly two-dimensional analysis would be very helpful to better understand wave modes, reflections, resonances, etc. Nevertheless, some important conclusions can already be drawn from the work available.

3.1.1. Orb Webs

Orb webs are mechanically heterogeneous structures. They consist of different types of threads with different material properties and pretension (Denny 1976; Masters 1984a; Wirth and Barth 1992). The radii of the orb web, the spokes of the wheel, are much more important for the transmission of vibrations than the spiral and sticky threads. Accordingly, a spider ready for prey capture always contacts the radii with the tarsi of its legs. The main reason for the comparatively small attenuation of vibrations transmitted through radial threads is their Young's modulus of elasticity, which is higher than in the other types of thread (radii: $3–20 \times 10^9 \, \text{N/m}^2$; spiral thread: $0.05 \times 10^9 \, \text{N/m}^2$; Denny 1976; Masters 1984a). In addition, the density of the material as well as the number of crossings with other threads influence the transmission properties. Mechanical tension of the thread, however, is not a critical parameter for the transmission of longitudinal vibrations (Frohlich and Buskirk 1982), although the misled intuition of a biologist might think otherwise.

Another striking difference between a radius and framework silk, on the one hand, and viscid silk, on the other hand, is the much larger breaking strain (*strain* is the ratio of increase of length under load and original length) of the latter. Denny (1976) reports for the orb web of *Araneus sericatus* values close to 3.0 for viscid silk but only 1.2 for framework silk. The shock-absorbing function of viscid silk is underlined both by its stress–strain curve, which exhibits a very low initial Young's modulus, and by its impressive extensibility. The sticky spiral of *A. diadematus* can be extended reversibly by 517% of its initial length, and that of *Meta reticulata* by 1400% (de Wilde 1943; Lucas 1964; Work 1976).

For the reception of vibratory signals, the particular significance of the hub as a geometrical and mechanical center in the orb web derives from the convergence of radii in this area. If an orb weaver was not already sitting in the hub (as it normally does), it first returns to it upon vibration of the web and then orients toward the source of vibration. The alternative is to sit in a niche or corner of the web instead of in the hub. In such a case the tarsus of a front leg is placed on a signal thread, which transmits the vibrations from the hub and thus from the entire web. In the sector spider web (*Zygiella x-notata*) a single signal thread runs through a sector of the web devoid of radii and spiral threads. It links the hub to the spider's retreat in the web periphery (Fig. 7.1F). Upon vibrations of the signal thread, the spider darts to the hub along the signal thread, turns toward the source of

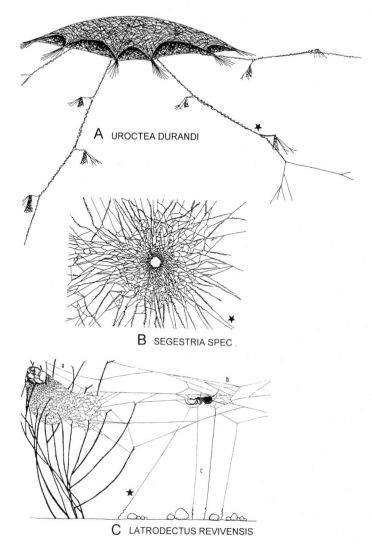

FIGURE 7.1. *Spider webs and vibrations.* A few examples demonstrating the variety of ways in which spiders use web structures to receive information on sources of vibrations such as prey animals. Asterisks point to signal threads. (A from Kullmann et al. 1975, with permission. B from Grassé 1949, with permission. C from Konigswald, Lubin, and Ward 1990, with permission. D from Henschel and Lubin 1992, with permission. E from Burgess 1976, with permission. All figures modified from originals.) For further explanation see p. 235.

vibration, and runs toward it along the proper radius (Liesenfeld 1956; Klärner and Barth 1982). Many non-araneid spiders show a similar behavior (review, Barth 1982). Other spiders with completely different types of webs use signal threads as well.

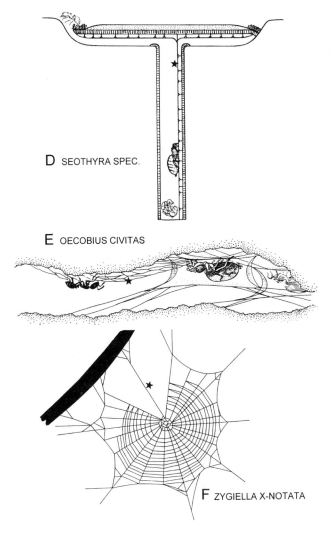

D SEOTHYRA SPEC.

E OECOBIUS CIVITAS

F ZYGIELLA X-NOTATA

FIGURE 7.1. *Continued.*

At the level of *individual radii* there are longitudinal, transverse, and torsional vibrations, according to the direction of movement relative to the direction of propagation. Transverse waves are further subdivided into waves with movement perpendicular to the plane of the web and lateral vibrations with movement within the plane of the web (Masters 1984a). This distinction is important because the strands crossing a radius will affect transverse and lateral vibrations differently (i.e., vibration parallel or perpendicular to cross-strands). Torsional vibrations have so far been neglected. They are hard to measure and are believed not to be important because torsional motion of a thread of such small diameter (a few micrometers) would result in negligible torque on an object (Masters 1984a). The estimated attenuation of torsional vibrations is high in the web of the golden silk spider *Nephila* (20 dB/10 cm; Landolfa and Barth 1996). Clearly, measurements are needed.

Data obtained with adequate noncontact methods on the vibration transmission in orb webs are available for *Nuctenea* and *Nephila*. The web of *Nephila clavipes* is particularly large (diameter up to ca. 1.5 m) and strong, and differs from that of the orb weaver *Nuctenea sclopetaria* by having a nonsticky auxiliary spiral thread in addition to the sticky spiral. The auxiliary spiral serves as a scaffolding during web construction in orb weavers. It is removed by *Nuctenea* when finishing its web.

Among the main conclusions from the studies on the orb webs of *Nuctenea* and *Nephila*, respectively, are the following (Masters 1984a; Landolfa and Barth 1996; see also *Zygiella*: Liesenfeld 1956, 1961; Graeser 1973; Klärner and Barth 1982):

1. Most of the energy of insect-generated *prey spectra* is below 100 Hz, with peaks between 5 and 50 Hz for nonbuzzing but struggling insects, and additional peaks around 100 to 300 Hz for buzzing flies and bees. Peak amplitudes of prey-induced vibrations span a range of up to 50 dB, when, for instance, comparing a fruitfly with a cricket. Although there is much overlap of the peak amplitudes and frequencies of the vibrations produced by different insects struggling in the web, there are also differences in detail that are faithfully transmitted to the vibration receptors.

2. *Attenuation* of longitudinal vibrations along individual radii is considerably smaller than that of transverse and lateral vibrations.

3. Upon single-point stimulation, vibrations measured in the *Nephila* web around the hub and at the spider's tarsi exhibit amplitude *gradients* of 20 to 30 dB. These gradients are likely to be used by the spider as indicators of stimulus direction. In contrast, vibration propagation velocities result in time-of-arrival differences at the spider's tarsi of less than 1 ms, which may be too brief to determine stimulus direction.

4. *Structural differences* between the webs of *Nuctenea* and *Nephila* significantly influence vibration transmission. The greater interconnectivity in a *Nephila* web leads to much higher attenuation. Vibrations may spread from a stimulated radius to neighboring radii via the auxiliary spiral (which is not taken down by *Nephila* when completing the web) and the greater number of sticky spiral threads.

5. If maximizing vibration transmission efficiency were an evolutionary priority over the other mechanical needs of a web, *Nephila* should use less spiral threads and remove the auxiliary spiral from its completed web, as indeed many other orb weavers do. Instead, *Nephila* adds mechanical redundancy to its web at the expense of a degradation of vibration transmission and gains a longevity of its web, which is outstanding among orb weavers.

3.1.2. Other Webs

Many spiders build webs very different from the orb webs of *Nuctenea* and *Nephila*, or *Zygiella* (Shear 1986). Very little is known about their vibration transmission properties, although these may be even more interesting than those of the rather regular orb web. From observations of many species of spiders, however, it is obvious that a wealth of nice tricks is applied by them in order to gain vibratory information.

Figure 7.1 illustrates this with a few examples. *Uroctea durandi* (Urocteidae) builds a tent-roof web. Signal threads lead from the floor and considerably extend the effective prey-catching area (diameter up to 0.5 m). Interestingly, the signal thread is attached to silken "telephone poles" instead of touching the ground. This most likely improves signal transmission to the spider's retreat under its silken roof (Fig. 7.1A; Kullmann et al. 1975). *Segestria* (Segestriidae) stretches signal threads radially on the ground from the opening of its tubular retreat to enlarge its detection and capture area. Other spiders do the same (Fig. 7.1B; Buchli 1969; Kullmann 1972). Some trapdoor spiders, such as *Anidiops villosus* (Ctenizidae), even arrange slender twigs instead of silken threads radially outward from the opening of their burrow for the same purpose (Main 1957).

Like other members of its family (Theridiidae), the widow *Latrodectus pallidus*, feared for its poisonous bite, catches prey with sticky threads arranged vertically between the ground and the catching platform of its web (Fig. 7.1C). These threads entangle the prey and transmit prey vibrations to the spider's retreat. When they break the spider comes and pulls them in together with its victim (Wiehle 1931; Szlep 1965). *Seothyra* (Eresidae) builds webs in sand dunes, which consist of a horizontal chamber on the sand surface and a vertical silk-lined burrow (Fig. 7.1D). Insects get caught by capture silk on the outside of the chamber edges. Signal threads attached to the inside of both the chamber and the burrow alert the spider to the presence of prey (Henschel and Lubin 1992). *Oecobius civitas* (Oecobiidae) is a gregarious spider living in large numbers in communal webs. When prey such as an ant disturbs the alarm web (Fig. 7.1E), the spider receives a vibratory signal in its retreat and catches its victim (Burgess 1976).

Unfortunately, none of these or other similar cases have been studied with regard to vibration transmission and vibration sense. There is a treasure mine for future research using current technologies.

3.2. Plants

The web and the ability to weave are among the distinctive features of spiders, and web vibrations are the most obvious ones compared with those in other media. There are many spiders, however, that do not use a web for prey capture nor as an arena for courtship and mating. Many of these spiders, including a large number of nocturnal species, are guided by substrate vibrations in plants, and even on the water surface, or in soil, and on rocks. Here we first concentrate on plants and the first question to be answered is: What kind of vibration can we expect in plants? The spiders we now have particularly in mind belong to the genus *Cupiennius* (Ctenidae). They live in close association with particular plants, mainly monocots such as bromeliads. Their vibratory communication and prey-capture behavior have been examined in some detail (Barth 1985a, 1993, 1997).

Structure-borne sound is the subject of a highly specialized and complicated field of physics. When trying to understand the generation and propagation of motions in plants, as biologists we would be better off to find a shortcut to the essentials instead of dealing with all the details that are found in the textbooks of physics (Morse 1948; Sommerfeld 1970; Skudrzyk 1971; Cremer, Heckl, and Ungar 1973). There are two basic types of waves in solids, longitudinal and transverse, with particle motion parallel and perpendicular to the wave propagation direction, respectively.

3.2.1. Longitudinal Waves

True longitudinal waves, with localized compression and rarefaction of the medium and movement exclusively in the direction of wave propagation (similar to sound waves), are of no concern here because they occur only in solids much larger than the wavelength of the vibration in all three dimensions. Theoretically, *quasi-longitudinal* waves might occur. These are characterized by an additional local transverse change of the diameter of the structure considered according to its Poisson ratio (i.e., the ratio of sideways motion to compression). Many plant leaves fulfill the size condition for the occurrence of such waves, which is that the structure should be small with respect to wavelength in one or two directions. When traveling along a rod, particles are displaced not only in the primarily longitudinal but also in the lateral (x,y) directions due to the (very small) crosscontraction. The ratio of the greatest lateral displacement to the greatest longitudinal displacement is roughly equal to the ratio of the rod's thickness to the wavelength, that is, very small (see propagation speed later; Cremer, Heckl, and Ungar 1973).

3.2.2. Transverse Waves

Transverse waves do not propagate in air or water. However, particle motion at a right angle to wave propagation and in the plane of the structure's surface does occur in structures large compared with wavelength in all three dimensions. It also occurs in flat plates of uniform thickness, which are able to transmit elastic-shearing forces. Because wavelengths are in the range of several meters to decimeters, plant leaves or stems are unlikely to support such waves.

Torsional waves may be considered as a special case of transverse waves. They are found in structures that are long compared with their thickness, that is, narrow beams excited by torque. In this case cross sections rotate about the axis of the beam, with circumferential displacement as a consequence.

3.2.3. Bending Waves

Bending or flexural waves are the most important ones in plants in our context (Fig. 7.2). They only occur in stiff material and not in an orb web because the silken threads lack bending stiffness. Bending waves go along with rather large lateral deflections. They are not transverse waves, however. Despite particle motion in a plane perpendicular both to the direction of wave propagation and to the surface, bending waves differ from transverse waves in various aspects. Thus, there are stresses and strains in the longitudinal direction. In addition, propagation is dispersive, which means that the higher frequency components propagate with a higher velocity than the lower frequency components.

Phase velocity C_{ph} (the velocity of movement necessary to remain at the same phase of a sinusoidal wave motion) has to be distinguished from the group velocity of the carrier wave envelope (wave "group"; Cremer, Heckl, and Ungar 1973). In rods with a diameter $d < \lambda/6$ (λ, wavelength)

$$C_{ph} = \sqrt[4]{\frac{B}{m'}} \sqrt{\omega} = \frac{1}{2} C_g, \tag{1}$$

where B is bending stiffness ($E \cdot I$, where E is Young's modulus of elasticity and I is the axial moment of inertia), m' is mass per unit length ($\varrho \cdot s$, where ϱ is density and s is length), and $\omega = 2\pi f$ (f, frequency). Because the propagation velocity of bending waves is a function of the fourth root of bending stiffness and the inverse value of mass per unit length, it varies only little with the structure's mechanical properties. Because I is proportional to r^4 (r, beam radius) and m' to r^2, propagation velocity is proportional to $\sqrt{r}$ and $\sqrt{f}$. Note that the propagation velocity of both longitudinal waves (C_L)

$$\text{rod: } C_L = \sqrt{\frac{E}{\varrho}}; \tag{2}$$

$$\text{plate: } C_L = \sqrt{\frac{E}{\varrho(1 - \mu^2)}} \tag{3}$$

(E, Young's modulus; ϱ, density; μ, Poisson ratio) and transverse waves (C_T)

$$C_T = \sqrt{\frac{G}{\varrho}} \tag{4}$$

[shear modulus $G = E/2 \cdot (1 + \mu)$] is independent of frequency.

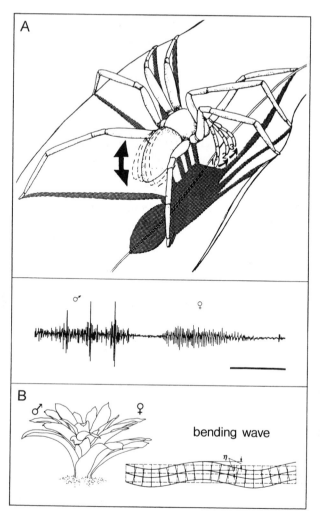

Figure 7.2. *Bending waves.* **A**: Courting male of the wandering spider *Cupiennius getazi*. Vibrations are introduced into the plant by up-and-down movements of the opisthosoma (without touching the plant) and are transmitted into the plant through the legs. The oscillogram shows the corresponding plant vibration. Scratching and drumming on the plant with the pedipalps also generates plant vibrations. The scale bar below the oscillogram represents 1 s. **B**: Schematic representation of a bending wave as it travels through a plant, such as a bromeliad, during reciprocal vibratory courtship communication of *Cupiennius*. (Adapted from Barth 1993, with permission.)

3.2.3.1. Actual Propagation Velocities

Measured propagation velocities of bending waves are considerably lower than the velocity of a longitudinal wave. In addition to the displacement perpendicular to the surface, the low propagation velocity is indeed the key property for the identification of a bending wave. In a pioneering study, Michelsen and colleagues (1982) have related the issues outlined earlier (Cremer, Heckl, and Ungar 1973) to the measurement of the transmission properties of plants used by "small cicadas" (planthoppers and leafhoppers, Cicadina) and cydnid bugs for their vibratory communication. Their laser vibrometry data, in particular the agreement between calculated and measured propagation velocities and their increase with the square root of frequency, clearly pointed to the existence of bending waves. The propagation velocity at a particular frequency was found to be largely independent of the plant's mechanical properties (soft bean plant, stiff reeds, and maples; Table 7.1).

Similar measurements were done on banana plants (Barth and Bohnenberger, unpublished; Barth 1985a). These are among the favorite dwelling plants of *Cupiennius*, on which this spider emits and receives vibrations. Both in the leaves and the pseudostem, sinusoidal signals of 100 Hz and 500 Hz were transmitted at velocities smaller than 50 m/s. Measurements on agavae (*Agave americana*) also supported the concept of bending waves (Wirth 1984; Barth 1993). Again, velocities increased with frequency, demonstrating the dispersive nature of the underlying wave type. In addition, propagation velocity of the same frequency was larger in the basal parts of a leaf than in its apical parts. This is in good agreement with bending-wave propagation speeds calculated from the mechanical properties of the leaf, which is thicker and stiffer at its base than apically. Thus, group velocity at 30 Hz (a frequency representative of the female courtship signal) in an *Agave* leaf 50 cm long and 12 cm broad was 4.4 m/s in the apical third of the leaf and 35.7 m/s in its basal region. In another similar leaf, propagation velocities were 8 m/s at 30 Hz, 10 m/s at 50 Hz, and 18 m/s at 80 Hz. At 200 Hz and 2000 Hz, values were 26 m/s and 80 m/s in the median region of the leaf and 87 m/s and 278 m/s in its basal region.

TABLE 7.1. Propagation velocities of vibrations on plants used by "small cicadas" and cydnid bugs for communication.

Plant	Propagation speed (C_g; m/s) calculated (measured)	
	At 0.2 kHz	At 2.0 kHz
Vicia faba (broad bean)	39 (36)	122 (120)
Galeobdolon vulgare (yellow archangel)	42 (45)	132 (143)

From Michelsen et al. (1982), with permission.

Such measurements are not only made to identify the underlying wave type. They are also needed to identify the differences in time of arrival of vibrations at different spider legs, which may be used as cues for orientation (see later). Adding an analysis of the vibration propagation in plants (blackberry, chamomile, rush, grass) used by stridulating bushcrickets (Kämper and Kühne 1983), we conclude that vibrations in plants used as communication channels by spiders and insects are primarily carried by bending waves. From the propagation velocity values, it can also be followed that the cross-sectional dimensions of these plants are small compared with wavelength, which is a condition for pure bending waves, for example, in the case of the *Agave*

$$\lambda = \frac{v}{f} = \frac{4.4 \, \text{m/s}}{30 \, \text{Hz}} = 0.14 \text{m or } \frac{26 \, \text{m/s}}{200 \, \text{Hz}} = 0.13 \text{m}.$$

3.2.3.2. Attenuation and Frequency Filtering

The attenuation and frequency filtering of a vibration on its way from the source to the receiver was studied in structurally simple monocotyledonous plants (*Musa sapientium*, banana plant; *Agave americana*) and in spiders of the genus *Cupiennius* in relation to their vibratory sensitivity and communication behavior (Wirth 1984; Barth 1985a, 1993). The attenuation values found are surprisingly small. At the dominant frequency component of the male courtship signal (ca. 75 Hz), average attenuation was only 0.3 dB/cm on a banana leaf. Even when introducing the vibration into one leaf and measuring it on another leaf, or on the pseudostem, average attenuation values typically were 0.3 to 0.4 dB/cm (Barth 1985a). At the dominant frequency of the female vibratory courtship signal (ca. 30 Hz), attenuation is even smaller. Such values will later be used to calculate the active range of these signals. Irregularities of attenuation were concentrated between 150 Hz and 1 kHz. Even at 5 kHz attenuation was only 0.35 dB/cm when measured for a signal that had traveled about 1 m through the plant (from one leaf to another or down the pseudostem). In *Agave americana* the attenuation was up to about 0.4 dB/cm for frequencies up to 85 Hz, which includes the main components of the courtship signals of *Cupiennius* (Wirth 1984).

Michelsen et al. (1982) support the idea that the branched dicotyledonous plants examined by them (see Table 7.1) are considerably more heterogenous mechanically than monocot "*Cupiennius* plants." Songs of the cydnid bugs and small cicadas often were more intense on the top of the plant or on a leaf far away than close to the singing insect on the stem. Reflections, frequency-dependent standing-wave patterns, and changes in mechanical impedance complicate the pattern of vibration. The attenuation of purely sinusoidal vibrations changed by up to 30 dB by changing the frequency or the position of the animal. It therefore is much to the advantage of the animals to use broad-band signals (see later).

Even in the agave, however, amplitude does not always decrease linearly with distance from the signal source. Propagation speed is larger not only basally, but also along the midline of a leaf as compared with its apical and marginal regions. It is not known how the spider deals with these heterogeneities when using vibrational cues for orientation (see later). A note of caution may therefore be in order here. The signal receivers on plants have to cope with serious signal distortions. There will be distortions not only due to the mechanical heterogeneity of plants but also due to reflections from the ends of the plant or parts of it, and, as a consequence, due to interference and superposition. In this regard it is surprising to see how well the vibratory courtship signal of *Cupiennius* is preserved as it reaches the receiver after having traveled distances of 2 m or more (see later). It may well be that the preference of *Cupiennius* for monocotyledonous plants as dwelling plants (Barth et al. 1988a) not only has to do with its need for a retreat but also with the predictability of vibration transmission.

More extended two- or even three-dimensional analyses of vibration transmission by plants considering complex wave fields are still needed but are difficult to carry out. Another interesting question for further research is how different modes of introducing vibrations into the relevant plants affect vibration transmission. *Cupiennius* introduces its abdominal vibrations into the plant by its eight legs (see later), that is, not through one point source but by eight regularly spaced sources. Again, this is a field for further study, which will not be easy but should be rewarding.

3.2.4. Surface Waves

Free surface waves combining longitudinal and transverse waves are called *Rayleigh waves* (Rayleigh 1885). Close to the surface compression leads to a transverse displacement (amplitude depending on the material's Poisson ratio μ). Consequently, particle motion is intermediate between purely longitudinal and transversal, respectively, and is similar to a gravity wave in deep water (Cremer, Heckl, and Ungar 1973). The propagation velocity (phase velocity) of Rayleigh waves is slightly less than that of longitudinal and transverse waves[1] but considerably higher than that of bending waves. It can be calculated from the following equation:

$$C_R = \frac{\omega}{S_R},\qquad(5)$$

where $\omega = 2\pi f$ and S_R is the trace wave number ($S = k_x/k_T$; $k = \dfrac{2\pi}{\lambda}$; indices

x and T refer to the actual trace wave numbers and those of transverse waves, respectively; Cremer, Heckl, and Ungar 1973).

---

[1] The ratio of propagation velocities, C_R/C_T, depends only moderately on the Poisson ratio μ of the material: at $\mu = 0$ it is 0.87, at $\mu = 0.25$ it is 0.92, and at $\mu = 0.5$ it is 0.95.

In summary, the research available on biologically relevant plants underlines the primary importance of bending waves as carriers of arthropod vibrations. The involvement of other types of waves cannot be generally ruled out yet, however.

3.3. Water Surface

Not many spiders live on the water surface. The best example for *semi-aquatic spiders* are the fishing spiders of the genus *Dolomedes* (Pisauridae; Carico 1973). They share the water surface with water striders and water bugs. Alerted by the surface ripples generated by insects that have fallen onto the water, *Dolomedes* darts toward its victim (Bleckmann and Barth 1984; Bleckmann 1994). Water surface waves are also generated by male *Dolomedes* during courtship and most likely are used for sexual communication (Bleckmann and Bender 1987). A wolf spider, *Pirata piraticus* (Lycosidae), behaves similarly to *Dolomedes* (Berestyńska-Wilczek 1962). There is only one spider known to actually live *in* the water, the water spider *Argyroneta aquatica* (Argyronetidae). This spider takes some of its terrestrial environment with it under water by collecting air in a web, which keeps it from bubbling to the water surface. Prey that hits the suspension threads of the web cause vibrations. The alerted spider swims through the water to catch its prey, much like a spider living in an aerial web. The effects of the surrounding water on signal transmission and reception have not been studied.

What *kind of vibrations* do we find on a water surface? Particle movement is in a plane perpendicular to the undisturbed surface and, depending on water depth, along a circle or ellipse (water depth is small compared with wavelength; Fig. 7.3). The propagation of water surface waves is not due to the elasticity of the medium, as in the case of the waves in the other media considered so far, but to gravity and surface tension. It is dispersive, like the propagation of bending waves in plants. There are additional complications (Sommerfeld 1970; Lighthill 1980; Bleckmann 1994). These result from a difference in the relationship of propagation speed and frequency in gravity waves (»13 Hz) and capillary waves («13 Hz), respectively (Fig. 7.3C). Shorter waves travel faster than longer ones in capillary waves but slower in gravity waves. As a result the phase velocity C_{ph} is minimal at 13 Hz (wavelength 1.7 cm; at 13 Hz $C_{ph} = C_g$), where it measures 23.1 cm/s, and larger at both lower and higher frequencies.

Natural signals will hardly ever be purely sinusoidal but contain many frequency components. Very low frequency components will travel ahead of components between about 5 Hz and 20 Hz, as will higher frequencies. Group velocity is minimal at about 6 Hz ($C_g = 17.5$ cm/s) and increases towards both lower and higher frequencies (at 100 Hz $C_g = 55$ cm/s). For water depths much larger than wavelength λ and with typical water density and surface tension,

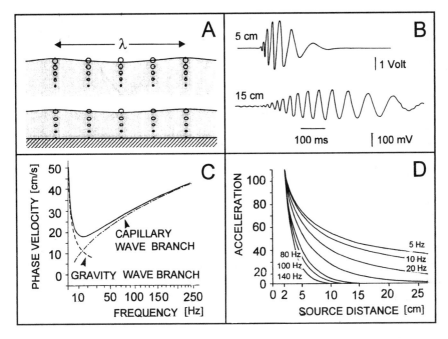

FIGURE 7.3. *Water surface waves.* **A**: When a sinusoidal wave (λ, wavelength) travels along the water surface, the motion of water particles depends on the depth of the water body. Whereas in deep water particles move in circles (radius equal to maximum surface elevation), their motion describes ellipses in shallow water. Dots indicate instantaneous position of particle. (From Lighthill 1980, with permission.) **B**: A click stimulus after having traveled distances of 5 cm and 15 cm, respectively. Note massive changes in waveform. **C**: The phase velocity of water surface waves as a function of frequency (solid line). Phase velocity is smallest at 13 Hz, at which frequency the values for pure gravity waves and pure capillary waves are identical. It is larger at both lower and higher frequencies. **D**: Calculated attenuation of surface waves as a function of wave frequency and distance from the wave source. (Adapted from Bleckmann 1994, with permission.)

$$C_{\mathrm{ph}} \approx \frac{\omega}{k}, \tag{6}$$

where $\omega = 2\pi f$ (f, frequency) and $k = 2\pi\lambda^{-1}$. A water surface wave like that produced by a click stimulus consists of a group of waves of different wavelengths after having traveled for some distance (Fig. 7.3B). Frequency modulation depends on the distance traveled and the propagation speed of the wave packet. Theoretically, source distance D can be calculated from the local frequency ω and the frequency modulation $\dot{\omega}$ around it (Bleckmann 1994);

$$D = -4.5 C_{\mathrm{ph}}\left(\omega/\dot{\omega}\right). \tag{7}$$

Attenuation of water surface waves along with their propagation is much stronger (several dB/cm) than that of plant-borne bending waves. Because it is also highly frequency dependent, a water surface vibration changes its spectral composition enormously with distance from the wave source. At distances of a few centimeters, frequencies higher than 200 Hz will already be gone (Fig. 7.3D). In other words, the water surface is a low-pass filter. The slope of the attenuation curve decreases with distance already traveled, which implies that the signal is attenuated most strongly close to its source.

4. Vibration Receptors

4.1. Types

In technology airborne sound usually is measured as acoustic pressure, and velocity is derived from pressure gradient measurements. Regarding structure-borne vibrations, the situation is different. Here kinematic variables, such as displacement, velocity, or acceleration, are measured, whereas stresses (pressure) and forces are mostly deduced from such measurements. Another difference in the measurements of airborne sound and structure-borne vibrations is that the first is measured well in the interior of the stimulus field, whereas the second usually is only accessible at the exterior of the vibrating solid structure.

Spiders have sensory devices to measure at the boundary of vibrating structures that is at the solid–air or water–air interface. These sensors are located on the legs and are either hair sensilla or slit sensilla (Barth and Blickhan 1984; Barth 1985b; Fig. 7.4). The most important vibration sensor is the *metatarsal lyriform organ*, a compound slit sense organ that is well

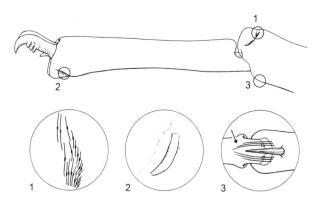

FIGURE 7.4. *Vibration receptors.* Different cuticular spider sensilla known to be vibration sensitive. The most prominent and sensitive vibration receptor is the metatarsal lyriform organ (1). (2) Tarsal single-slit sensillum. (3) Cuticular hairs bridging the tarsus–metatarsus joint. (From Speck-Hergenröder and Barth 1988, with permission.)

adapted to its exteroreceptive function. It is found mid-dorsally at the distal ends of all eight walking leg metatarsi and is made up by slits arranged in parallel and at a right angle to the long axis of the leg (see inserts in Fig. 7.7). Both of these features are exceptional among the slit sensilla and are particularly important for the organ's vibration sensitivity: Whenever the tarsus is moving up and down or sideways due to vibrations of the substrate, the slits are compressed and thus adequately stimulated. A deep furrow in the cuticle at both lateral sides of the organ increases the organ's deformability and focuses the compressional forces transmitted by the tarsus onto it (Barth 1972a,b; Barth and Geethabali 1982).

In *Cupiennius salei* the metatarsal organ is made up of 21 slits. These vary in length from about 20 to 120 µm (Barth 1971). In other spiders the number of slits is smaller, but the largest slits always follow a common pattern of arrangement and can be identified individually. Variation is mainly seen with the shorter slits, which vary in number from 0 to 10. Examples for total numbers of slits in the metatarsal organ are as follows: *Salticus scenicus* (Salticidae) 11 slits, *Zygiella x-notata* and *Nephila clavipes* (Araneidae) 20 slits, *Tegenaria larva* (European house spider, Agelenidae) 16 slits, and *Achaearanea tepidariorum* (American house spider, Theridiidae) 8 to 10 slits.

The metatarsal organ is the most obvious vibration receptor and the most sensitive one. It is not the only vibration-sensitive organ of spiders, however (see Fig. 7.4). (a) There are two *single slits* (each ca. 45 µm long in *Cupiennius*) on the pretarsus directly behind the claws (Barth and Libera 1970), which are not only stimulated proprioceptively by active pretarsal movements but also by vibrations of the substrate. Their sensitivity (measured between 0.01 Hz and 1 kHz) is lower by a factor of about 100 than that of the metatarsal organ but shows a similar high-pass characteristic (Speck and Barth 1982). (b) In addition, there are *cuticular hairs* (16 in *Cupiennius salei*) bridging the metatarsus–tarsus joint ventrally. The tips of these hairs touch the tarsus and are moved together with it by substrate vibrations. According to recordings from interneurons onto which the sensory cells of these hairs converge, sensitivity again is lower by at least two powers of 10 than in the metatarsal organ, with a maximum at 70 Hz and 150 Hz. (c) Finally, the very sensitive slits with a highly phasic response characteristic of *lyriform organs* more proximally on the leg may respond to substrate vibrations as well (Barth and Bohnenberger 1978; Bohnenberger 1981). Direct experimental evidence for this is lacking but, aside from the slits' very high sensitivity, the rather efficient transmission of vibrations through the spider leg (Fig. 7.5) supports the idea.

4.2. Sensitivity

We have determined threshold curves for 10 individual slits of the metatarsal organ by recording extracellularly from them while vibrating the tarsus sinusoidally (Barth and Geethabali 1982). In *Cupiennius salei* all 10 of the

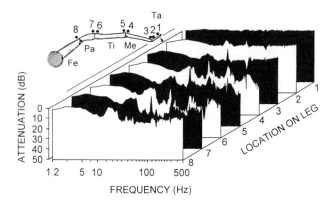

FIGURE 7.5. *Vibration transmission through spider leg (Cupiennius salei).* Attenuation of vibrations introduced at the tarsus (arrow) on their way to the femur. The velocity of movement was measured perpendicular to the cuticular surface with a laser vibrometer at locations 1 to 8. *Ta*, tarsus; *Me*, metatarsus; *Ti*, tibia; *Pa*, patella; *Fe*, femur. (From Barth and Bohnenberger, unpublished, with permission.)

21 slits of the metatarsal organ behaved like high-pass filters (measured between 0.1 Hz and up to 3 kHz). Their sensitivity at threshold is 10^{-3} to 10^{-2} cm up to about 10 to 40 Hz. Displacement values decrease rapidly at higher frequencies by up to 40 dB/decade and reach 10^{-6} to 10^{-7} cm at 1 kHz (Fig. 7.6). At low frequencies threshold curves roughly follow constant displacement of the tarsus, whereas at higher frequencies (beyond the bend in the curve) they follow constant acceleration. This seems to make sense. When measuring vibrations with a technical transducer, one would measure displacement (d) rather than acceleration (a) at very low frequencies because of the small acceleration values to be expected. At high frequencies, however, acceleration is high, even at very small displacement values, and would be the preferred parameter to be measured ($d = a/4\pi^2 f^2$, where f is frequency).

At least some of the slits not only respond to up and down movements but also to lateral movements of the tarsus with a very similar threshold curve (Barth and Geethabali 1982). The spider is therefore capable not only of picking up transverse waves (movement perpendicular to plane of substrate) but also longitudinal waves. This may be important in orb webs, in which both longitudinal and transverse vibrations occur (Landolfa and Barth 1996).

Vibration sensitivities of different ecotypes of spiders do not obviously differ from each other. Threshold curves of spiders living in an orb web (*Zygiella x-notata*, Liesenfeld 1961), in a densely woven sheet web (*Tegenaria*, Liesenfeld 1961), and on the water surface (*Dolomedes triton*, Bleckmann and Barth 1984) are all very similar (Barth and Geethabali 1982). Threshold curves are also largely independent of the stiffness of the

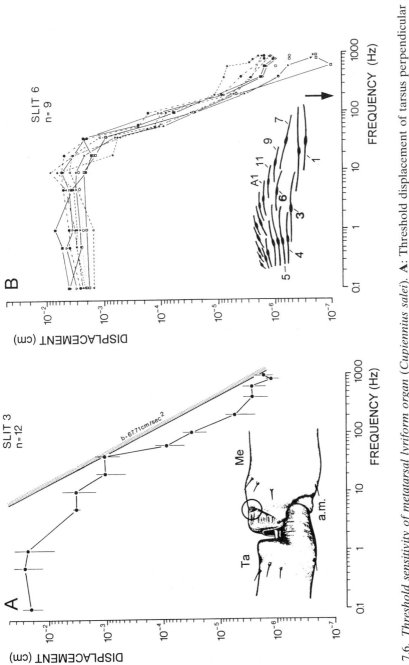

FIGURE 7.6. *Threshold sensitivity of metatarsal lyriform organ (Cupiennius salei)*. **A**: Threshold displacement of tarsus perpendicular to substrate surface needed to elicit a spike response in slit 3. Straight line gives displacement at constant acceleration of 67.7 cm/s². Inset: position of vibration receptor on leg. *Ta*, tarsus; *Me*, metatarsus; a.m., articular membrane. **B**: Threshold curve of slit 6, determined for nine different preparations. Inset: Arrangement of slits within metatarsal organ (dorsal view) of *C. salei*. Arrow points toward the leg tip (tarsus). (Adapted from Barth and Geethabali 1982, with permission.)

metatarsus–tarsus joint, which differs considerably among these spiders (van de Roemer 1980).

When using band-limited noise stimuli with a bandwidth of one-third octave ($Q = 0.35$)[2] instead of sinusoidal stimulation of the tarsus, the threshold values of metatarsal slits decrease by up to 10 dB in *Cupiennius salei*. This finding is paralleled by a similar increase of sensitivity of behavioral reactions and of the response of vibration-sensitive interneurons (Barth 1985a, 1993; Speck-Hergenröder and Barth 1987). Because natural vibrations, in particular prey signals, are never purely sinusoidal, this finding is of great behavioral significance (see later).

A comparison of the absolute vibration sensitivity of *Cupiennius* and other spiders (see earlier) with that of other animal groups known to be particularly vibration sensitive demonstrates that spiders do very well. Recall that thresholds for tarsal *displacement* are as low as 10^{-6} to 10^{-7} cm at stimulus frequencies between 400 Hz and 1000 Hz, and 10^{-2} to 10^{-3} cm at low frequencies between 0.1 Hz and about 40 Hz. Thresholds for tarsal *acceleration* are lower than 0.8 cm/s^2 (at about 100 Hz, i.e., the dominant frequencies of the male courtship signal; Barth and Geethabali 1982; Baurecht and Barth 1992). This is also the behavioral threshold for the female response to the male vibration (Schüch and Barth 1990).

For comparison, scorpion metatarsal organs respond to square displacement pulses of less than 10^{-7} cm (10 Å; Brownell and Farley 1979). In leaf-cutting ants, again a displacement threshold of 10^{-7} cm (acceleration 2 to 3 cm/s^2 at 100 Hz) was found to be necessary to still get responses from vibration-sensitive cells (Markl 1969). Grasshoppers and cydnid bugs respond to accelerations of about 1 cm/s^2 at the receptor level (Devetak, Gogala, and Čokl 1978; Kalmring, Lewis, and Eichendorf 1978) and ghost crabs with vibration receptors differing from both arachnid metatarsal organs and insect subgenual organs to 0.1 cm/s^2 (at frequencies between 1 and 3 kHz; Horch and Salmon 1969). More examples are found in reviews (e.g., Markl 1973).

According to Autrum (1941) and Schnorbus (1971), the real champions are orthopterans such as cockroaches and tettigoniids. Their subgenual organs responded to displacements of the substrate as small as 10^{-9} cm and even 4×10^{-10} cm (0.04 Å). This is roughly 100 times more sensitive than was found in scorpions and spiders. However, this low threshold recently has been revised upward by two powers of 10 (Shaw 1994). The value now

---

[2] Q values are determined by dividing the maximally effective value of a given parameter (in the given case, the frequency peak of the frequency spectrum) by the 3-dB bandwidth. The 3-dB bandwidth is the range of values along the x-axis that is received by drawing a horizontal line at a distance of 3 dB from the maximum of the curve. Hence the Q value is without dimension. It is used in engineering to describe the steepness of electronic filters.

postulated is in the same range as that measured for the vibration-sensitive subgenual organ of a cricket hindleg (Dambach 1989). It is still remarkably low, considering that it is very similar to the basilar membrane displacement needed to stimulate a vertebrate cochlear hair cell and, indeed, the hair-bundle deflection at threshold (ca. 0.4 nm or 4 Å; Sellik, Patuzzi, and Johnstone 1982; Pickles and Corey 1992). This threshold is limited by the mechanical noise caused by Brownian motion (Sivian and White 1933; Hudspeth 1985, 1989) and thus at the limit of the physically possible.

Among the vertebrates, white-lipped frogs (*Leptodactylus albilabris*), which communicate with seismic signals, are particularly vibration sensitive. Accelerations of the sustratum (main frequency components between 20 and 70 Hz) at levels as low as $0.001 \, cm/s^2$ effectively elicit responses in the auditory vestibular nerve. The corresponding displacement value is ca. $1 \times 10^{-6} \, cm$ (100 Å). Single fibers originating in the saccule saturate at whole-animals displacements of only 10 Å peak to peak (Narins and Lewis 1984; Lewis and Narins 1985; Narins 1995). This is much lower than the values known of warm-blooded vertebrates: Acceleration thresholds at best frequencies are, for example, $40 \, cm/s^2$ at 300 Hz for ducks (Dorward and McIntyre 1971), $55 \, cm/s^2$ at 400 Hz for the bullfinch (Schwarzkopff 1948), and $4 \, cm/s^2$ at 100 Hz for the human (Keidel 1956). In cases in which the somatosensory system provides "seismic" sensitivity, displacement values at threshold in the best frequency range are comparatively high as well. Examples are lamellated corpuscles in the kangaroo leg (0.4 to 6 μm, 250 to 400 Hz; Gregory, McIntyre, and Proske 1986), Herbst corpuscles of the pigeon wing (0.1 μm, ca. 800 Hz; Höster 1990), and Pacinian corpuscles of humans, which need a compression of 0.5 μm, even when stimulated directly with a rapidly increasing stimulus (most effective frequencies 200 to 300 Hz; McIntyre 1980).

The overall conclusion from the above-mentioned comparison is that the spider metatarsal organ, together with hair-cell–based organs in vertebrates and insect subgenual organs, is among the most sensitive vibration receptors known, and that the threshold values of these are rather similar.

5. Signal Production

Spiders produce courtship and rival "songs" like those of famous singers such as crickets (Schmitt, Schuster, and Barth 1992, 1994; Barth 1993). There is true communication in the sense of reciprocal signaling. Mechanisms of signal production are highly developed in spiders and include stridulation. Unfortunately, there is not much known about these mechanisms, which is in striking contrast to the many hints to obviously interesting phenomena in the literature. We shall first consider the only case of a more detailed experimental analysis and then add a glimpse at other ways by which spiders produce vibrations.

5.1. Courtship Vibrations of Cupiennius

The elaborate male courtship signal of *Cupiennius*, with its syllables reminiscent of cricket song, is due to vibrations of the opisthosoma (Rovner and Barth 1981; see Fig. 7.2). When vibrating, the opisthosoma does not hit the plant. Instead, the vibrations are introduced into the plant through the spider's legs. The same is true for many other spiders, and indeed other arthropods, although actual drumming seems to be more widespread among insects and crustaceans (Markl 1969, 1983; Barth 1982, 1985a; Uetz and Stratton 1982; Gogala 1985; Ewing 1989). Whereas the signals produced in this way were described several times, there is no detailed study on the mechanisms involved in their production, except for the spider case (Dierkes and Barth 1995).

Males of *Cupiennius getazi* produce opisthosomal signals much like *C. salei* (Barth 1993). The main frequency component of these signals, which are essential for species recognition and come in series of syllables, is around 80 Hz. According to high-speed video analysis, the bobbing movements of the opisthosoma producing the syllables are dorsoventral around an axis within the pedicel. The opisthosomal deflection angle is only about 2° in the beginning of a series and increases to about 30° (displacement at spinnerets 6 mm) during the terminal three to four syllables. These terminal syllables effectively elicit the female's response (Schmitt, Schuster, and Barth 1992) and were therefore further analyzed using laser vibrometry.

Two frequency components characterize the bobbing movements of the opisthosoma: (a) a "low-frequency component" of 10 to 20 Hz, and (b) a "main frequency component" of low amplitude (ca. 0.5 mm, 2.5°) and about 80 Hz superimposed onto the last upward motion of the opisthosoma at the end of every syllable (Fig. 7.7). It is this 80-Hz component which leads to the maxima of the substrate vibration. Accordingly, not only the temporal pattern of the opisthosomal movement, but also its frequency spectrum is very similar to that of the plant movement (acceleration). The plant does not resonate in the frequency range of the male courtship signal. When measuring the transfer function of the spider's own body by transmitting plant vibrations through its body, resonances were found between 30 and 100 Hz, with peaks at ca. 50 Hz and 80 Hz, that is, in the frequency range of the courtship vibration. These resonances could amplify the main frequency component of the signal before it is transmitted into the plant. Frequencies above 200 Hz are strongly attenuated by the spider body.

The 80 Hz component in the opisthosomal movement originates from the activity of the opisthosoma depressor muscle 85, one of 36 muscles attached to the pedicel (for details, see Dierkes and Barth 1995). This muscle is comparatively large and is active only during courtship. Its activity closely matches the motion of the opisthosoma. Each syllable is accompanied by several muscle potentials (see Fig. 7.7), culminating in a discharge frequency of about 200 Hz and the tetanizing of the muscle for about 40 ms.

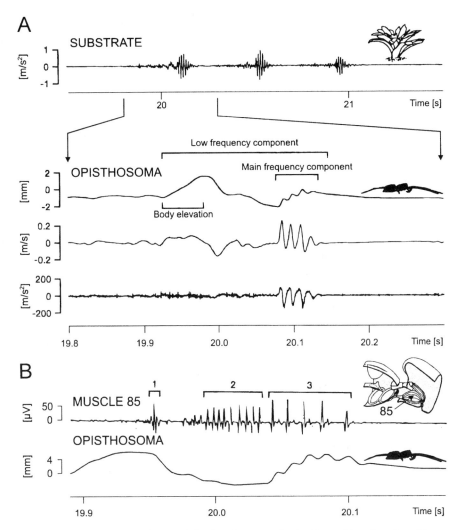

FIGURE 7.7. *Signal production* (*Cupiennius getazi*). **A**: Courtship vibration (three syllables shown) on bromeliad (upper trace) and on the spider opisthosoma itself (lower three traces, representing displacement, velocity, and acceleration, respectively). **B**: Spike activity of opisthosoma depressor muscle 85 (upper trace) and corresponding movement of opisthosoma. Muscle activity: *1*, single action potential; *2*, action potential frequency about 200 Hz; *3*, action potential frequency at 80 to 90 Hz, which corresponds to the main frequency component of male courtship vibration. (Adapted from Dierkes and Barth 1995, with permission.)

When the opisthosoma is moved up again due to the activity of a lavator muscle (not yet identified), the depressor, muscle 85, contracts again three to five times at intervals of 11 to 12 ms, that is, at a frequency of 80 to 90 Hz. These contractions, then, lead to the main frequency component of the male courtship vibration. There is no special mechanism, such as stridulation, involved in frequency multiplication.

5.2. Other Ways to Vibrate

There are many other ways to produce vibrations in arthropods. Ewing (1989) distinguished (a) percussion of body parts against each other or the substrate, (b) tymbal or click mechanisms, (c) stridulation, (d) air expulsion, and (e) vibration of the body or its appendages. In spiders leg oscillation (Rovner 1980; Schüch and Barth 1985), leg rubbing, and leg waving have been observed in addition to pedipalpal and opisthosomal movements. All of these vibration-producing mechanisms may be found in the same species (Roth 1986). There are no tymbal nor air-expulsion mechanisms known for spiders but many cases of percussion and stridulation. The best known drummers are lycosids (wolf spiders). They use their pedipalps to produce sound during courtship (not unlike *Cupiennius*). One of them, *Lycosa gulosa*, is referred to as the "purring spider" because the sound it produces is audible to the human ear over distances of several meters (Kaston 1936; Harrison 1969). Similarly, *Hygrolycosa rubrifasciata*, a wolf spider (Lycosidae), drums on leaf litter with a sclerotized cuticular plate on the ventral surface of its opisthosoma (Köhler and Tembrock 1987), and the resulting sound can be heard by the human observer. Rovner (1975) demonstrated that the palpal drumming of many lycosid males belonging to the genera *Lycosa* and *Schizocosa*, and including *L. gulosa*, is more complicated than previously thought and is better defined as substratum-coupled stridulation. There is a stridulatory organ at the tibio-tarsal joint of the palps of these species, with the file on the tibia and the scraper on the tarsus. The palps do not drum but oscillate anterio-posteriorly, with flexions and extensions in the tibio-tarsal joint. Vibrations are conducted into the substratum by stout spines on the palpal tarsus. The behavioral significance of these vibrations in courtship and in reproductive isolation is well documented (Rovner 1967; Uetz and Stratton 1982; Stratton and Uetz 1983).

Stridulation is not only known for wolf spiders but also for many other species belonging to a total of 22 families (Legendre 1963; Kronestedt 1973; Uetz and Stratton 1982; Starck 1985). The Theridiidae are another group in which stridulatory courtship behavior (in the web) has received attention (Meyer 1928; Braun 1955; Gwinner-Hanke 1970). In some large mygalomorph spiders (among them bird and trapdoor spiders, Ctenizidae and Theraphosidae), stridulation seems to be a threat behavior but has not been studied in detail (review, Starck 1985; Marshall, Thoms, and Uetz 1995). In addition to the palpal type of stridulatory organ, there are many others involving different parts of the body. Starck (1985) distinguishes 12 morphological types of stridulatory organs, elaborating on the types summarized by Legendre (1963). The prosoma, opisthosoma, chelicerae, pedipalp and walking leg coxae, and the petiolus serve as pars stridens in different species and may also carry the plectron.

According to the sparse literature, substrate-borne vibrations are the most likely signals produced by stridulation in spiders. This does not rule out the significance of airborne signals in specific cases. Hearing in the sense of far-field pressure reception is the exception in most cases of stridulation, whereas air particle movement in the nearfield is a rather likely stimulus for the trichobothria in many cases (Barth et al. 1993).

6. Signals and Noise

Several basic questions regarding the vibration sense of spiders can only be answered with some knowledge of the vibrations spiders are exposed to under biologically relevant conditions. Such questions are related to problems of signal recognition and distinction, to the orientation toward sources of vibration, and to the spatial range covered by the vibration sense.

6.1. Types of Vibrations

Figure 7.8 gives representative examples of vibrations that spiders will be exposed to on plants (e.g., *Cupiennius*), on the water surface (e.g., *Dolomedes*), and in the orb web (e.g., *Nephila*). On the plant and on the water surface, the most obvious difference between abiotic background vibrations due to wind- and prey-generated signals is the much broader spectrum of the latter and their high-frequency content. Courtship vibrations, on the other hand, have a prominent temporal structure, whereas their frequencies are intermediate between vibrations induced by wind and prey. In the orb web (Masters and Markl 1981; Masters 1984b; Landolfa and Barth 1996) struggling prey produced vibrations, with most of the energy at frequencies below 50 Hz. Components above 100 Hz become important in the case of buzzing insects. Vibrations due to wind show narrower spectra and energy peaks below 10 Hz. It seems to be a rule then, applicable to web-borne vibrations as well, that low frequencies and small bandwidths are characteristic of the vibratory background, whereas broad spectra having high-frequency components are characteristic of prey signals. The amplitudes of all these vibrations vary to a large extent, depending on the intensity of the signal emitted and the distance and geometrical configuration between the sender and the receiver.

6.2. Discrimination

There is good behavioral evidence that spiders can indeed discriminate among the types of vibrations described in the preceding paragraph. Noisy signals, even if limited to a small frequency band (one-third octave; $Q = 0.35$; see footnote 2), are significantly more effective in eliciting a prey-capture response in *Cupiennius salei* than are sinusoidal vibrations of the

PLANTS

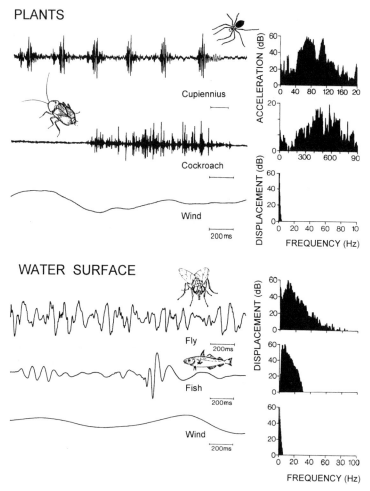

WATER SURFACE

ORB WEB

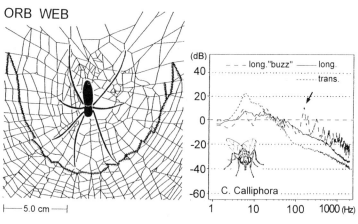

same center frequency (Hergenröder and Barth 1983a). Likewise, the fishing spider (*Dolomedes triton*) much more often responds to broad-band wave stimuli than to narrow-band stimuli (Bleckmann 1994). Experiments with stimuli of varying frequency and constant amplitude, and vice versa, provided behavioral evidence that *Cupiennius salei* is capable of frequency and amplitude discrimination (Hergenröder and Barth 1983a). The hypothesis that nonsinusoidal signals, made up of a wide range of frequencies and irregular in the time domain, are recognized as prey signals is supported by two more observations.

1. Watching *Cupiennius* in the field, we have noticed that sometimes otherwise attractive prey passes by the spider as if unnoticed, even when almost touching its legs (Barth et al. 1988b). The vibrations produced in such cases have low amplitude and, more importantly, do not contain any high-frequency components. As a matter of fact, the potential prey animals seem to effectively avoid transients by a smooth, slow, and regular "sinusoidal" gait. The same habit has paved the way for some insects and spider kleptoparasites into the webs of other spiders (review Barth 1982). Vollrath (1979a,b) even found that *Argyrodes* (Theridiidae, cobweb weavers) (a spider intruding the orb web of *Nephila*) guards against a sudden release of tension when it cuts out the stolen prey from the host's web, thus avoiding fast transients (high-frequency vibrations). Low frequencies and the lack of fast transients may also be one of the reasons why young spiderlings tolerate each other or feed as commensals in their mother's web without being eaten (Tretzel 1961) and why *Dolomedes* detects prey signals of only a few micrometers on top of wind-induced surface waves of more than 1 cm.

2. The second additional argument supporting our hypothesis takes us back to neurobiology. According to their threshold curve, the slits of the metatarsal organ are displacement receptors at low frequencies (i.e., up to ca. 10 to 30 Hz) and are very insensitive in this range (see Fig. 7.6; Barth and Geethabali 1982). We conclude that, as a consequence, much of the low-frequency background vibrations are kept out of the vibration processing system. This is not the case for the higher frequency ranges representing

---

FIGURE 7.8. *Natural vibrations.* Above: *On plants* (from above to below): male courtship vibrations (acceleration) of *Cupiennius salei*, vibrations (acceleration) on banana plant pseudostem due to crawling cockroach, and wind-induced vibrations (displacement of banana plant). Middle: *On water surface*: Vibrations (displacement) generated by a fly entrapped in the water surface, by a fish touching the surface from below, and by wind. Below: *In orb web* (*Nephila clavipes*): a fly, *Calliphora vicina*, caught in the web and struggling to free itself; *long*, longitudinal and *trans*, transverse vibrations (velocity) of a radius measured with laser vibrometer at a distance of 16 cm and 13.5 cm from the fly, respectively. *Long* "*buzz*", fly beating with its wings during measurement (arrow). (A and B adapted from Barth 1985a, with permission. C from Landolfa and Barth 1996, with permission.)

courtship and prey signals. Recordings from interneurons of the *Cupiennius* central nervous system that are sensitive to substrate vibration have demonstrated a consistent difference from the primary receptor cells. Whereas the latter show high-pass characteristics (with regard to displacement), threshold curves of the interneurons show band-pass characteristics. At least some of them have best frequencies in the frequency ranges of courtship and prey signals, respectively. On the other hand, all interneurons so far recorded from (a difficult undertaking in spiders) were very insensitive at the low frequencies representing background vibrations (Fig. 7.9) (Barth 1986; Speck-Hergenröder and Barth 1987).

The current concept is thus a two-step filter. The first component (primary sensory cells) keeps out low-frequency components, and the second component (interneurons) picks out biologically important frequencies. Both the sensory neurons and the interneurons are significantly more sensitive to "noisy" signals than they are to purely sinusoidal stimulation, which supports these ideas. Sensitivity increases up to 20 dB by applying a frequency band that is only one-third octave broad ($Q = 0.35$) compared with monofrequency stimulation (Speck-Hergenröder and Barth 1987).

Courtship vibrations clearly form a class of their own. Various species of *Cupiennius* were used in a detailed case study of their significance in the spiders' precopulatory behavior (review Barth 1993). The male vibration sent through the dwelling plant to the female is important in ensuring species isolation. The female is tuned to various parameters contained in the male signal. This was examined extensively by using synthetic and systematically altered male signals and by observing the female's behavioral response to them (Schüch and Barth 1990). The female's readiness to respond with her own vibration can be taken as a measure of the attractiveness of variously modified male signals. The female's releasing mechanism has been shown to be rather narrowly tuned, both to the carrier frequency and to several of the temporal parameters contained in the male signal. It is indeed the temporal structure which most obviously distinguishes the male signals of different species of *Cupiennius* from each other.

Using 3-dB ranges and Q values (see footnote 2 for definitions) to quantify the narrowness of the tuning, the durations of the male syllable (SD) and the silent pause (PD) in between them were found to be the most influential parameters, apart from the carrier frequency. For the signals to be maximally effective, the values of SD and PD should form particular combinations, expressed by the duty cycle (DC), which represents the relative share (as a percentage) of SD of SD + PD (Fig. 7.10A). Not surprisingly, the effective values of the various parameters used in the synthetic signals are close to those in the natural signals. Likewise, the limited tolerance of the female to variation in the male signal is reflected by the difference in SD and PD values among different species of *Cupiennius* (Fig. 7.10B). Both the SD and the PD (and therefore also the duty cycle) are

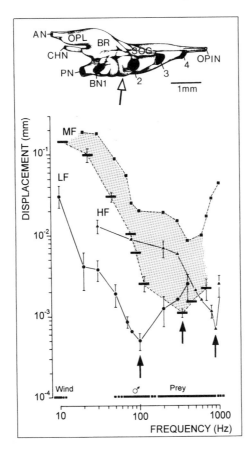

FIGURE 7.9. *Interneurons*. Threshold curves of vibration-sensitive interneurons recorded in the subesophageal ganglionic mass of the central nervous system of *Cupiennius salei*. Note that all cells have an optimum frequency (arrows), which is found in different spectral ranges. Applying small frequency bands (■■, one-third octave) instead of sinusoidal vibrations (■, same neuron) to the tarsus shifts thresholds to considerably lower values (see shaded area). *LF, MF, HF*: low, medium, and high frequency interneuron. Bars below threshold curves indicate frequency ranges of naturally occurring vibrations. The inset above shows the central nervous system of *Cupiennius salei* and the area (arrow) recorded from. *AN*, optic nerve; *CHN*, cheliceral nerve; *PN*, pedipalpal nerve; *BN1*, nerve of first walking leg; *OPIN*, opisthosomal nerve; *OPL*, optic lobe, *BR*, brain; *SOG*, subesophageal ganglionic mass. (Adapted from Barth 1986, with permission.)

largely independent of ambient temperature (Shimizu and Barth 1996). This finding underlines the importance of these parameters in the species-recognition process accomplished by the female. Both the temporal pattern and the carrier frequency of the male courtship signal (the opisthosomal signal, see later) remain largely unchanged on their way through the plant,

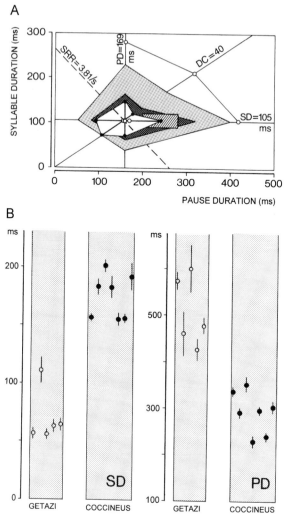

FIGURE 7.10. *Effective signal parameters.* Synthetic male courtship signals. (*Cupiennius salei*), such as those seen on Figures 7.8A and 7.11, were systematically varied to determine which parameters are the most influential in eliciting a female vibratory response. **A**: Combinations of syllable duration and pause duration are most effective if in the innermost white area (70.7% to 100% responses), and less so in the more peripheral areas indicated by different shading (50% to 70.7%; 10% to 50%). The rectangular hatched area includes the combinations of syllable and pause duration typically found in natural signals. *SRR*, syllable repetition rate; *PD*, pause duration; *DC*, duty cycle; *SD*, syllable duration. (From Schüch and Barth 1990, with permission.) **B**: Difference in the values for syllable duration (SD) and pause duration (PD) in the male vibratory signal of two sympatric species (*Cupiennius getazi*; *C. coccineus*). Each value represents the average taken from five (*C. getazi*) or seven (*C. coccineus*) courtships of four animals of each species. Vertical bars give standard errors. (Adapted from Barth 1993, with permission.)

although both amplitude and overall frequency contents of the male vibration change considerably (Baurecht and Barth 1992).

According to electrophysiological experiments on the metatarsal vibration receptors with both natural and synthetic male signals, much of the female filtering occurs in the sensory periphery (Baurecht and Barth 1992, 1993). Examples are the following:

1. The courting male produces two kinds of vibrations: by scratching and drumming against the leaves of the dwelling plant with his pedipalps and by vibrating his opisthosoma (Schüch and Barth 1985). It has been known since the work of Schüch and Barth (1990) that opisthosomal signals are both necessary and sufficient to elicit a response of a female *Cupiennius salei*. The pedipalpal signal contains high-frequency components (>1 kHz). These are, however, attenuated very strongly (>30 dB) on their way through the plant, whereas low-frequency (<200 Hz) components are well represented, even when having traveled to the tip of a contralateral leaf (Fig. 7.11). A possible function of the pedipalpal signal is to inform the female about the distance to the male, which is indicated by the presence or absence of high-frequency components.

The slits of the metatarsal lyriform vibration receptor process these two types of signals in parallel. Whereas pedipalpal signals elicit responses from all slits, opisthosomal signals mainly elicit responses from long distal slits (Baurecht and Barth 1992). Parallel processing continues in the central nervous system (Friedel and Barth 1995; Fig. 7.12).

2. Typically, the syllables of male courtship signals of *Cupiennius salei* come in a series. At least three consecutive syllables are needed to elicit a vibratory response in more than 50% of the females, and the effectiveness increases up to 12 syllables/series (Schüch and Barth 1985, 1990). Electrophysiological studies suggest that one of the advantages gained by the repetition of syllables is the reduction of the signal-to-noise ratio in the receptor response. In addition, the receptor response to a series of syllables is marked by a poststimulus depression (Baurecht and Barth 1992). This finding reflects the importance of breaking up long trains of syllables into shorter series in behavioral experiments: a train of 1000 consecutive syllables never elicited more than two female responses (Schüch and Barth 1990).

3. *Signal amplitude* is one of the less important parameters of the male courtship signal with regard to its effect on the number of female behavioral responses. This is not surprising if one considers the variability of signal amplitudes under natural conditions. The logarithmic relation between stimulus level and the number of action potentials elicited in long slits of the metatarsal organ (see earlier) enables the receptor to represent nearly the whole natural range of courtship signal amplitudes. At the same time, however, it reduces the ability to discriminate different acceleration amplitudes (Baurecht and Barth 1993). In order to elicit a female behavioral response, acceleration amplitude not only has to be above a lower threshold

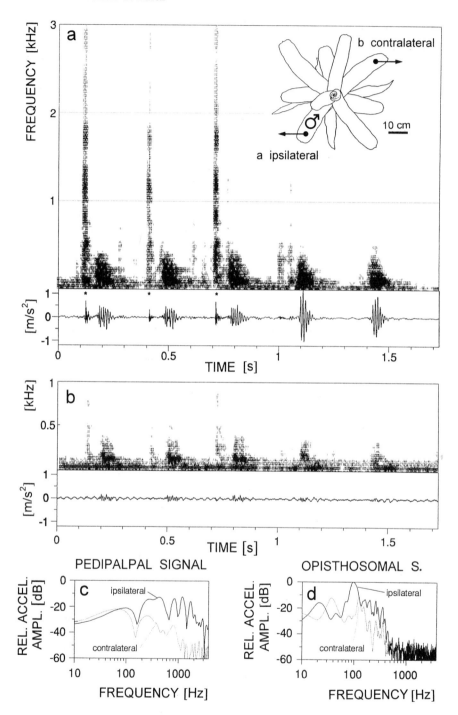

PEDIPALPAL SIGNAL

OPISTHOSOMAL S.

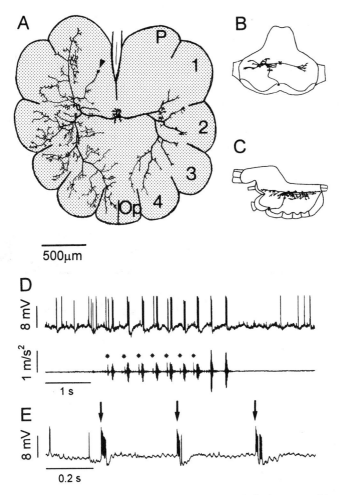

FIGURE 7.12. *Interneurons.* Response of female bilateral plurisegmental interneuron in subesophageal ganglionic mass to male courtship vibration (*Cupiennius salei*) (**D**) and to airpuffs (arrows in **E**). **A, B,** and **C** show the dorsal, frontal, and sagittal views of a cobalt-filled neuron. Note that the neuron responds reliably (upper trace in D) only to the low-frequency syllables, which are caused by oscillations of the opisthosoma but not reliably to the high-frequency pedipalpal signals. (*, lower trace of D showing the vibratory signal.) *1–4*, leg neuromeres; *P*, pedipalpal neuromere; *Op*, opisthosomal neuromere. Scale bar in A, 500 µm. (From Friedel and Barth 1995, with permission.)

FIGURE 7.11. *Signal attenuation on plant.* **a**: "Sonagram" and oscillogram (accelera-tion) of courtship vibrations (*Cupiennius salei*) introduced by a male into a brome-liad at *a* (see inset, ipsilateral leaf). Vibrations produced by pedipalps are marked with *. **b**: The same vibrations recorded on the contralateral leaf at *b* after having traveled through the plant. **c**: and **d**: Damping of pedipalpal and opisthosomal signals. Note enormous damping of pedipalpal signals, which contain high-frequency components only on the ipsilateral leaf as seen in the FFTs (fast fourier transforms). 0 dB corresponds to the highest amplitude in the FFT of the opisthosomal signal. (From Baurecht and Barth 1992, with permission.)

value (ca. $8\,\text{mm/s}^2$, Schüch and Barth 1990) but also below an upper limit, which is indeed the limit of the natural range ($>1000\,\text{mm/s}^2$). When receiving signals larger than this, spike activity occurs in between the syllables (decrease of synchronization factor) in the receptor cells of the metatarsal organ and precise information on the syllable and pause durations is no longer available. The rapid decrease of the behavioral response probability at amplitudes above the natural range may therefore simply be due to the misrepresentation of the temporal structure of the male signal in the sensory periphery (Baurecht and Barth 1993; Fig. 7.13).

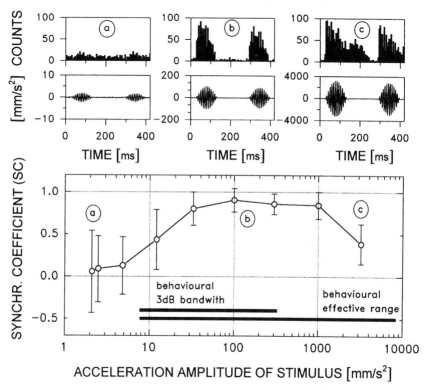

FIGURE 7.13. *Metatarsal vibration receptor and signal amplitude.* Response of female vibration receptor (slit 2, see inset in Fig. 7.6B) to synthetic male courtship vibrations of increasing acceleration amplitude. Uppermost row of graphs: peri–stimulus-time histograms of spike response. Counts refer to 10 stimulus series, each consisting of 10 male syllables, 2 of which are shown in the second row of graphs. The lower graph shows the synchronization coefficient SC of the response. SC = 1 indicates perfect copying of syllable and pause durations and SC = 0 indicates no copying. Note the decrease in SC with very low and very high values of stimulus amplitude. SC = $(ns/ds - np/dp)/(ns/ds + np/dp)$; ns (np), number of spikes during syllable (pause); ds (dp), duration of syllable (pause). (From Baurecht and Barth 1991, with permission.)

6.3. Orientation

6.3.1. Angle

When a hungry spider is exposed to a vibratory prey signal, it turns toward the source of stimulation. Obviously, it is able to determine the target angle. This has been demonstrated for all three spider substrates: web, plant, and water surface (Klärner and Barth 1982; Hergenröder and Barth 1983b; Bleckmann and Barth 1984). The case studies available tell us how *angular orientation* may be brought about.

6.3.1.1. Orb Web

Zygiella x-notata, the sector spider, is an orb weaver. It localizes a vibratory stimulus source with a mean error angle (stimulus angle minus turning angle) of only 3.6° (±7.7°). *Nephila clavipes*, another orb weaver, is rather precise as well (7.0° ± 8.2°) (Klärner and Barth 1982). When turning toward the stimulus in the hub, the movements of both these species contain only a small translatory component: they turn and then run in a straight line toward the vibration source. Although there is a moderate increase in error angle with stimulus angle in *Nephila* (stimulus angle 0° to 50°: 3.9° ± 5.6°; stimulus angle 50° to 100°: 10.1° ± 9.1°), the angular precision of the turn is high over the whole range of stimulus angles tested (280° for *Zygiella*; 200° for *Nephila*).

Which clues may the spiders use to determine the angle toward a single-point stimulus in their webs? The vibrations are not confined to the stimulated radii but spread laterally across the web. As a consequence, both amplitude and temporal gradients develop in the web, which may contain the information used by the spider. We have measured this in some detail in the *Nephila* web (Landolfa and Barth 1996). Attenuation is least in the stimulated radius and increases progressively in more distant radii. The same applies to all three vibration types (longitudinal, transverse, lateral). The resulting gradients, however, are steepest for longitudinal waves, whereas they are similar for the other two wave types (see also Masters 1984a). This can be quantitatively expressed by the angular space constants (Fig. 7.14). In order to know exactly what gradient is available to the spider, a radius was vibrated longitudinally and the vibration of the tarsi of different legs was measured directly. The gradients are on the order of 20 dB and more and are therefore considered the primary candidates for being the cues used in orientation.

The use of *time-of-arrival* and/or *phase cues* is much more uncertain in web spiders because, in contrast to the water surface and plants (see later), the propagation speed of vibrations is high. Using rectangular pulses of 1 ms duration, we measured 986 ± 390 m/s for longitudinal vibrations and 129 ± 7 m/s and 207 ± 56 m/s for transverse and lateral vibrations, respectively, in different threads. Whereas the first value is only about one-half the

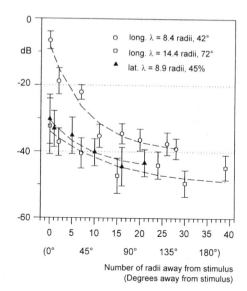

FIGURE 7.14. *Vibration gradients in the web.* Comparison of gradients among longitudinal, transverse, and lateral vibration in the web of *Nephila clavipes*, the golden silk spider. λ, angular space constant, or the number of radii (angular extent) over which levels of vibration fall to 1/e, or 37%, of their initial values. (From Landolfa and Barth 1996, with permission.)

velocity predicted for isolated strings (Frohlich and Buskirk 1982), the values for transverse and lateral vibrations are within the ranges of the calculated values. The conversion of these velocities into maximum time-of-arrival differences at the tarsi of a *Nephila* with a leg span of 10 cm yields values of 0.1 ms (longitudinal), 0.78 ms (transversal), and 0.48 ms (lateral). It is not known yet whether *Nephila* is able to resolve such small time intervals. For *Nephila* spiderlings or the small adults of *Zygiella*, the situation would be even more demanding because time-of-arrival differences are correspondingly smaller (by a factor of 0.1 and less). A further complication may be the high variability of the propagation velocities within the same and across different threads, even if the spider determines stimulus direction only by determining which of its legs was stimulated first.

Active localization is used by orb weavers such as *Zygiella* and *Nephila* to locate objects like prey or dirt hanging motionless in their web. The spiders vibrate the radii with their forelegs and are believed to use information derived from the vibratory echo but not from thread tension. *Zygiella* was shown to detect a 0.4 mg particle 6 cm away from the hub in 75% of the cases; its record was a particle of only 0.05 mg (Klärner and Barth 1982).

6.3.1.2. Plant

When *Cupiennius salei*, the wandering spider, is standing with one or several of its legs on a platform that vibrates, it turns toward the source of vibration if the stimulus resembles prey signals. Its turning angle much depends on the leg or combination of legs stimulated. When the forelegs are stimulated, the error angle is considerably smaller than when the hindlegs are stimulated. By systematically varying the combination of legs stimulated (simultaneously and with same stimulus amplitude), the following rules for the central nervous interaction of the inputs from the eight legs in determining the turning angle could be found (Hergenröder and Barth 1983b; Fig. 7.15): (a) The stimulus angles (angle between long axis of spider and the line connecting the stimulus site and the center of the prosoma) are multiplied by a constant weighting factor F, which is largest for the first leg (1.2) and smallest for the fourth leg (0.4). (b) There is additive ipsilateral inhibition: stimulation of the first leg reduces F of all ipsilateral posterior legs by 0.1. (c) There is also contralateral inhibition when legs on both sides are stimulated. This inhibition acts only from front to back and is multiplicative, reducing F by a factor of 0.5 in all cases. The connectivity diagram developed for *Cupiennius* is thus an inhibitory network. It allows a quantitative prediction of the turning angle upon stimulation of a leg or of leg combinations with substrate vibrations.

Both ipsilateral and contralateral anatomical connections of the sensory neurons found in the central nervous system support the model (Babu, Barth, and Strausfeld 1985; Babu and Barth 1989; Anton and Barth 1993). For a comparison of this model with similar networks postulated to represent the central nervous interactions of the sensory inputs from the legs in a scorpion (*Paruroctonus*) and a backswimmer (*Notonecta*), see Hergenröder and Barth (1983b). The main message from these results is that angular orientation is possible by the connectivity of the inputs from the different legs, even without differences in amplitude or time of arrival.

Under natural conditions, however, both amplitude and time differences are to be expected and *Cupiennius* is able to make use of them for angular orientation (Hergenröder and Barth 1983b). This was demonstrated by the simultaneous stimulation of different legs with different vibration amplitudes (Δd), or their stimulation with identical amplitude but a small time shift (Δt). Values of 4 ms for Δt and of 10 dB (smaller differences not tested yet) for Δd had marked effects: The spider turns as if only the leg that received the stronger or earlier stimulus had been stimulated. Even at a Δt of 2 ms, most turns are toward the leg stimulated earlier. There are interneurons in the subesophageal ganglionic mass that receive sensory input from several legs and show reactions to Δt and Δd similar to the spider's behavior (Speck-Hergenröder 1984; Wirth 1984; Speck-Hergenröder and Barth 1987).

How do these values relate to what a *Cupiennius* is likely to experience in the field? The time-of-arrival differences at the eight legs much depend on stimulus frequency, due to the dispersive nature of signal propagation, and on the specific location of the spider on, let us say, an agave or bromeliad leaf. With the female courtship signal particularly in mind, which guides the male to the female and has its main frequency peak close to 30 Hz, we have

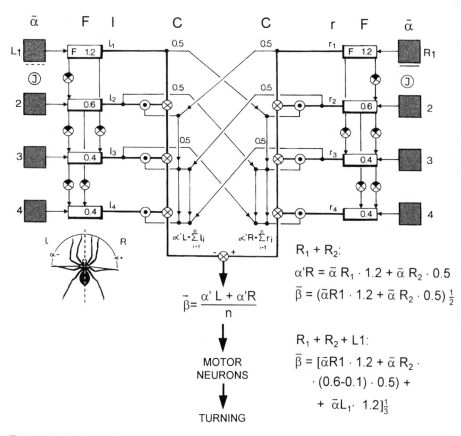

FIGURE 7.15. *Central nervous interaction.* Behavioral studies in which the turning angle of the wandering spider *Cupiennius salei,* in response to substrate vibrations at individual or selected groups of legs, was determined reveal some formal rules governing the central nervous interaction of the inputs from the vibration receptors of the eight legs. ⊗, Addition; ⊗, subtraction; ⊙, multiplication; $\bar{\alpha}$, mean stimulus angle; F, factor; r, product of stimulus angle and F; I, additive, ipsilateral inhibition; C, multiplicative, contralateral inhibition. The stimulus angles α of the right ($R_1 - R_4$) and left ($L_1 - L_4$) legs are multiplied by the weighting factors F (1.2, 0.6, or 0.4), the products ($r_1 \ldots r_4$, $l_1 \ldots l_4$) are added to give $\alpha'R$ and $\alpha'L$, and the result is divided by the number of stimulated legs (n). The value so obtained is the turning angle $\bar{\beta}$. *Thin lines,* direction of contralateral or ipsilateral inhibition; *thick lines,* processing of stimulus angle $\bar{\alpha}$ from the resting position of individual legs to the stage at which it is used to calculate the mean turning angle $\bar{\beta}$. (From Hergenröder and Barth 1983b, with permission.)

found a 10-ms time-of-arrival difference for the tarsi most distant from each other and 2 to 3 ms for the two front legs. At 80 Hz, the main peak of the male courtship signal, these values reduce to about one half but are still in the range known to be effective as orientation cues. For still higher frequen-

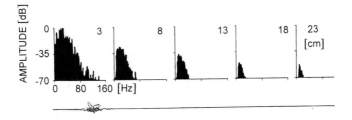

FIGURE 7.16. *Water surface.* The surface waves generated by a fly caught on the surface and struggling changes drastically with the distance traveled away from its source. Note, in particular, the loss of the high-frequency components. 0 dB is equal to the highest amplitude value measured 3 cm away from source. Numbers above spectra give the distance from the wave source. (From Bleckmann 1994, with permission.)

cies, such as those contained in prey signals, values for Δt decrease to less than 1 ms. The same applies if we consider smaller spiders instead of adult *Cupiennius salei* with a leg span of 10 cm. In the behavioral studies available to date, such small values for Δt did not elicit orientation toward the leg first stimulated and we do not know yet whether the vibration receptors can resolve them.

From the values determined for signal attenuation on the way through the plant, we conclude that Δd values in the behaviorally effective range do occur (Barth 1986). The question, however, of how the spider deals with the heterogeneity of signal propagation through the plant is still unanswered (Wirth 1984). Phase shifts as well as changes in frequency composition are additional parameters theoretically available to the spider, but their evaluation appears to be even more complicated.

6.3.1.3. Water Surface

The low propagation speed of water surface waves makes time differences a good candidate parameter for use in the determination of the direction toward a wave source. It is indeed known to be used by *Dolomedes*, the fishing spider (Bleckmann et al. 1994), and also serves the back swimmer *Notonecta* (Murphey 1973; Wiese 1974) for the same purpose.

Like temporal differences, amplitude differences might well be used by semiaquatic spiders for angular orientation. Because attenuation is strongest at high frequencies (see Section 3.3; Fig. 7.16), high-frequency waves should be the best guide, provided the spider is close to the source where they are still present. There is evidence, however, that amplitude gradients are not important in the given context: neither decreasing the distance to a wave source nor increasing wave frequency (monofrequency stimulation) increases the accuracy of the turning angle in *Dolomedes* (Bleckmann et al. 1994).

6.3.2. Distance

Evidence for successful distance determination is scarce for spiders. An orb weaver may not even be in need of such information. Once it has found the radius with the largest longitudinal vibrations, it simply has to follow the "track" to eventually reach an entangled insect. On the plant, and even more so on the water surface, there are no such "tracks." The possibility of at least a rough estimation of distance on the plant was already suggested: a polyfrequency signal such as a courtship or a prey vibration will contain more high-frequency components close to the sender than further away (see Section 3.2 on dispersive transmission of bending waves; see Fig. 7.11). Whether *Cupiennius* or any other wandering spider really makes use of this possibility still needs to be shown experimentally.

Dolomedes, the fishing spider, clearly does determine the distance to a source of concentric water surface waves up to a distance of 15 cm (Bleckmann 1988; Bleckmann et al. 1994). It runs to the source of stimulation, even after this source has been removed shortly after the application of a signal. Among the various parameters theoretically allowing for such a behavior, it is the curvature of the wave stimulus which tells *Dolomedes* the distance to its target. When stimulated by an uncurved (nonconcentric) wave, *Dolomedes* does the job significantly worse (Bleckmann 1994).

7. Active Space

Information gathered using the vibration sense usually arises from nearby sources. The vibration sense may therefore not impress us as much as far distance senses, such as eyes detecting stars which are light years away or ears detecting thunder from many kilometers away. There are particular virtues of vibratory signals and vibration sensing, however, that must have contributed to the remarkable evolutionary success of spiders. Thus vibratory signals, unlike optical ones (except light production), are available to nocturnal animals such as many spiders. Short transmission distances give vibratory courtship signals a rather private character, keeping both enemies and competitors from being informed. Compared with the spreading of odors and airborne acoustic signals, the transmission of vibrations along the leaves of a bromeliad or the threads of a spider web is less diffuse and the signals are not blown away by wind or blocked by visual obstacles.

With knowledge of the signals, their transmission, and the receiver's vibration sensitivity now at hand, we can quantitatively estimate the signal ranges on the various spider substrates.

Vibrations generated by entrapped insects in the orb web differ in magnitude by more than 50 dB, not considering the initial insect–web impact (e.g., fruitfly, *Drosophila* 0.03 mm/s; cricket, *Acheta* 10 mm/s). Since *Nephila* does not respond to all entrapped struggling *Drosophila*, the vibrations generated by fruitfly-sized insects seem to be near the behavioral threshold (Landolfa and Barth 1996). Thresholds determined with sinusoidal transverse vibrations are considerably higher (Klärner and Barth 1982).

Because the transmission of longitudinal vibrations through the *Nephila* web is very efficient (attenuation ca. 0.3 dB/cm), this component of prey vibrations will reach the spider from anywhere in the web with less than 20-dB attenuation, and much less from areas close to the hub. Transverse (and lateral) vibrations are attenuated nearly six times as much (ca. 1.7 dB/cm). Their strong representation in the insect vibration received by the spider suggests that the insects generate transverse vibrational components more strongly than longitudinal ones (Landolfa and Barth 1996).

Individual slits of the metatarsal vibration receptor of *Nephila* respond to longitudinal vibrations down to 0.05 to 0.8 mm/s. Vibrations from flies larger than *Drosophila*, such as *Calliphora*, and from crickets and similar-sized insects, will reach the spider from all over its web and still be above threshold for the metatarsal vibration receptor (Landolfa and Barth, in preparation). Thus, the full size of the *Nephila* web may be considered the active space of the spider's vibrational sense. The same applies to the orb web of *Nuctenea*, which transmits vibrations even more efficiently (Masters 1984a).

For the *plant* case, the signal range was calculated for *Cupiennius salei*, in which female courtship vibrations function to guide the male to its mate (Barth 1993). The largest acceleration amplitudes known of the female courtship signal measured ca. 1.6 m/s^2, about 10 cm away from the spider (Schüch and Barth 1990). The thresholds of the slits of the metatarsal vibration receptor are as small as ca. 7 mm/s^2 in the range of the dominant frequency components of the female signal (Barth and Geethabali 1982; Baurecht and Barth 1993). The average attenuation of the signal on its way through the plant is 0.3 dB/cm under favorable conditions (Barth 1985a; Barth et al. 1988b). The active range calculated from these data is roughly 150 cm, or close to 200 cm if one considers more naturalistic band-limited noise (ca. one-third octave, $Q = 0.35$) instead of purely sinusoidal stimuli (Barth 1985a). Behavioral studies in the field in Costa Rica have shown that the female of *C. coccineus* still responds to a courting male when 3.5 m away from it on a banana plant (Barth 1993).

The active space for the reception of *water surface* waves is rather limited because of the strong attenuation (see Fig. 7.16) of the stimulus on its way to the receiver. *Dolomedes* responds to an insect falling onto the water surface with prey-capture behavior up to a distance of ca. 40 cm. The threshold displacements necessary to elicit a behavioral response with sinusoidal stimuli are smallest between 40 Hz (ca. 2 μm) and 70 Hz (ca. 1 μm), and increase toward both higher and lower frequencies (Bleckmann and Barth 1984). The electrophysiologically determined thresholds for single slits of the metatarsal lyriform organ are about 10 μm at 40 Hz and about 5 μm at 70 Hz. Unlike the behavioral threshold curve (but identical to the slits of the metatarsal organs of both web and wandering spiders), the slits of *Dolomedes* behave like high-pass filters. Thresholds are high (ca. 75 μm) up to 10 to 30 Hz and decrease continuously at higher frequencies, down to about 10^{-5} mm (0.01 μm) at 1 kHz (Bleckmann and Barth 1984).

From the study of Lang (1980) we know that prey-generated water surface waves range from about 2 to 80 μm. Taking a stimulus frequency of 40 Hz, the original amplitude will be reduced to 10% after 30 cm (see Fig. 7.3). Even when starting from 20 μm, the metatarsal slits would be barely able to detect such a stimulus. Figure 7.16 illustrates how quickly a fly-generated surface wave signal declines with distance.

8. Summary

In order to make sense of any sensory system, the analysis has to consider several different questions and levels of organization. These include the physics of the adequate and the biologically relevant stimuli, the primary processes at the receptor cell level, the interneurons in the central nervous system, and the actual role played by the particular information provided by the sensory system in behavior. We should ask not only proximate but also ultimate questions. By the time we have accomplished all this, we will have learned that studying a sensory system necessitates a broad systemic approach and an in-depth exercise in many fields of biology.

We are still a long way from such a goal. However, we do already enjoy a number of interesting vistas and may ask where further digging seems to be particularly necessary and promising.

1. At the *receptor level*, we do have some idea about stimulus transformation in slit sensilla in general and lyriform organs in particular (Barth 1985b). Intracellular recording and patch-clamp analysis, however, has just recently been started in a lyriform organ on the spider leg patella (Seyfarth and French 1994). It would certainly be worthwhile to extend such studies to the metatarsal vibration receptor.

Apart from general problems of membrane physiology, the origin of the frequency selectivity of different slits in the metatarsal organ when exposed to suprathreshold natural stimuli and their "tuning" to the temporal properties of courtship vibrations (Baurecht and Barth 1992) still has to be clarified. At least part of the answer is likely to be found in the micromechanics of the slit assembly of the organ, which has to be analyzed under dynamic conditions. This will be a difficult task, mainly because of the high spatial resolution needed to measure the deformation of the individual slits.

2. At the level of the *central nervous system*, spiders still are rather hard to study. As has been shown, however, intracellular recording from

vibration-sensitive interneurons is possible (Friedel and Barth 1995). Clearly, more data are needed to see how the central nervous system handles temporal and intensity information contained in the stimuli and how this relates to orientation behavior. Other aspects still to be treated more extensively are the multimodal properties of interneurons (Speck-Hergenröder and Barth 1987) and their relation to the multimodal interaction, both known from behavioral experiments (Hergenröder and Barth 1983a,b) and suggested by neuroanatomical studies (Babu and Barth 1989; Anton and Barth 1993). At a more general level, it seems intriguing that according to all of our neuroanatomical work, there are only few mechanosensory projection pathways from the subesophageal ganglionic mass to the supraesophageal central nervous system, that is, the brain. We still do not know if and to what extent the brain is participating in shaping the behavior guided by vibrational stimuli. Evidence at hand suggests only a minor role. It would be valuable to know where the vibration sense is represented in the brain and what the relation of its representation to the other senses, mainly to air movement detection and vision (Strausfeld and Barth 1993; Barth, Nakagawa, and Eguchi 1993), is like.

3. At the *behavioral level*, we note that stridulatory mechanisms have never been studied in spiders, although stridulation is widespread in this animal group (and in most cases is used as a source of substrate-borne vibrations instead of airborne sound). Considering the wealth of different spider webs and their significance as tools picking up vibrations and transmitting them to the web owner, a gold mine for future research is ready to be exploited for those interested in a combination of biomechanics and behavioral studies.

References

Anton S, Barth FG (1993) Central nervous projection patterns of trichobothria and other cuticular sensilla in the wandering spider *Cupiennius salei* (Arachnida, Araneae). Zoomorphology 113:21–32.

Autrum H (1941) Über Gehör- und Erschütterungssinn bei Locustiden. Z Vergl Physiol 28:580–637.

Babu KS, Barth FG (1989) Central nervous projections of mechanoreceptors in the spider *Cupiennius salei* KEYS. Cell Tissue Res 258:69–82.

Babu KS, Barth FG, Strausfeld NJ (1985) Intersegmental sensory tracts and contralateral motor neurons in the leg ganglia of the spider *Cupiennius salei* KEYS. Cell Tissue Res 241:53–57.

Barth FG (1971) Der sensorische Apparat der Spaltsinnesorgane (*Cupiennius salei* Keys., Araneae). Z Zellforsch 112:212–246.

Barth FG (1972a) Die Physiologie der Spaltsinnesorgane. I. Modellversuche zur Rolle des cuticularen Spaltes beim Reiztransport. J Comp Physiol A 78:315–336.

Barth FG (1972b) Die Physiologie der Spaltsinnesorgane. II. Funktionelle Morphologie eines Mechanoreceptors. J Comp Physiol A 81:159–186.

Barth FG (1982) Spiders and vibratory signals: Sensory reception and behavioral significance. In: Witt PN, Rovner JS (eds) Spider Communication: Mechanisms and Ecological Significance. Princeton, NJ: Princeton University Press, pp. 67–122.

Barth FG (1985a) Neuroethology of the spider vibration sense. In: Barth FG (ed) Neurobiology of Arachnids. Berlin: Springer-Verlag, pp. 203–229.

Barth FG (1985b) Slit sensilla and the measurement of cuticular strains. In: Barth FG (ed) Neurobiology of Arachnids. Berlin: Springer-Verlag, pp. 162–188.

Barth FG (1986) Vibrationssinn und vibratorische Umwelt von Spinnen. Naturwissenschaften 73:519–530.

Barth FG (1993) Sensory guidance in spider pre-copulatory behavior. Comp Biochem Physiol 104A:717–733.

Barth FG (1997) Vibratory communication in spiders: Adaptation and compromise at many levels. In: Lehrer M (ed) Orientation and Communication in Arthropods. Basel: Birkhäuser, pp. 247–272.

Barth FG, Blickhan R (1984) Mechanoreception. In: Bereiter-Hahn J, Matoltsy AG, Richards KS (eds) Biology of the Integument I. Invertebrates. Berlin: Springer-Verlag, pp. 544–582.

Barth FG, Bohnenberger J (1978) Lyriform slit sense organ: threshold and stimulus amplitude ranges in a multi-unit mechanoreceptor. J Comp Physiol A 125:37–43.

Barth FG, Geethabali (1982) Spider vibration receptors. Threshold curves of individual slits in the metatarsal lyriform organ. J Comp Physiol A 148:175–185.

Barth FG, Libera W (1970) Ein Atlas der Spaltsinnesorgane von Cupiennius salei Keys., Chelicerata (Araneae). Z Morph Tiere 68:343–369.

Barth FG, Schmitt A (1991) Species recognition and species isolation in wandering spiders (Cupiennius spp.; Ctenidae). Behav Ecol Sociobiol 29:333–339.

Barth FG, Seyfarth E-A, Bleckmann H, Schüch W (1988a) Spiders of the genus Cupiennius Simon 1891 (Araneae, Ctenidae). I. Range distribution, dwelling plants, and climatic characteristics of the habitats. Oecologia 77:187–193.

Barth FG, Bleckmann H, Bohnenberger J, Seyfarth EA (1988b) Spiders of the genus Cupiennius Simon 1891 (Araneae, Ctenidae). II. On the vibratory environment of a wandering spider. Oecologia 77:194–201.

Barth FG, Nakagawa T, Eguchi E (1993) Vision in the ctenid spider Cupiennius salei: spectral range and absolute sensitivity (ERG). J Exp Biol 181:63–79.

Barth FG, Wastl U, Humphrey JAC, Devarakonda R (1993) Dynamics of arthropod filiform hairs. II. Mechanical properties of spider trichobothria (Cupiennius salei Keys.). Philos Trans R Soc Lond B 340:445–461.

Baurecht D, Barth FG (1991) Vibratory communication in spiders: receptor response to synthetic vibratory signals. In: Elsner N, Penzlin H (eds) Proceedings of the 19th Göttingen Neurobiology Conference. Stuttgart: Thieme, p. 133.

Baurecht D, Barth FG (1992) Vibratory communication in spiders. I. Representation of male courtship signals by female vibration receptor. J Comp Physiol A 171:231–243.

Baurecht D, Barth FG (1993) Vibratory communication in spiders. II. Representation of parameters contained in synthetic male courtship signals by female vibration receptor. J Comp Physiol A 173:309–319.

Berestyńska-Wilczek M (1962) Investigations of the sensitivity of the spider Pirata piraticus (Clerck) to vibrations of the water surface. Acta Biol Cracowiensia 5:263–277.

Bleckmann H (1988) Prey identification and prey localization in surface-feeding fish and fishing spider. In: Atema J, Fay RR, Popper AN, Tavolga WN (eds) Sensory Biology of Aquatic Animals. New York: Springer-Verlag, pp. 619–642.

Bleckmann H (1994) Reception of Hydrodynamic Stimuli in Aquatic and Semi-aquatic Animals. Stuttgart: G Fischer.

Bleckmann H, Barth FG (1984) Sensory ecology of a semiaquatic spider (*Dolomedes triton*). II. The release of predatory behavior by water surface waves. Behav Ecol Sociobiol 14:303–312.

Bleckmann H, Bender M (1987) Water surface waves generated by the male pisaurid spider *Dolomedes triton* during courtship behavior. J Arachnol 15:363–369.

Bleckmann H, Borchardt M, Horn P, Görner P (1994) Stimulus discrimination and wave source localization in fishing spiders (*Dolomedes triton* and *D. okefinokensis*). J Comp Physiol A 174:305–316.

Blest AD (1985) The fine structure of spider photoreceptors in relation to function. In: Barth FG (ed) Neurobiology of Arachnids. Berlin: Springer-Verlag, pp. 79–102.

Bohnenberger I (1981) Matched transfer characteristics of single units in a compound slit sense organ. J Comp Physiol A 142:391–401.

Braun R (1955) Zur Biologie von *Teutana triangulosa*. Z Wiss Zool 159:255–318.

Brownell P, Farley RD (1979) Detection of vibrations in sand by tarsal sense organs of the nocturnal scorpion, *Paruroctonus mesaensis*. J Comp Physiol A 131:23–30.

Buchli HHR (1969) Hunting behavior in the Ctenizidae. Am Zool 9:175–193.

Burgess W (1976) Social spiders. Sci Amer 234:100–106.

Carico JE (1973) The nearctic species of the genus *Dolomedes* (Araneae: Pisauridae) Bull Mus Comp Zool 144:435–488.

Cremer L, Heckl M, Ungar EE (1973) Structure-Borne Sound. Berlin: Springer-Verlag.

Dambach M (1989) Vibrational responses. In: Huber F, Moore TE, Loher W (eds) Cricket Behavior and Neurobiology. Ithaca, NY: Cornell University Press, pp. 178–197.

Denny M (1976) The physical properties of spiders' silk and their role in the design of orb-webs. J Exp Biol 65:483–506.

Devetak D, Gogala M, Čokl A (1978) A contribution to the physiology of vibration receptors in the bugs of the family Cydnidae (Heteroptera). Biol Vestn 26:131–139.

Dierkes S, Barth FG (1995) Mechanism of signal production in the vibratory communication of the wandering spider *Cupiennius getazi* (Arachnida, Araneae). J Comp Physiol A 176:31–44.

Dorward PK, McIntyre AK (1971) Responses of vibration-sensitive receptors on the interosseus region of the duck's hind limb. J Physiol 219:77–87.

Ewing AW (1989) Arthropod Bioacoustics. Neurobiology and Behavior. Ithaca, NY: Cornell University Press.

Forster L (1985) Target discrimination in jumping spiders (Araneae: Salticidae). In: Barth FG (ed) Neurobiology of Arachnids. Berlin: Springer-Verlag, pp. 249–274.

Friedel T, Barth FG (1995) Responses of female interneurons to male courtship vibrations in a spider (*Cupiennius salei* Keys., Ctenidae). J Comp Physiol A 177:159–171.

Frohlich C, Buskirk RE (1982) Transmission and attenuation of vibration in orb spider webs. J Theor Biol 95:13–36.

Gogala M (1985) Vibrational communication in insects (biophysical and behavioral aspects). In: Kalmring K, Elsner N (eds) Acoustic and Vibrational Communication in Insects. Berlin: Parey, pp. 117–126.

Graeser K (1973) Die Übertragungseigenschaften des Netzes von *Zygiella x-notata* (Clerck) für transversale Sinusschwingungen im niederen Frequenzbereich und Frequenzanalyse beutetiererregter Netzvibrationen. Diplomarbeit, Universität Frankfurt, Frankfurt am Main.

Grassé P (ed) (1949) Traité de Zoologie. Tome VI. Paris: Masson et Cie.

Gregory JE, McIntyre AK, Proske U (1986) Vibration-evoked responses from lamellated corpuscles in the legs of kangaroos. Exp Brain Res 62:648–653.

Gwinner-Hanke H (1970) Zum Verhalten zweier stridulierender Spinnen, *Steatoda bipunctata* Linné und *Teutana grossa* Koch (Theridiidae, Araneae), unter besonderer Berücksichtigung des Fortpflanzungsverhaltens. Z Tierpsychol 27:649–678.

Harrison JB (1969) Acoustic behavior of a wolf spider, *Lycosa gulosa*. Anim Behav 17:14–16.

Henschel JR, Lubin YD (1992) Environmental factors affecting the web and activity of a psammophilous spider in the Namib Desert. J Arid Environ 22:173–189.

Hergenröder R, Barth FG (1983a) The release of attack and escape behavior by vibratory stimuli in a wandering spider (*Cupiennius salei* Keys.). J Comp Physiol A 152:347–358.

Hergenröder R, Barth FG (1983b) Vibratory signals and spider behavior: How do the sensory inputs from the eight legs interact in orientation? J Comp Physiol A 152:361–371.

Horch KW, Salmon M (1969) Production, perception and reception of acoustic stimuli by semiterrestrial crabs (genus *Ocypode* and *Uca*, family Ocypodidae). Forma et Functio 1:1–25.

Höster W (1990) Vibrational sensitivity of the wing of the pigeon (*Columba livia*) — a study using heart rate conditioning. J Comp Physiol A 167:545–549.

Hudspeth AJ (1985) The cellular basis of hearing: the biophysics of hair cells. Science 230:745–752.

Hudspeth AJ (1989) How the ear's works work. Nature 341:397–404.

Kalmring K, Elsner N (1985) Acoustic and Vibrational Communication in Insects. Berlin: Parey.

Kalmring K, Lewis B, Eichendorf A (1978) The physiological characteristics of primary sensory neurons of the complex tibial organ of *Decticus verrucivorus* L. (Orthoptera, Tettigonioidae). J Comp Physiol A 127:109–121.

Kämper A, Kühne R (1983) The acoustic behavior of the bushcricket *Tettigonia cantans*. II. Transmission of airborne sound and vibration signals in the biotope. Behav Proc 8:125–145.

Kaston BJ (1936) The senses involved in the courtship of some vagabond spiders. Entomol Am 16:97–167.

Keidel WD (1956) Vibrationsreception. Der Erschütterungssinn des Menschen. Erlangen Forschungen, Reihe B: Naturwissenschaften, Bd. 2. Erlangen, Univ. Bibliothek.

Klärner D, Barth FG (1982) Vibratory signals and prey capture in orb-weaving spiders (*Zygiella x-notata*, *Nephila clavipes*; Araneidae). J Comp Physiol A 148:445–455.

Köhler D, Tembrock G (1987) Akustische Signale der Wolfsspinne *Hygrolycosa rubrofasciata* (Arachnida: Lycosidae). Zool Anz 219:147–153.

Konigswald A, Lubin Y, Ward D (1990) The effectiveness of the nest of a desert widow spider, *Latrodectus revivensis* in predator deterrence. Psyche 97:75–80.

Kronestedt T (1973) Study of a stridulatory apparatus in *Pardosa fulvipes* (Collett) (Araneae, Lycosidae) by scanning electron microscopy. Zool Scripta 2:43–47.

Kullmann E (1972) The convergent development of orb-webs in cribellate and ecribellate spiders. Am Zool 12:395–405.

Kullmann E, Otto F, Braun T, Raccanello R (1975) Fundamentals and classification — A survey of spider net constructions. In: IL 8, Netze in Natur und Technik. Stuttgart: Mitt Inst Leichte Flächentragwerke.

Land MF (1985) The morphology and optics of spider eyes. In: Barth FG (ed) Neurobiology of Arachnids. Berlin: Springer-Verlag, pp. 53–78.

Landolfa MA, Barth FG (1996) Vibrations in the orb web of the spider *Nephila clavipes*. Cues for discrimination and orientation. J Comp Physiol A, 179:493–508.

Lang HH (1980) Surface wave discrimination between prey and nonprey by the backswimmer *Notonecta glauca* L (Hemiptera, Heteroptera). Behav Ecol Sociobiol 6:233–246.

Legendre R (1963) L'audition et l'emission de sons chez les Aranéides. Ann Biol 2:371–390.

Lewis ER, Narins PM (1985) Do frogs communicate with seismic signals? Science 227:187–189.

Liesenfeld FJ (1956) Untersuchungen am Netz und über den Erschütterungssinn von *Zygiella x-notata* (Cl.) (Araneidae). Z Vergl Physiol 38:563–592.

Liesenfeld FJ (1961) Über Leistung und Sitz des Erschütterungssinnes von Netzspinnen. Biol Zbl 80:465–475.

Lighthill J (1980) Waves in Fluids. Cambridge, UK: Cambridge University Press.

Lucas F (1964) Spiders and their silks. Discovery 25:20–26.

Main BY (1957) Adaptive radiation of trapdoor spiders. Austral Mus Mag 12:160–163.

Markl H (1969) Verständigung durch Vibrationssignale bei Arthropoden. Naturwissenschaften 56:499–505.

Markl H (1973) Leistungen des Vibrationssinnes bei wirbellosen Tieren. Fortschritte d Zool 21:100–120.

Markl H (1983) Vibratorial communication. In: Huber F, Markl H (eds) Neuroethology and Behavioral Physiology. Heidelberg: Springer-Verlag, pp. 332–353.

Marshall SD, Thoms EM, Uetz GW (1995) Setal entanglement: an undescribed method of stridulation by a neotropical tarantula (Araneae: Theraphosidae). J Zool Lond 235:587–595.

Masters WM (1984a) Vibrations in the orb webs of *Nuctenea sclopetaria* (Araneidae). I. Transmission through the web. Behav Ecol Sociobiol 15:207–215.

Masters WM (1984b) Vibrations in the orb webs of *Nuctenea sclopetaria* (Araneidae). II. Prey and wind signals and the spider's response threshold. Behav Ecol Sociobiol 15:217–223.

Masters WM, Markl H (1981) Vibration signal transmission in spider orb-webs. Science 213:363–365.

Masters WM, Markl HS, Moffat AJM (1986) Transmission of vibration in a spider's web. In: Shear WA (ed) Spiders: Webs, Behavior and Evolution. Palo Alto, CA: Stanford University Press, pp. 49–69.

McIntyre AK (1980) Biological seismography. Trends Neurosci Sept:202–205.

Meyer E (1928) Neue Sinnesbiologische Beobachtungen an Spinnen. Z Morph Ökol Tiere 12:1–69.

Michelsen A, Fink F, Gogala M, Traue D (1982) Plants as transmission channels for insect vibrational songs. Behav Ecol Sociobiol 2:269–281.

Morse PM (1948) Vibration and Sound. New York: McGraw-Hill.

Murphey RK (1973) Mutual inhibition and the organisation of a nonvisual orientation in *Notonecta*. J Comp Physiol A 84:31–69.

Narins P (1995) Frog communication. Sci Amer 273:62–67.

Narins P, Lewis ER (1984) The vertebrate ear as an exquisite seismic sensor. J Acoust Soc Am 76:1384–1387.

Pickles JO, Corey DP (1992) Mechanotransduction by hair cells. Trends Neurosci 15:254–259.

Rayleigh Lord DCL (1885) On waves propagated along the plane surface of an elastic solid. Proc Math Soc Lond 17:4–11.

Roemer van de A (1980) Eine vergleichende morphologische Untersuchung an dem für die Vibrationswahrnehmung wichtigen Distalbereich des Spinnenbeins. Diplomarbeit, Universität. Frankfurt, Frankfurt am Main.

Roth E (1986) Vibratorische Balz von *Cupiennius coccineus*: zur ethologischen Bedeutung der Signale. Diplomarbeit, Universität. Frankfurt, Frankfurt am Main.

Rovner JS (1967) Acoustic communication in a lycosid spider (*Lycosa rabida* Walckenaer) Anim Behav 15:273–281.

Rovner JS (1975) Sound production by Nearctic wolf spiders: a substratum-coupled stridulatory mechanism. Science 190:1309–1310.

Rovner JS (1980) Vibration in *Heteropoda venatoria* (Sparassidae): a third method of sound production in spiders. J Arachnol 8:193–200.

Rovner JS, Barth FG (1981) Vibratory communication through living plants by a tropical wandering spider. Science 214:464–466.

Schmitt A, Schuster M, Barth FG (1992) Male competition in a wandering spider (*Cupiennius getazi*, Ctenidae). Ethology 90:293–306.

Schmitt A, Schuster M, Barth FG (1994) Vibratory communication in a wandering spider (*Cupiennius getazi*, Ctenidae): female and male preferences for various features of the conspecific male's releaser. Anim Behav 48:1155–1171.

Schnorbus H (1971) Die subgenualen Sinnesorgane von *Periplaneta americana*: Histologie und Virationsschwellen. Z Vergl Physiol 71:14–48.

Schüch W, Barth FG (1985) Temporal patterns in the vibratory courtship signals of the wandering spider *Cupiennius salei* Keys. Behav Ecol Sociobiol 16:263–271.

Schüch W, Barth FG (1990) Vibratory communication in a spider: female responses to synthetic male vibrations. J Comp Physiol A 166:817–826.

Schwartzkopff J (1948) Der Vibrationssinn der Vögel. Naturwissenschaften 35:318–319.

Sellick PM, Patuzzi R, Johnstone BM (1982) Measurement of basilar membrane motion in the guinea pig using the Mössbauer technique. J Acoust Soc Am 72:131–141.

Seyfarth E-A, French AS (1994) Intracellular characterization of identified sensory cells in a new spider mechanoreceptor preparation. J Neurophysiol 71:1422–1427.

Shaw SR (1994) Re-evaluation of the absolute threshold and response mode of the most sensitive known "vibration" detector, the cockroach's subgenual organ: a cochlea-like displacement threshold and a direct response to sound. J Neurobiol 25:1167–1185.

Shear WA (1986) (ed) Spiders: Webs, Behavior, and Evolution. Palo Alto, CA: Stanford University Press.

Shimizu I, Barth FG (1996) The effect of temperature on the temporal structure of the vibratory courtship signal of a spider (*Cupiennius salei* Keys.). J Comp Physiol A 179:363–370.

Sivian LJ, White SD (1933) On minimum audible sound fields. J Acoust Soc Am 4:288–321.

Skudrzyk E (1971) The Foundations of Acoustics. Vienna: Springer-Verlag.

Sommerfeld A (1970) Vorlesungen über theoretische Physik II. Mechanik der deformierbaren Medien. Leipzig: Akad Verlagsges.

Speck J, Barth FG (1982) Vibration sensitivity of pretarsal slit sensilla in the spider leg. J Comp Physiol A 148:187–194.

Speck-Hergenröder J (1984) Vibrationsempfindliche Interneurone im Zentralnervensystem der Spinne *Cupiennius salei* Keys. Ph.D. thesis, Universität Frankfurt, Frankfurt am Main.

Speck-Hergenröder J, Barth FG (1987) Tuning of vibration sensitive neurons in the central nervous system of a wandering spider, *Cupiennius salei* Keys. J Comp Physiol A 160:467–475.

Speck-Hergenröder J, Barth FG (1988) Vibration sensitive hairs on the spider leg. Experientia 44:13–14.

Starck JM (1985) Stridulationsapparate einiger Spinnen — Morphologie und evolutionsbiologische Aspekte. Z Zool Syst Evolutionforsch 23:115–135.

Stratton GE, Uetz GW (1983) Communication via substratum — coupled stridulation and reproductive isolation in wolf spiders (Araneae: Lycosidae). Anim Behav 31:164–172.

Strausfeld NJ, Barth FG (1993) Two visual systems in one brain: neuropils serving the secondary eyes of the spider *Cupiennius salei*. J Comp Neurol 328:43–62.

Szlep R (1965) The web spinning process and web structure of *Latrodectus tredecimguttatus*, *L. pallidus* and *L. revivensis*. Proc Zool Soc Lond 145:75–89.

Tautz J (1979) Reception of particle oscillation in a medium. An unorthodox sensory capacity. Naturwissenschaften 66:452–461.

Tautz J (1989) Medienbewegung in der Sinneswelt der Arthropoden. Fallstudien zu einer Sinnesökologie. Stuttgart: G Fischer.

Tretzel E (1961) Biologie, Ökologie und Brutpflege von *Coelotes terrestris* (Wider) (Araneae: Agelenidae), II. Brutpflege. Z Morphol Ökol Tiere 50:375–524.

Uetz GW, Stratton GE (1982) Acoustic communication and reproductive isolation in spiders. In: Witt PN, Rovner JS (eds) Spider Communication: Mechanisms and Ecological Significance. Princeton, NJ: Princeton University Press, pp. 123–159.

Vollrath I (1979a) Behavior of the kleptoparasitic spider *Argyrodes elevatus* (Araneae, Theridiidae). Anim Behav 27:515–521.

Vollrath F (1979b) Vibrations: their signal function for a spider kleptoparasite. Science 205:1149–1151.

Wiehle H (1931) Neue Beiträge zur Kenntnis des Fanggewebes der Spinnen aus den Familien Argiopidae, Uloboridae und Theridiidae. Z Morph Ökol Tiere 23:349–400.

Wiese K (1974) The mechanoreceptive system of prey localization in *Notonecta*. J Comp Physiol A 92:317–325.

Wilde J de (1943) Some physical properties of the spinning threads of *Aranea diademata*. L Arch Neerl Physiol 27:117–132.

Wirth E (1984) Die Bedeutung von Zeit- und Amplitudenunterschieden für die Orientierung nach vibratorischen Signalen bei Spinnen. Diplomarbeit, Universität Frankfurt, Frankfurt am Main.

Wirth E, Barth FG (1992) Forces in the spider orb web. J Comp Physiol A 171:359–371.

Work RW (1976) The force elongation behavior of web fibers and silks forcibly obtained from orb-web-spinning spiders. Textile Res J 46:485–492.

8
The Sensory Coevolution of Moths and Bats

JAMES H. FULLARD

1. Introduction

Coevolution is the accumulation of reciprocal adaptations that phylogeneti-cally distant species undergo as a result of their interactions over evolution-ary time. Although the classic definition of coevolution requires highly specialized, one-to-one relationships between the participants (e.g., the proboscises of bees and the corollas of orchids, Darwin 1862), today we recognize coevolutionary relationships in a global perspective that does not involve localized interactions (Thompson 1994). Predator-prey relation-ships have formed the basis for many coevolutionary studies, most using the morphology or behavior of the organisms involved to examine the presence of reciprocal adaptations (e.g., the defenses of cladocerans, Dodson 1988; Riessen and Sprules 1990). Coevolutionary processes between predators and prey can be difficult to observe because of the multiple functions that most animal structures serve, and it can be difficult to determine which parts of an organism's phenotype have been subjected to the specific changes induced by coevolutionary effects (e.g., are the horns of male deer directed against predators or other males?). Vertebrates, in particular, present special problems in the study of coevolution because of their rela-tively long life span and the complexities of their interactions with other organisms, both predators and competitors.

Insect sensory systems are useful for the study of predator-prey coevolu-tion because of their simple neural organization and the fact that the short reproductive life span of insects (usually one season) amplifies the effects of other organisms on their life histories. There have been recent advances in our understanding of the evolution of animals using phylogenetically based methods of comparative biology (Brooks and McLennan 1991), which pro-vide templates onto which we can map our observations of extant species to interpret their evolutionary past. As inviting as they seem at first for evolu-tionary studies, most insect sensory systems retain the problem of how to witness the specific effects of coevolutionary pressures in the presence of multiple functions. An ideal sensory system for studies of coevolution would be one that was both neurologically simple and had functions limited to the selection pressures exerted by a small number of coevolutionary participants.

One such insect sensory system is the bat-detecting moth ear. Kenneth D. Roeder was the first to recognize its potential as a model to understand the neural control of natural behavior and, with his colleagues, he conducted studies that also demonstrated the value of the moth ear for our understanding of insect auditory systems and their evolution. In experiments that persist as models for complete neuroethological research, Roeder demonstrated how noctuid moths detect the echolocation signals of hunting bats in time to initiate evasive flight maneuvers and to avoid ending up as dinner (summarized in Roeder 1967). By observing the responses of wild moths, Roeder and Treat (1962) estimated that moths that respond to approaching bats have a 29% to 58% advantage compared with moths that do not respond. This sensory ability has given the moths that possess it a tool of profound selective advantage.

Although the adaptiveness of the moth ear is clear, what is debatable (Jones 1992) is whether it has actually coevolved with bats or, more precisely, with the acoustic characteristics of bat sonar calls. To demonstrate a coevolutionary process, it is necessary to show that, first, the moth's ear is designed to detect those bats that exert selection pressure and, second, that some bats have designed their calls to counteract this ability. Critical to the argument for coevolution is the assumption that the primary purpose of the moth ear is to detect the echolocation calls of bats, a difficult premise to accept considering the many functions served by our own auditory system. Of the estimated 200,000 species of Lepidoptera in the world (Holloway, Bradley, and Carter 1987; Common 1990), about one half belong to families that possess ears (assuming, for example, that all Noctuidae are eared). Of these 92,000 species of eared moths, only seven have been demonstrated to use sounds for social communication (e.g., Surlykke and Gogola 1986; Conner 1987; Alcock, Gwynne, and Dadour 1989; Sanderford and Conner 1990; Alcock and Bailey 1995).

As fascinating as singing moths are, their use of ears for this purpose is likely to be a secondary adaptation in the evolution of moth auditory system. Consider that, of the 9702 species of birds in the world (Monroe and Sibley 1993), 17 use their wings for swimming and, although this percentage is higher than that of singing moths, we accept the derived state of swimming in the evolution of wings. Swimming is a remarkable and *exceptional* adaptation for some birds and, as such, cannot be used to understand the general evolution of the wing. In turn, the social use for ears in moths is the *exception* for this sensory system and may not explain its evolutionary origins or the present-day selection pressure maintaining it in most moths. Accordingly, this chapter makes the assumption that the majority of moths use their ears only to listen for bats, and the echolocation calls of these predators have served as the selective force acting upon their design.

2. The Moth Ear: Peripheral Nervous System

The moth ear has attracted the attention of entomologists since the late 19th century (Swinton 1877; White 1877). Studies in the first half of the 20th century examined the tympanal structures (Forbes 1916; Richards 1932) and the auditory peripheral nervous system (Eggers 1919, 1925) of moths and demonstrated that their ears contain one (Notodontidae), two (e.g., Noctuidae), or four (e.g., Geometridae) sensory neurons (Fig. 8.1). In noctuids, two auditory receptors (A1 and A2; Roeder and Treat 1957; Roeder 1967; Ghiradella 1971) monitor the tympanic membrane, which faces obliquely backwards from the metathoracic segment. The tympanic membrane appears to act as a high-pass filter, favoring frequencies above 20 kHz to stimulate the receptors, whereas the receptors themselves may be intrinsically tuned (Oldfield 1985) at 2, 14, and 60 kHz (Adams 1972).

The simplicity of its receptor cell–membrane association renders the moth ear a useful tool for biophysical studies of how acoustic energy is transformed into neural information (Adams 1971; Adams and Belcher 1974; Schiolten, Larsen, and Michelsen 1981; Coro and Pérez 1984, 1990; Pérez and Coro 1984, 1986; Surlykke, Larsen, and Michelsen 1988; Coro, Pérez, and Machado 1994; Waters 1996; Tougaard 1996). The spiking responses of the closely aligned receptor cells within the auditory chordotonal organ has led some authors to suggest that inhibitory interactions exist between these cells (Coro and Pérez 1983, 1984; Pérez and Coro 1986).

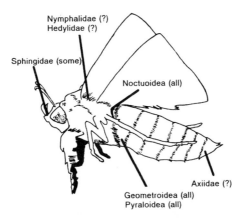

FIGURE 8.1. The places where ears occur on moths with examples of families that use those sites. All of the proposed sites, including the nontympanate ears of some Sphingidae, have been included in this figure, although some (question marked) have only been described morphologically. (Adapted from Yack and Fullard 1990, Wiley-Liss Publishers, with permission.)

For noctuid moths, the A1 and A2 cells possess similar tuning but are separated in their threshold sensitivities, with the A2 cell being about 20 dB less sensitive than the A1 cell (Fig. 8.2). The fact that both receptors reveal similarly shaped audiograms indicates that the noctuid ear is tone-deaf (Roeder 1967) and treats all similarly structured sounds as the same but of greater or lesser intensity. While the two-celled ear of the noctuids prohibits frequency discrimination for these moths, the four-celled peripheral arrangement of pyralids (Pérez and Zhantiev 1976) may allow these moths to identify sounds by their frequencies (Spangler 1988). In noctuids, the axons of the two auditory receptors pass through the tympanic air sac, where they meet with the B cell, a nonchordotonal receptor whose function remains unknown (Treat and Roeder 1959; but see Lechtenberg 1971). Axons of the A1, A2, and B cell form the auditory nerve (IIIN1b; Nüesch 1957) and enter the metathoracic neuromere of the fused mesothoracic and metathoracic ganglia via the hind-wing nerve.

The simplicity of the moth's two-celled peripheral auditory nervous system invites the belief that an understanding of the response patterns of these cells may provide enough information to decipher the internal

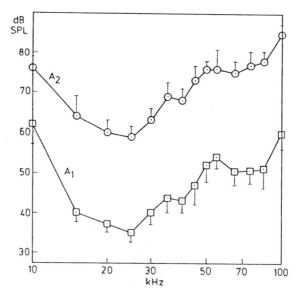

FIGURE 8.2. The auditory threshold curves (audiograms) of the two auditory receptor cells (A1 and A2) in the noctuid moth, *Baratha brassicae*. The curves are similarly tuned but are separated by approximately 20 dB, indicating that this two-celled peripheral auditory system, although tone-deaf, can estimate the relative distance of an approaching bat. (From Madsen and Miller 1987, Springer-Verlag Publishers, with permission.)

wiring responsible for this behavior or at least build testable hypotheses about those circuits. Roeder (1967, 1974) proposed the "bimodal" theory of the noctuid's auditory-evoked bat evasion whereby the most sensitive cell, A1, evokes the first stage of evasive behavior, coordinated negative phonotaxis, while the less sensitive A2 cell causes the moth to undergo erratic looping and diving responses to very close (i.e., intense) bats. This theory has been tested by Surlykke (1984), who used notodontids, moths that have a single auditory receptor (Eggers 1919) in each ear. According to Roeder's theory, these moths should express a unimodal response to bat calls, but Surlykke's (1984) study suggested that notodontids also express a two-stage evasive response. The bimodal behavioral response, if it exists, may be governed by differential levels of only A1 activity, rendering the neural control system even simpler than originally imagined by Roeder.

Following Roeder's (1966) suggestion that the A1 cell signals a condition of "no-bat" by means of spontaneous, single-fire discharge, Fullard (1987a) proposed that some minimal interspike interval must be encoded by A1 before the moth's central nervous system recognizes the condition of "bat." Because the natural environment is noisy, it is expected that, with its short reproductive adult life span, a moth must be able to distinguish the calls of hunting bats from harmless sounds. The short sounds of crackling branches and the long sounds of chorusing bushcrickets should be ignored by a flying moth searching for pheromone trails or oviposition sites. The moth's auditory receptors are therefore adapted to not only bats but to their total acoustic environment so as to minimize any time wasted in unnecessary avoidance flight. These discriminatory abilities do not reside within the moth's peripheral nervous system, however, because jingling keys, hand slaps, and ultrasonic television channel changers can evoke respectably batlike receptor responses in moth auditory preparations. It seems that although the moth's peripheral nervous system provides for ideas about how it deals with its complex acoustic environment, it is to this insect's central nervous system that we must look for the ways in which it recognizes signals.

3. The Moth Ear: Central Nervous System

Compared with the studies of the moth ear's peripheral nervous system, there are few studies of its central nervous system (CNS) processing (Roeder 1966; Paul 1973; 1974; Agee and Orona 1988; Coro and Alonso 1989), and only five that have used intracellular techniques (Boyan and Fullard 1986, 1988; Madsen and Miller 1987; Boyan, Williams, and Fullard 1990; Boyan and Miller 1991). At the receptor level, only the noctuid A1 cell has been intracellularly recorded within the CNS and morphologically identified (Fig. 8.3). This cell projects ipsilaterally into the metathoracic, mesothoracic, and prothoracic ganglia, resembling the auditory receptors of

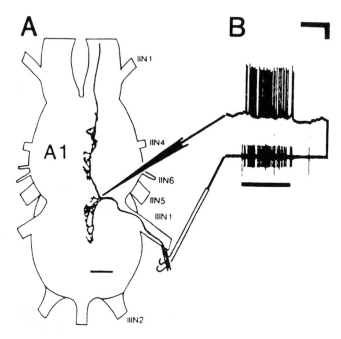

FIGURE 8.3. The central nervous system projection of the A1 receptor cell in the noctuid moth, *Agrotis infusa*. A: The morphology of the cell within the metathoracic and mesothoracic ganglion. B: Simultaneous extracellular (bottom trace) and intracellular (top trace) recordings of the A1 response to a white-noise sound burst (solid line). Scale bars: vertical (intracellular only) 7.5 mV; horizontal 26 ms. (From Boyan and Fullard 1988, Springer-Verlag Publishers, with permission.)

acridid Orthoptera. The morphology of the A1 axon suggests that synapses occur between it and interneurons in all three thoracic ganglia and possibly in more anterior centers. As with other insect auditory fibers, the noctuid A1 cell projects into various regions of the ring tract, suggesting an evolutionary conservatism of auditory processing circuitry (Boyan, Williams, and Fullard 1990; Boyan 1993; Boyan, Chapter). The CNS projection patterns of A2 have not been ascertained from intracellular recordings and its anatomy, which has been described as restricted to the metathoracic ganglion (Surlykke and Miller 1982), should be considered tentative. The nonauditory B cell enters the pterothoracic ganglion on the same nerve trunk that carries the two auditory cells, but its CNS anatomy, like that of the A2 cell, has only been inferred from whole nerve staining (Agee and Orona 1988). These studies imply that the A1 and B cells project into the ventral intermediate tract, whereas the A2 cell projects anteriorly into the dorsal intermediate tract of the CNS (Boyan, Williams, and Fullard 1990). Because the B cell is a nonchordotonal cell, it is unlikely that its axon would project into the same neuropilar region as that of the chordotonal A1 cell, while this cell's partner, A2, enters a different tract. Investigations of receptor morphologies of other insect mechanosensory systems (e.g., cricket cerci, Murphey et al. 1983) indicate that particular receptor cell types have corresponding CNS destinations (Boyan 1989, 1993). It is possible that the A2 and B cell CNS morphologies have been misidentified and that it is the B cell that is restricted to the metathoracic ganglion.

Early studies of acoustically activated interneurons (Roeder 1966; Paul 1974) revealed two responses: repeater cells and pulse-marker cells. Repeaters exhibit a tonic spiking pattern, while pulse markers show a phasic, single-spike response (Coro and Alonso 1989). Boyan and Fullard (1986) intracellularly identified seven interneurons activated by the A1 receptor in the noctuid, *Heliothis virescens*, one of which, 501 (Fig. 8.4), may function as a noise filter, allowing the moth to ignore long-duration bouts of ultrasound, such as choruses of singing bushcrickets (Boyan and Fullard 1988). Boyan and Miller (1991) extended these findings by examining the responses of two interneurons, 501 and 504, to the prerecorded echolocation calls of attacking bats. Both cells exhibit postsynaptic summation to the rapidly repeated echolocation calls typical of the terminal phase of a bat's attack, suggesting that certain interneurons may provide specific information about the position of the approaching bat.

Although a critical part of the moth avoidance response circuit is the motor output, there has been only one intracellular study of ultrasound-evoked activity in moth flight motoneurons (Madsen and Miller 1987). Using artificial echolocation pulses as stimuli to the noctuid, *Barathra brassicae*, they examined the somal responses of 28 unidentified dorsal longitudinal flight motoneurons. Twenty-three of the cells probed were quiescent until acoustically stimulated, whereupon they responded with either depolarizing or hyperpolarizing responses (never both) that summated to increasing stimulus intensity and pulse repetition rates. Five cells were spontaneously active and could be excited or inhibited by acoustic stimulation but only when in fictive flight, suggesting that these cell's auditory responses are triggered only in the presence of a flight generator's activity and play a role in powered evasive flight responses to close-bat stimuli.

4. Moth Ear Origins

A polyphyletic origin for moth ears is suggested by the diversity of locations in which they presently reside. With some exceptions, the position of the ear of a moth can be predicted by which superfamily it belongs to (e.g., Noctuoidea, metathorax; Geometroidea, abdomen). The systematic ordering of the moth ear within the taxonomic structure of the Lepidoptera suggests that it evolved rapidly and simultaneously from some strong

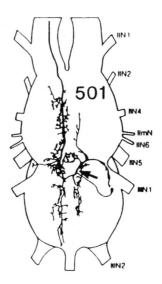

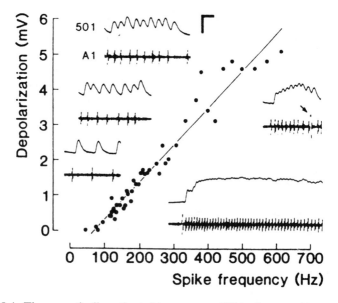

FIGURE 8.4. The acoustically activated interneuron 501 in the noctuid moth, *Agrotis infusa*. Top: This cell extends from the posterior margin of the metathoracic ganglion to at least the prothoracic ganglion. Bottom: Simultaneous extracellular recordings of A1 receptor (insets, bottom traces) and intracellular recordings of 501 (insets, top traces) indicate that the very short excitatory postsynaptic potentials of 501 do not summate to A1 firing levels less than 100 Hz and may act as a noise filter. (From Boyan and Fullard 1988, Springer-Verlag Publishers, with permission.)

selection pressure (e.g., echolocating bats). It should not be assumed, however, that all moth ears have arisen along predictable lines. Moth ears appear in a variety of diverse and unexpected taxa, and there are many examples that have not received attention since they were first reported. Roeder, Treat, and Vandeberg (1968) discovered the pilifer ears on the faces of certain hawkmoths, but the biophysics and the nervous system of this ear remains unstudied, as have the purported ears of certain axiids, tineids, dudgeoneids, thyridids, and hedylids (reviewed in Scoble 1992).

4.1. Interspecific Homology

A discussion of the evolution of moth ears requires definitions. *Eared* species have adaptively functional auditory systems that participate in some element of the insect's natural life (Yack and Fullard 1993a); *deaf* species are taxa whose ancestors had ears but lack them, partially or completely, today; and *earless* species have never had ears throughout their evolutionary history.

It has been suggested that insect tympanal ears evolved from specialized proprioceptor organs made of chordotonal sensilla (von Kennel and Eggers 1933; Pumphrey 1940; Haskell 1961). Certain chordotonal organs (COs) detect low-amplitude vibrations (e.g., subgenual organs that sense the substrate-borne vibrations of predators or conspecifics, Markl 1983) and appear preadapted as auditory receptors (Scoble 1992). Proprioceptor COs are scattered over the insect body but appear to be most concentrated on those segments associated with membranes (e.g., leg joints). These membranes provide attachment points for the COs and are themselves preadapted to function as tympana. The ears of present-day insects reflect this membrane affinity and morphological similarity to the ancestral CO proprioceptors from which they descended (Fig. 8.5). Boyan (1993, Chapter) summarized the evidence supporting the theory that insect auditory systems evolved as specialized chordotonal receptor networks already connected to preexisting interneurons. These data come from a diversity of insect orders and reflect the antiquity of these evolutionary adaptations.

Moths can provide considerable interspecific information about the adaptations of peripheral receptors that have led to the specializations of ears. Using anatomical, ultrastructural, and physiological results, Yack and her associates (Yack and Fullard 1990; Yack 1992; Yack and Roots 1992) proposed homology between the peripheral nervous system of earless moths (bombycoids and sphingoids) and the auditory nervous system of noctuoid moths (Fig. 8.6). These studies describe the peripheral branching patterns of the auditory nerve of noctuoids (IIIN1b) and its homologue in the earless species, *Actias luna* (luna moth) and *Manduca sexta* (tobacco hornworm moth). In addition to a striking similarity in destination points for the IIIN1b in the two auditory states, a three-celled CO originates at the

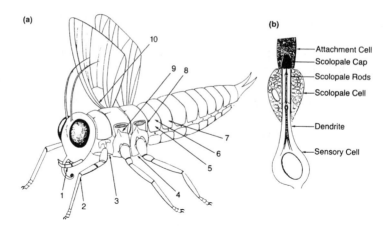

FIGURE 8.5. a: Locations where tympanate ears occur on the generalized insect body: 1, Lepidoptera (Sphingidae, not technically tympanate); 2, Orthoptera (Ensifera); 3, Diptera (Tachinidae); 4, Mantodea (Mantidae); 5, Lepidoptera (Geometroidea, Pyraloidea); 6, Orthoptera (Acrididae); 7, Hemiptera (Cicadidae); 8, Lepidoptera (Noctuoidea); 9, Hemiptera (Corixidae); 10, Neuroptera (Chrysopidae). b: The operational unit of the insect ear, the chordotonal sensillum. (From Fullard and Yack 1993, Elsevier Publishers, with permission.)

IIIN1b1 nervelet of earless moths in the homologous position as the two-celled auditory CO of eared moths.

The CO of earless moths is anchored by a long attachment strand to a membranous area at the base of the hind wing and appears to serve as a stretch receptor organ, monitoring its position. The anatomy and physiology of the hind-wing CO of earless moths (Yack and Fullard, 1990) suggests that this organ represents the preauditory, plesiomorphic condition of the present-day moth ear and provides phylogenetic evidence supporting the proprioceptor theory of insect ear origins. Interestingly, the CO of the luna moth responds to intense, low-frequency sounds, a trait shared by other chordotonal proprioceptors (e.g., the locust hind-wing CO; Pearson, Hedwig, and Wolf 1989), and it is possible to derive an "audiogram" of the hind-wing CO (Fig. 8.7). The extraordinary sensitivity of this CO suggests that in ancestral prebat moths it was already preadapted as an auditory CO for those moth lineages that eventually evolved ears (Yack and Fullard 1990; Hasenfuss 1997). The epiphenomenon of acoustic responsiveness in proprioceptors also serves as a warning to the mislabeling of ears in insects (Yack and Fullard 1993a).

Another element of the noctuoid ear is the B cell, a nonchordotonal, multipolar sensory cell that Roeder and Treat (1957) described to be unaffected by sounds. Yack (1992) and Yack and Fullard (1993b) proposed that a nonchordotonal cell associated with the hind-wing CO in the earless *M.*

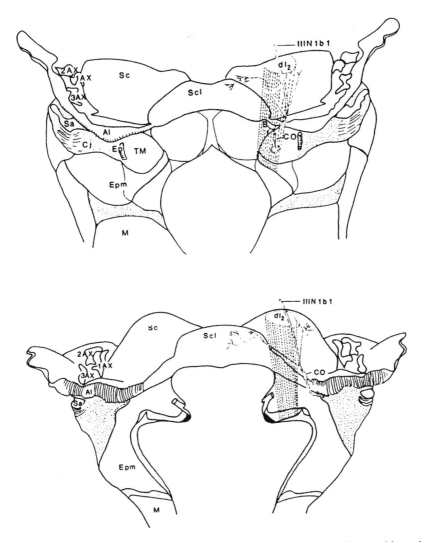

FIGURE 8.6. Comparison of the auditory metathorax of the eared noctuid moth, *Feltia heralis* (top), and the homologous, nonauditory metathorax of the earless saturniid moth, *Actias luna* (bottom). In both moths, a branch (IIIN1b1) of the hind-wing nerve travels past the dorsal longitudinal muscle (dl2) to the posterior margin of the metathoracic segment, but whereas in *F. heralis* this nerve ends in the auditory chordotonal organ (CO), it terminates in a homologous, nonauditory CO attached to the hind-wing membrane of *A. luna*. Sc, scutum; Scl, scutellum; AX, articular sclerites: Sa, subalar plate; Cj, conjunctivum; Ep, epaulette; Al, hind-wing alula; TM, tympanic membrane; Epm, epimeron; M, meron. (From Yack and Fullard 1990, Wiley-Liss Publishers, with permission.)

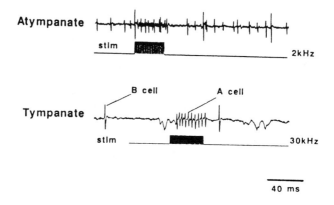

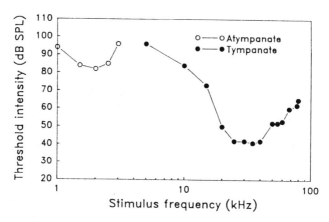

FIGURE 8.7. Top: An indication of the preadapted condition of the nonauditory Chordotonal organ (CO) in an earless (atympanate) moth (*Actias luna*) is seen in its receptor cells' response to low-frequency sounds. The ear of the eared (tympanate) arctiid moth, *Halysidota tessellaris*, responds to sounds that simulate echolocating bats. Bottom: The two COs also exhibit sensitivity curves, although it is the auditory CO of *H. tessellaris* that possesses the necessary sensitivity and tuning characteristics to detect the echolocation calls of hunting bats. (From Yack and Fullard 1990, Wiley-Liss Publishers, with permission.)

sexta is the homologue of the noctuoid B cell and demonstrated its proprioceptive qualities in monitoring the position of the hind wing (Fig. 8.8). Yack (1992) points out that a similar (perhaps homologous) organization exists in the wing-hinge receptor complex of locusts. The B cell of eared noctuids may therefore be an evolutionary leftover from a time when moths used the ancestor of this receptor for monitoring their hind-wing position. Whether this cell serves any function in present-day moths remains unknown.

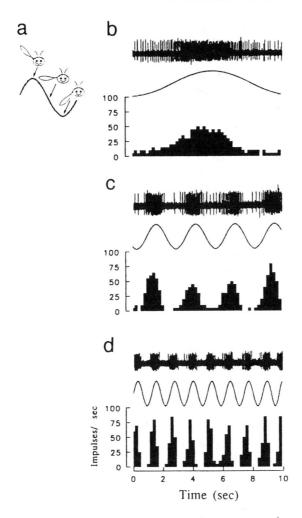

FIGURE 8.8. The nonauditory B cell of eared moths possesses a homologue in the earless sphingid moth, *Manduca sexta*. By moving the hind wing (a), this cell fires to tonically monitor its position through slow (b) or rapid (c, d) cyclic movements. (From Yack and Fullard 1993b, Springer-Verlag Publishers, with permission.)

4.2. Ontogenetic Homology

In addition to its interspecific auditory diversity, the Lepidoptera offers an ontogenetic route for testing the proprioceptor theory of ear origins. Moths are holometabolous insects with two very different life histories, a larval stage, which is earless to ultrasound in all species, and a flighted adult stage, which is eared in some species and earless in others. Caterpillars, untroubled by bats, do not bother listening for them, but the anatomy and physiology of their nervous systems provide clues as to the evolutionary origins of the auditory organs in eared species and their homologues in earless species.

Lewis and Fullard (1996) compared the metathoracic nervous systems of larval and adult gypsy moths (*Lymantria dispar*) with those of an earless species, the tent caterpillar moth, *Malacosoma disstria*. Neural homology, interspecific and ontogenetic, exists for these distantly related species at both the gross anatomical and cellular levels (Fig. 8.9). In the caterpillar, a branch of the larval IIIN1b nerve in both the eared and earless species ends in a three-celled proprioceptive CO and a single nonchordotonal, multipolar cell. Following metamorphosis, the CO and multipolar cell persists in both species, whereas other sensory structures serviced by the IIIN1b nerve (e.g., hair plates) are lost (Fig. 8.10). It appears that (1) the multipolar cell is retained in both species as the B cell, (2) the larval IIIN1b CO remains as a proprioceptor CO in the tent caterpillar moth, and, (3) the same organ becomes the auditory CO in the gypsy moth. Figure 8.11 further illustrates that peripheral nervous system homology is maintained in the receptor CNS projection patterns of the two species before and following metamorphosis. These results have recently been supported by Husenfuss (1997) with moths from different families.

These comparative results allow us to propose a pathway for the evolution of the bat-detecting moth ear. Ancestral moths, living in prebat times, were all earless, and the metathoracic IIIN1b CO was conserved as a proprioceptor in the larva to the adult, where it continued to serve this function. Once bats had evolved echolocation calls, the sensitivity of this CO preadapted it to the role of an ultrasonic ear. The function of the hind-wing CO in present-day earless moths suggests that it was also preadapted for evoking evasive flight behavior as an ear. The hind-wing COs of the locust, *Locusta migratoria*, although not true ears, can be activated by sounds and project to auditory centers of the CNS (Pearson, Hedwig, and Wolf 1989). Although they do not affect the wing beat frequency of this insect, the COs appear to function as vibration proprioceptors, influencing the steering behavior during flight. If ancestral moths were partially sensitive to sounds, the intense echolocation calls of early bats (Fenton et al. 1995) might have activated the hind-wing CO to affect flight in an preadaptive avoidance manner. Alternatively, the first auditory function of the hind-wing CO might have been to detect the sounds of approaching terrestrial predators, causing sudden jumping or flying escape responses similar to those seen today in acoustically activated startle responses (Hoy, Nolen, and Broadfueher 1989; Libersat and Hoy 1991). Increasing the sensitivity of the CO to achieve the status of an ear would not have required CNS rewiring to retain its effect on flight centers causing a bat-avoidance response. The existence of a proprioceptor auditory precursor in the caterpillar suggests that metamorphosis, with its massive tissue rearrangement, may have been the stage at which random changes in the sensillar/tympanal attachment occurred that allowed the

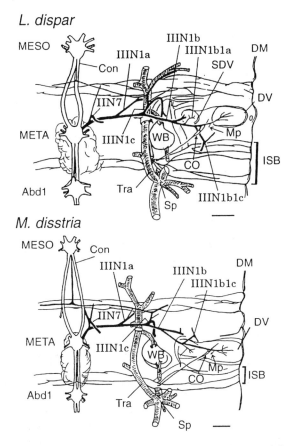

FIGURE 8.9. The larval metathoracic nervous system of the eared (as adult) gypsy moth, *Lymantria dispar*, and the earless (as larva and adult) tent caterpillar moth, *Malacosoma disstria*. Although these two species are not closely related, their peripheral nervous systems are almost identical. The premetamorphic auditory nerve (IIIN1b) in *L. dispar* and its homologue in *M. disstria* end in a chordotonal organ (CO) that attaches itself to the intersegmental membrane (ISB) of the meta-abdominal boundary. MESO, mesothoracic ganglion; META, metathoracic ganglion; Abd1, first abdominal ganglion; Con, ganglionic connective; WB, wing bud; SDV, subdorsal verruca/verricle; DM, dorsal midline; DV, dorsal verruca/verricle; Tra, tracheae; Sp, spiracle; Mp, multipolar cell. (From Lewis and Fullard 1996, John Wiley & Sons Publishers, with permission.)

appearance of auditory receptors and a new defense against the acoustic attacks of bats.

One problem with this scenario is the discrepancy of the 2 to 4 kHz "best frequency" of the preauditory tympana (see Fig. 8.7) and the 20 to 50 kHz calls of bats. If bats began echolocating with ultrasonic calls, it is difficult to envision a protoear with even slight sensitivity at those frequencies to begin the process of selective response. One possibility is that prebat Lepidoptera

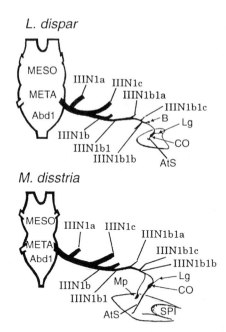

FIGURE 8.10. Following metamorphosis, the metathoracic nervous systems of the eared *L. dispar* resembles that of the earless *M. disstria*, except that the termination point of the IIIN1b1b nerve in the gypsy moth is the auditory chordotonal organ (CO), whereas in the forest tent caterpillar moth it is a putative proprioceptor. The nonauditory multipolar (Mp) B cell of the gypsy moth is represented in *M. disstria* by a homologue attached to the hind-wing base membrane. MESO, mesothoracic ganglion; META, metathoracic ganglion; Abd1, first abdominal ganglion; Lg, ligament; SPI, spiracle; AtS, attachment strand. (From Lewis and Fullard 1996, John Wiley & Sons Publishers, with permission.)

were *not* all earless, with some already ultrasonically tuned for listening to conspecific social signals (e.g., Surlykke and Fullard 1989). Although this explanation is not supported phylogenetically, considering the present-day paucity and unordered taxonomic appearance of singing moths the intense selective advantage of detecting bats could have overwhelmed the primitive social uses of early ears.

5. Moth Ears and Bat Calls: Coevolution or Coincidence?

While the moth ear has the advantage of representing a "simple" auditory nervous system, even simple systems are exposed to complex selection pressures. Since Roeder and Treat's studies in the 1950s, there have been many discoveries about the hearing and hunting abilities of bats (Fenton,

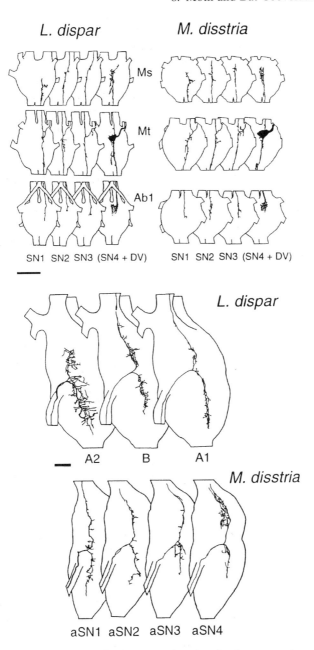

FIGURE 8.11. In larval *L. dispar* and *M. disstria*, the four receptor cells of the metathoracic chordotonal organs (CO) project into the same areas of the central nervous system. After metamorphosis, *M. disstria* retains all four cells projecting to the same ganglionic destinations, but *L. dispar* loses one and shrinks the projection patterns of another to the metathoracic ganglion. (From Lewis and Fullard 1996, John Wiley & Sons Publishers, with permission.)

Racey, and Rayner 1987; Popper and Fay 1995), and these adaptations have placed diverse selection pressures on the defensive systems of sympatric moths. To examine how ears have assisted moths in surviving bats and whether coevolution exists between these two participants first requires an appreciation of the different ways bats forage for their dinners.

5.1. Bat Foraging: Aerial Hawking versus Substrate Gleaning

The best understood type of foraging bat is the aerial hawker, which catches flying insects on the wing using a sophisticated sonar system to determine the position, distance, and perhaps identity of its prey. Using North American species (*Eptesicus fuscus* and *Myotis lucifugus*) as models, Kick and Simmons (1984) describe the phases that aerial hawking bats exhibit as they approach and capture a target. They propose four sequences (Fig. 8.12): (1)

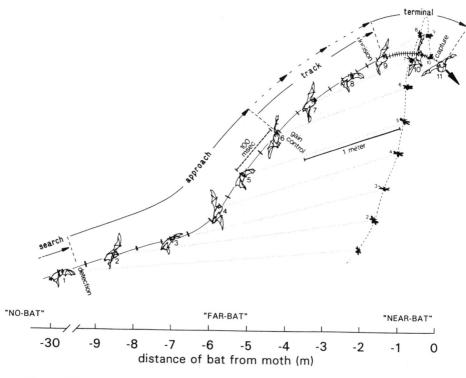

FIGURE 8.12. As a bat attacks a fleeing moth, it progresses through echolocation phases that present cues to the moth as to its proximity and intent. Moths perform evasive behaviors based on their estimation of the distance of the approaching bat from these cues. (Adapted from Kick and Simmons 1984, Society for Neuroscience Publishers, with permission.)

search (>3 m from target), in which the bat is emitting echolocation signals but receiving no echoes; (2) approach (3 to 1.5 m from target), in which the bat has detected the echo of a potential prey; (3) tracking 1.5 to 0.5 m from target), in which the bat makes a decision whether or not to attack the target; and (4) terminal (<0.5 m), in which the bat attacks its target. Common to all aerial hawkers is a reduction of their echolocation interpulse intervals as they enter the terminal phase. This cue is encoded by a moth's auditory cells and appears to be important in enabling it (and other eared insects) to distinguish between situations of "no-bat," "far-bat," and "near-bat" (Roeder 1967; Miller and Olesen 1979; Fullard 1987a; Yager and May 1990).

Evidence now suggests that it was to aerially hawking bats that the moth ear originally evolved and to which it now maintains its physiological design, but these bats are not the only threats that a moth faces at night. Substrate-gleaning bats hunt by using vision and/or by listening for the sounds created by insects (e.g., singing, leaf-litter rustling) and rely upon their echolocation more as a navigational tool than as a target detector. As they search for their prey, gleaners forage near to the ground or surrounding vegetation (Tuttle and Ryan 1981; Belwood 1988), using flight that presents orientation tasks requiring echolocation calls adapted to dealing with the problems of acoustic clutter (i.e., multiple echoes) and self-deafening (Fenton et al. 1995). As a result, the calls of gleaners tend to be high frequency, low intensity, of short duration, and emitted with a low duty cycle (some gleaners completely turn off their echolocation as they approach their target; reviewed in Neuweiler 1990; Faure and Barclay 1994). These call characteristics severely reduce their acoustic conspicuousness ("apparency," Waters and Jones 1996) and could allow gleaners to enjoy a significant foraging advantage over eared insects (Faure, Fullard, and Barclay 1990; Faure, Fullard, and Dawson 1993).

The result of these varied foraging tactics is that bats end up in all of the places used by moths. Aerial hawkers favor "open" habitats (e.g., above the forest canopy or in clearings), and substrate gleaners spend their time in "closed" habitats, zones close to vegetation (Fig. 8.13). Most bats probably fall between these two extremes and hunt in "edge" zones, foraging near to trees but occasionally entering the underbrush to pursue fleeing insects (Neuweiler 1984; Fenton 1990). For the moth, this diversity of bat-hunting strategies results in a mosaic of predation pressures, and it is necessary for us to understand the types of selection pressures that exist between these two combatants in the natural world.

Figure 8.14 illustrates three levels of selection pressures that hunters exert upon the hunted as levels of overlap that the two experience in their life histories (e.g., activity patterns, geographic distributions). In Figure 8.14A, the life histories of potential predators and potential prey are completely separated and experience no coevolutionary potential (e.g., lions and penguins). Figure 8.14B describes a more obvious situation: a common

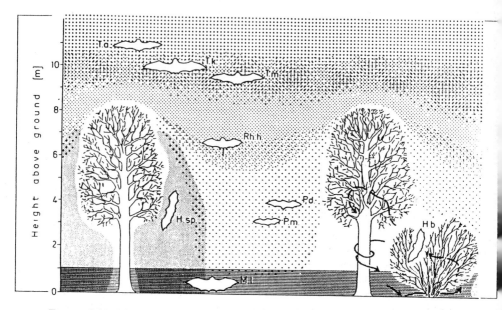

FIGURE 8.13. The foraging zones of an Indian community of bats. Whereas some bats hunt in the open zones above the tree canopy, others prefer edge habitats, and some restrict their foraging to very close to the vegetation, gleaning insects from the leaves or the ground. T.a., *Tadarida aegyptiaca*; T.k., *Taphozous kachhensis*; T.m., *Taphozous melanopogon*; Rh.h., *Rhinolophus hardwickei*; P.d., *Pipistrellus dormeri*; P.m., *Pipistrellus mimus*; H.sp., *Hipposideros speoris*; H.b., *Hipposideros bicolor*; M.l., *Megaderma lyra*. (From Neuweiler 1984, Springer-Verlag Publishers, with permission.)

predator in the same place and time as its intended prey (e.g., lions and gazelles). Here, we would expect a considerable degree of the prey's sensory abilities and defensive behavior directed toward that particular predator and coevolutionary adaptations would be specific and obvious. Figure 8.14C illustrates a more complicated situation but one that may describe much of the real world. This figure predicts that only a certain portion of a predator's life history will impact on its intended prey. If this overlap is large, animals will evolve adaptive antipredator behaviors, but if the overlap is negligible, these responses may never appear. An example of this has been the extinction of certain ground-nesting Hawaiian birds following their contact with introduced predators such as humans and mongooses (reviewed in Tomich 1986).

The fact that entire species can be driven to extinction because they have not had enough time to evolve defenses against novel predators demonstrates that prey can be constrained by their phylogenetic history and that these limitations can be exploited (even if by accident) by predators. In naturally coevolved populations, if the selection pressure of a rare predator

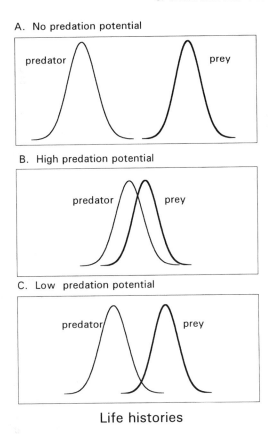

A. No predation potential

predator   prey

B. High predation potential

predator   prey

C. Low predation potential

predator   prey

Life histories

FIGURE 8.14. Three levels of predator–prey overlap. As potential predators increase their life-history overlap with that of prey, the prey will be increasingly influenced by the predator selection pressure and will evolve more and more specialized defenses against them.

remains low enough, it should be possible for it to remain in the community as a "cheater," exploiting the equilibrium between the primary predators and their intended prey. We will see how the overlap of hunter and hunted illustrated in Figure 8.14 is represented in the coevolution of moths and bats.

5.2. Evidence for Coevolution: Moths versus Bats

Roeder (1970) first proposed that moth ears are adaptively matched to the calls of sympatric bats. Audiograms of North American noctuid moths demonstrate a spectral tuning of these moths to frequencies between 20 and 50 kHz, the bandwidth commonly used by foraging bats in this area (Fig. 8.15). The broadness of the tuning curves of moths is a result of the diversity

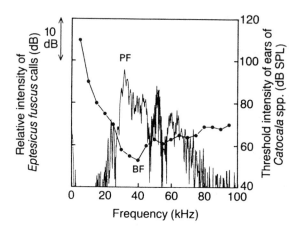

FIGURE 8.15. The sensitivity of moth has its best frequency (BF) matched to the peak frequency (PF) of the relevant bat predators with which it coevolves. In this figure, the moth audiogram is that of a population of Canadian noctuids, while the frequency spectrum is that of the echolocation calls of a sympatric moth-feeding bat, *Eptesicus fuscus*. (From Fullard and Yack 1993, Elsevier Publishers, with permission.)

of bats that these ears must detect. Bats emit species-specific echolocation calls (Novick 1977; Fenton and Bell 1981), and a community of these predators will emit a range of frequencies determined by the diversity of that community. Moth ears, being tone-deaf, cannot discriminate one bat from another and so, have evolved ears matched to the spectrum of the combined frequencies of all of the bats ("echolocation assemblage," Fullard 1982) that form significant predation pressures.

An echolocation assemblage has been described for the lowland rain forest of central Panamá (Fullard and Belwood 1988), where there are over 50 species of bats representing every known foraging strategy. The frequency spectra of the echolocation signals of 37 of these species (Fig. 8.16) suggests, at first, a daunting array of sounds for which moths need to listen. However, not all bats are insectivorous and the calls of some species (e.g., strict frugivores) should not influence the ears of sympatric moths. By combining the individual spectra of Figure 8.16, we can derive the spectrum of the total echolocation assemblage of this community (Fig. 8.17A). By comparing the audiograms of Panamanian notodontid moths with the assemblage spectrum, it appears that their ears are tuned to the calls of those bats that form only the relevant (i.e., moth-eating) predation potential (Fig. 8.17B–D).

That moths do not listen for bats that do not hunt them may seem a trivial observation, but the broad antibat tuning that has resulted from this selection pressure carries with it some costs. Figure 8.18 illustrates the echoloca-

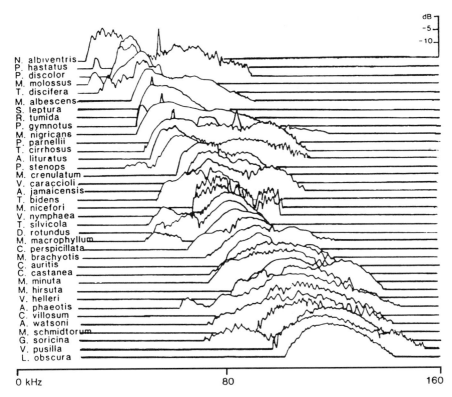

FIGURE 8.16. The frequency spectra of 37 species of bats from a lowland Panamanian rain forest ordered from the lowest to the highest peak frequency (PF). All of the bats recorded are included in this figure although many (e.g., nectar feeders) do not present a predation potential to the ears of moths. (From Fullard and Belwood 1988, Plenum.)

tion frequencies within (syntonic) the maximum sensitivity of the moth's ear and those outside (allotonic) the moth's best frequency bandwidth (Fullard 1987a) compared to the peak frequencies of a variety of syntonic and allotonic bats. The advantage of allotonic over syntonic bats to come closer to eared moths before they are detected depends upon the degree to which the peak frequencies of their calls are mismatched to the ears of the moths they are hunting. Roeder (1970) proposed that moths are deaf at allotonic frequencies (e.g., those used by singing frogs and crickets) because they have nothing to listen for at these bandwidths. We will find that Roeder was only partially correct in his idea about allotonic deafness in moths and that there are hidden dangers at these frequencies.

The first part of our examination of the coevolution of moth ears and bat calls requires a demonstration that moths have specifically and independently adapted their ears to the bats with which they are sympatric. This is

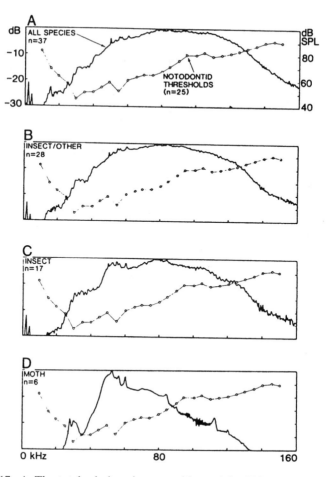

FIGURE 8.17. A: The total echolocation assemblage of the 37 bat species of Fig. 8.16 as one frequency spectrum plotted against the median audiogram of 25 sympatric notodontid moths. The frugivorous species have been removed from the assemblage B, only the insectivorous species are shown C and only those bats reported as feeding on moths are included D. (From Fullard and Belwood 1988, Plenum.)

possible for two convenient reasons: First, bat calls are species specific (Fenton and Bell 1981; Figs. 8.16 and 8.18), so the echolocation assemblage of an area will represent the bats (and their diets) that comprise it, and, second, the global diversity of bats and their calls allows us to compare moths' ears from substantially different echolocation communities. The echolocation assemblages of speciose African bat communities (e.g., Côte d'Ivoire and Zimbabwe) resemble those of Panamá, with many more species using allotonic bandwidths than those in North American sites (e.g., Ontario and Arizona; Fenton and Fullard 1979). Correspondingly,

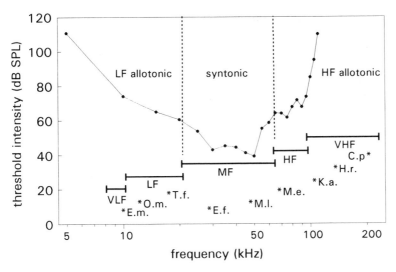

FIGURE 8.18. Terms used in this chapter regarding the echolocation frequencies used by bats as seen by the evolutionary perspective of the moths' ears that listen for them. Bats emit echolocation calls with peak frequencies that are very low frequency (VLF, e.g., E.m., Euderma maculatum); low frequency (LF, e.g., O.m., Otomops martiensseni, T.f., Tadarida fulminans); midfrequency (MF, e.g., E.f., Eptesicus fuscus, M.l., Myotis lucifugus); high frequency (HF, e.g., M.e., Myotis evotis); and very high frequency (VHF, e.g., K.a., Kerivoula argentata, H.r., Hipposideros ruber, C.p., Cloeotis percivali). Bat peak frequencies (asterisks) from Fenton and Bell (1981).

moths sympatric with these assemblages have ears that are more sensitive to high and low frequencies than those exposed to less diverse bat environments (Fullard 1982; Fullard, Fenton, and Furlonger 1983).

At first, the explanation seemed clear — allotonic bat calls produce allotonically sensitive moth ears. Qualifications to this have arisen in recent years, however, with our increased understanding of the natural biology of bats. It now appears that the very high frequency (VHF; see Fig. 8.18) bats may be substrate-gleaning species that do not contribute to the selection pressures acting on moths' ears (see next section). Nevertheless, it appears that there are enough high-frequency (HF; see Fig. 8.18) aerial hawkers to exert selection pressure that has resulted in increased allotonic sensitivities in sympatric moths. Two species of neotropical notodontid moths have pinna-like external auditory modifications that increase their high-frequency sensitivities, apparently to deal with the acoustic diversity of the predatory bat community with which these moths are sympatric (Fullard 1984a, 1987b).

At the other end of the frequency spectrum, a special case of low-frequency sensitivity exists in the ears of Hawaiian noctuids, moths exposed to the hunting pressure of only one species of bat, *Lasiurus cinereus semotus* (Belwood and Fullard 1984; Tomich 1986). The isolated evolutionary habitat offered by certain oceanic islands provides for a natural laboratory (Simon 1987) in which to observe coevolutionary events more easily than in more complex regions of the world. Endemic noctuids analyzed on Kaua'i (Fullard 1984b) possess ears that superficially resemble those of mainland populations, but they are significantly more sensitive in the very low frequency and low-frequency end of the spectrum (Fig. 8.19A). This increased sensitivity may exist as a result of these moths listening for the 10 kHz agonistic calls that *L. c. semotus* broadcasts to conspecifics as it forages around local insect concentrations. In the absence of other bats, the low-frequency social call of *L. c. semotus* provides for an unusual cue for detecting the whereabouts of the sole species that Hawaiian moths have evolved with.

Support for this is seen by comparing the audiograms of a pan-Pacific species, *Ascalapha odorata*, sampled in Kaua'i and Panamá (Fig. 8.19B). As with assemblages of moths, single species reveal similar adaptive increases in sensitivity that are appropriate to their predatory habitat. Although it could be argued that the taxonomy of *A. odorata* is poorly understood and we are actually comparing two separate species, the phenotypic changes accompanying this speciation probably include auditory sensitivities appropriate to the bat predation-potential levels of these two habitats.

5.3. Evidence for Coevolution: Bats versus Moths

Although it is apparent that moths evolved ears to detect bats, demonstrating coevolution requires showing that some bats have evolved countermaneuvers to circumvent these auditory defenses. The ears of moths are syntonic with the bandwidth of frequencies of the bats that form their heaviest predation potential, but allotonic frequencies will be more difficult to detect and bats that use them will appear further away to the moth. Figure 8.18 predicts what echolocation frequencies are theoretically suited as countermaneuvers against moth ears. Bats that use extremely allotonic echolocation frequencies (i.e., very low frequency and very high frequency, see Fig. 8.18) can exploit the moth's total deafness at these unusual frequencies as an advantage to allow them to approach them completely undetected.

5.3.1. High-Frequency Allotonic Echolocation

Some bats echolocate with extraordinarily high frequencies (e.g., *Cloeotis percivali*: 212 kHz; Fenton and Bell 1981), far beyond the auditory abilities of sympatric moths (Fullard 1982). The undetectable qualities of these sounds to moth ears gave rise to the idea that allotonic echolocation evolved as a method to circumvent the auditory defenses of moths (Fenton

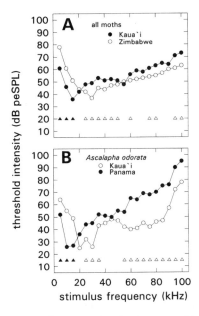

FIGURE 8.19. A: The mean audiogram of endemic noctuoid moths collected on the Hawaiian island of Kaua'i versus that of an African population sampled in Zimbabwe. The moths in the bat-reduced Pacific island have significantly lower sensitivities (indicated by open triangles) in the MF, HF and VHF bandwidths but have ears that are more sensitive in the VLF and LF bandwidths (indicated by filled triangles) because of the special acoustic cues provided to them by this island's single bat species, *Lasiurus cinereus semotus*. B: The mean audiograms of the noctuid *Ascalapha odorata* collected on Kaua'i compared to those measured in Panamá. The same differences witnessed between populations of moths are observed within a single species. (Adapted from Fullard 1984, Springer-Verlag.)

and Fullard 1979; Fullard and Fenton 1980). African allotonic bats approach closer to moths before their echolocation calls are detected (Fullard and Thomas 1981), and the fact that allotonic bats from a variety of locations take a high percentage of moths in their diets (Whitaker and Black 1976; Jones 1992) supports the theory of high-frequency countermaneuvering by bats.

Fullard and Thomas (1981) and Fullard (1987a) suggested that in addition to mismatching their call frequencies, bats could reduce the durations and/or intensities of their echolocation pulses to render themselves even more inconspicuous. Very short calls would evoke single-receptor spikes in the moth's ears, resembling spontaneous firing (e.g., "no-bat" condition, see Fig. 8.12), treated by interneuronal filters within the moth's CNS as extraneous noise and ignored (Boyan and Fullard 1988). Although screening out environmental noise would be advantageous for moths, Waters (1996) demonstrates that this filtering could be costly because it would allow short-pulsing bats to approach within echo-detecting distances before a moth begins evasive maneuvers.

Waters and Jones (1996) used the prerecorded calls of six species of syntonic and allotonic bats as stimuli to the ears of a sympatric noctuid moth and computed the predicted detection distances of the moth's ears to the bats. Whereas certain high-frequency allotonic bats (e.g., *Rhinolophus hipposideros*) elicit short detection distances comparable with those witnessed in Africa using wild bats (Fenton and Fullard 1979; Fullard and Thomas 1981), other supposedly allotonic bats are more detectable (e.g., *R. ferrumequinum*). Waters and Jones (1996) suggest that the inconspicuousness of the allotonic signals of constant frequency (CF) bats is compromised by their long durations and shortening them can reduce their detectability. Bats hunting in close quarters routinely shorten their echolocation pulses, presumably to avoid pulse-echo overlap (Fenton et al. 1995), but for allotonic bats a coevolved benefit of this could be an increase in their calls' inconspicuousness.

As mentioned earlier, another use for very high frequency echolocation signals is an increased resolution of small targets. Closed habitat foraging requires a gleaning bat to be aware of the many obstacles presented to it (e.g., branches), and many bats that hunt in these conditions use high-frequency echolocation (Neuweiler 1990). The long-eared bats, *Myotis evotis* and *M. septentrionalis*, are vespertilionids that glean in North American forests using short, high-frequency allotonic calls that are very undetectable to sympatric moths (Faure, Fullard, and Barclay 1990; Faure, Fullard, and Dawson 1993). Resting moths, their ears covered by their wings, are extremely vulnerable to gleaners, but the bats themselves are constrained by the clutter that surrounds their intended prey.

The question regarding coevolution is which came first: increased echo resolution or moth ear countermaneuver? An answer to this may lie in the dietary composition of very high frequency gleaners. If moth ear circumvention did not play a role in the evolution of very high frequency echolocation, we would expect to see a diverse array of insect types represented in the diets of these bats because many earless insects (e.g., beetles) also spend their time walking about on leaf surfaces. It appears, however, that moths form a very high proportion of the insects selected (Whitaker and Black 1976; Jones 1992), which suggests that although navigational functions for high-frequency echolocation presently exist, coevolution against the auditory defenses of moths was an important factor selecting for its evolution.

Although most bats support the prediction that allotonic echolocation allows for greater moth foraging efficiency (Jones 1992), the picture is not clear for all allotonic bats. Waters and Jones (1996) demonstrated that the leaf-nosed bat, *Rhinolophus ferrumequinum*, emits a 83-kHz call that is allotonic to sympatric moths and yet is readily detected by the ears of those moths. As previously described, this conspicuousness may arise from the long duration and high intensity of this bat's calls, qualities that compromise their allotonic nature. Although this bat feeds heavily on Lepidoptera (Jones 1990, 1992), the percentage of eared moths in their diet (which can be estimated only from identified moth remains; fecal pellets may contain earless species) is low relative to other insects throughout the summer (Jones 1990), so this bat may not eat as many eared moths as suspected. The North American and Hawaiian hoary bats, *Lasiurus cinereus cinereus* and *L. c. semotus*, are bats whose moth-eating habits (Black 1972; Belwood and Fullard 1984) also seem to be at odds with their syntonic echolocation. Assuming that the moths these bats eat can hear, it remains a puzzle why they take as many moths in their diets as they do. Ears, like helmets in wartime, only offer partial defense, and bats such as *R. ferrumequinum* and *L. cinereus* may have developed echolocation emission patterns (e.g., narrow-beam projection) or flight behaviors that compensate for their conspicuous calls and allow them to capture escaping moths. We should not forget that the process of coevolution predicts that for maneuvers there will be countermaneuvers, some of which may be as cryptic to us as they are to the moths.

5.3.2. Low-Frequency Allotonic Echolocation

Although testing the theory that high-frequency allotonic echolocation evolved as a countermaneuver against moth ears is confounded by the presence of the other acoustic functions for those frequencies, low-frequency allotonic calls (see Fig. 8.18) possess none of the advantages of high-frequency calls and, in fact, present the bat that uses them with significant problems. Low-frequency echolocation, although persisting longer in the atmosphere, returns poor resolution echoes from the items it encounters because of its long wavelengths (Griffin 1971; Lawrence and Simmons 1982). Low-resolution echoes increase the minimum size of the target that a bat can detect, and their ineffectiveness against closed-zone clutter constrains the bat to forage in open habitats where insects may not be as abundant. Eared moths have good odds for detecting hunting bats flying in these habitats if they emit syntonic calls, but if a bat could use low-frequency allotonic calls while flying in this habitat, it would increase its chances of drawing near to a moth before it detects its presence.

   Which bats echolocate with low-frequency allotonic calls? Many species of free-tailed bats (Molossidae) favor this form of echolocation (Fenton and Bell 1981). *Tadarida australis* emits a human-audible 12.6-kHz call in western Australia (Fullard et al. 1991), whereas the European species, *T. teniotis*, may use its 11 to 12 kHz echolocation (Zbinden and Zingg 1986) to come more closely to eared moths (Rydell and Arletazz 1994). In North America, the vespertilionid, *Euderma maculatum*, emits the lowest echolocation call yet described, with a peak frequency of 9 to 11 kHz (Woodsworth, Bell, and Fenton 1981; Leonard and Fenton 1984; Obrist

1995), and reports that *E. maculatum* feeds almost exclusively upon moths (Pouché 1981; Wai-Ping and Fenton 1989) make it reasonable to suppose that very low frequency echolocation and high moth dietary preference are related. Fullard and Dawson (1997) tested this idea by exposing the ears of moths to the prerecorded calls of hunting *E. maculatum* and observed that they were deafer (some completely so) to these calls than to the 25- to 35-kHz syntonic calls of another sympatric species, *E. fuscus* (Figs. 8.20 and 8.21). Very low frequency allotonic bats, like *T. teniotis* and *E. maculatum*, may be species whose unusual calls have coevolved in direct countermea-

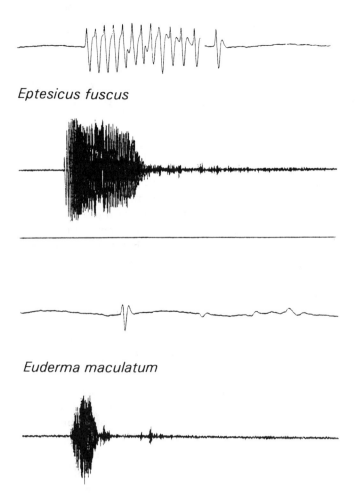

Figure 8.20. Auditory receptor responses of a western Canadian moth *Zotheca tranquilla* to the echolocation calls of the syntonic bat, *Eptesicus fuscus* and the low frequency allotonic bat, *Euderma maculatum*. (From Fullard and Dawson, 1997, Company of Biologists.)

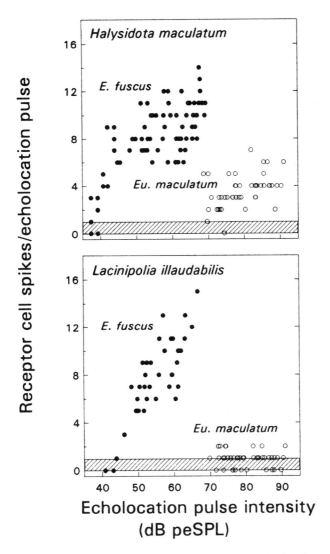

FIGURE 8.21. The auditory receptor responses of the moths, *Halysidota maculatum* and *Lacinipolia illaudabilis* to the prerecorded echolocation calls of *Euderma maculatum* (open circles) and *Eptesicus fuscus* (closed circles). The shaded band represents a firing pattern of one spike per echolocation call, responses that are probably rejected by the moth's CNS as noise. (From Fullard and Dawson 1997, Company of Biologists.)

sure to the auditory capabilities of moths because there seems to be little other adaptive value for this type of echolocation.

If allotonic bats are so successful at capturing eared moths, why haven't moths evolved better ears to detect them? As described in Figure 8.14, the selection pressure exerted by a rare predator may be insufficient to cause an evolutionary response in prey. The author suggests that allotonic bats, with their inconspicuous echolocation calls, form a small proportion of bat communities (Fullard et al. 1991; Fenton et al. 1995) and their selection pressures have never been strong enough to favor ears that can detect them. Because an all-sensitive ear, like a perfect microphone, is a physical impossibility, moths have been evolutionarily tuned to the major source of predation pressure in their lives: aerially hawking bats with their intense, syntonic calls.

An apparent paradox results from this situation: the most dangerous bats provide the least selection pressure. The paradox, however, is a "cannot see the forest for the trees" mistake. As fascinating as allotonic bats are, they are the exception (the "trees") compared with the majority of bats that echolocate in the air with syntonic calls (the "forest"). Faced with the need to protect moths against the bats that present the heaviest predation potential, evolution has left the ears of moths powerless against the few bats that exploit the coevolved status quo between most bats and most moths. We will see how this sensory deficit has selected for different defenses in earless moths.

6. Earless Ways of Dealing with Bats

Although most of the attention in the moth–bat story has focused on the ability of moths to detect the echolocation calls of hunting bats, it is worthwhile to examine the defenses of earless moths. These insects serve as a reminder that there are nonauditory ways of dealing with echolocating bats but that lacking ears may limit their evolutionary success. With some exceptions, moths in the superfamilies Sphingoidea (hawkmoths) and Bombycoidea (silkworm moths) are earless as adults and yet, at first glance, inhabit similar habitats as insectivorous bats.

Habitats, temporal and spatial, are composed of varied microhabitats, and it is in these that earless moths have found their protection. Yack (1988) demonstrated that earless moths emerge during the early parts of the summer in a Canadian site before most bats in this site arrive and before they shift their foraging activities from shoreline to overland (i.e., moth-inhabited) sites. Morrill and Fullard (1992) and Lewis, Fullard, and Morrill (1993) tested the hypothesis (Roeder 1974) that earless moths defend themselves by flight patterns that physically conceal themselves from bats. They found that, compared with eared moths, earless moths spend less time in the air and, when they do fly, remain closer to the ground and fly more erratically. For bats that spend most of their hunting time in open or edge (Neuweiler 1984; Fenton 1990) habitat zones, these flight patterns provide for an effective defense as it would keep those moths away from the bulk of bat "traffic," where the predation pressures are the highest. They would not, however, be as effective against closed-zone bats (e.g., Fig. 8.13), who hunt within the underbrush that these insects use. It appears that ears in moths are the norm when considering the total number of individuals flying at any one time, suggesting that earless defenses, although effective, may restrict the activities of the moths that use them and reduce their evolutionary "success."

7. Selection Relaxation

If moths have evolved auditory defenses specifically designed to deal with the predation potential of bats, we should expect to witness changes in the ears of moths that have evolved life histories that have relaxed or eliminated the selection pressures of bats. We will see how moths have done this by (1) migrating to bat-free habitats (spatial isolation), (2) moving into bat-free times (temporal isolation), and (3) adopting flight patterns that conceal them from hunting bats (behavioral isolation).

7.1. Spatial Isolation

Assuming that bats are the main selection force acting on the maintenance of moth ears, it is reasonable to predict that moths that have evolved in areas free of bats will show signs of deafness. This prediction carries with it two caveats. First, the habitat sampled to test this prediction must have been bat free for an evolutionary significant amount of time. Although bat-free habitats are rare (a testimony to the worldwide niche-filling abilities of bats), there are some places that satisfy this requirement. The second, and more difficult, condition to satisfy is that the moths sampled in these places must represent taxa that have resided there long enough for the effects of selection relaxation to be expressed. A moth that had accidentally arrived on an isolated oceanic island by an intercontinental flight a few days before is an obviously poor representative of the physiological condition of species that have resided thousands of years, but how can we determine the length of time that species have existed in particular habitats? Observations of the gradual changes in proteins and nucleic acids that occur over time allows an estimate of low long those taxa have been separated from their parent stock and thereby their time in genetic isolation (Avise 1994), but these techniques are young and still relatively field incompatible. In the meantime, assuming that taxonomy reflects phylogenetic relationships, if we sample from areas with endemic species, those that occur nowhere else, we can be somewhat confident that these taxa have evolved in isolation for some time.

With these precautions in mind, the bat-free habitats of oceanic islands have been used to search for auditory changes in moths released from bat predation pressure. The first study was on the Hawaiian island of Kaua'i, chosen because of this area's extraordinarily high level of insect endemism (Zimmerman 1958) and because of its single species of bat, *Lasiurus cinereus semotus* (see Section 5). We first thought that this island, with its small bat community, would contain little enough predation pressure that endemic moths would be completely deaf. True to Murphy's law, however, *L. c. semotus* turned out to be a voracious moth feeder, similar to its North American congener, *L. c. cinereus* (Whitaker and Tomich 1983). Field observations of foraging Hawaiian bats (Belwood and Fullard 1984) revealed that they persistently attack moths, which, in turn, respond with evasive flight maneuvers typical of moths in other parts of the world. Not surprisingly, these moths have maintained syntonic auditory sensitivities (see Fig. 8.19) comparable with those of African moths (Fullard 1984b), and it is apparent that the bat community of Hawai'i still exerts a sufficiently strong selective force to maintain functional ears in moths.

Although the ears of Hawaiian moths are sensitive at the frequencies they need to be, they reveal evolutionary changes that have been induced by their reduced predatory environment. Compared with African and North American moths, Kauaian moths exhibit significant deafness and increased threshold variability to frequencies above those used by *L. c. semotus* (see Fig. 8.19). Presumably, with no bats to listen for at these allotonic frequencies, the ears of Hawaiian moths have begun the evolutionary process of desensitization, beginning at the high-frequency end of the acoustic spectrum. Study of the endemic moths of Hawai'i reveals that degenerative changes in auditory systems occur at frequencies not selected upon but that even single species of bats can place a strong stabilizing evolutionary force on the ears of sympatric moths. To witness more profound sensory changes in moths, we must go to places where bats are completely absent.

Surlykke (1986) examined the ears of moths from the bat-free islands of the Faeroes in the North Atlantic. These moths, although completely released from bats, possess neurally responsive ears with best-frequency (BF) sensitivities comparable with those of mainland European populations. Surlykke (1986) concluded that these moths have retained their ears from a time when they were selectively maintained as antibat defenses, but Fullard (1987a) suggested that the moths on the Faeroe Islands are not genetically isolated from those of mainland individuals whose ears are maintained by bats.

To meet the criteria previously described, Fullard (1994) studied moths on the islands of French Polynesia, a region that has always been bat free and possesses a community (albeit shrinking from the pressures of human activities) of endemic moths. In this study, endemic species were compared with immigrant species, with the assumption that immigrants express greater genetic similarity with their mainland (i.e., bat-influenced) conspecifics. It was hoped that the endemic species of French Polynesia would have experienced enough evolutionary time freed from bats to show significant signs of bat deafness. Supporting this prediction was the observation of Clarke (1971) of the Polynesian pyralid moth, *Lathrotelis obscura*, having no ears whatsoever. This moth belongs to an otherwise eared family, and its complete deafness strengthened the case for island moth bat deafness.

Figure 8.22 summarizes the results from our studies with Polynesian moths that have expanded the picture of auditory degeneration first observed on Kaua'i. As on Kaua'i, the endemic noctuids of French Polynesia possess neurally responsive ears with sensitivity curves similar to immigrants in the 20 to 50 kHz region but with significant levels of high-frequency deafness to frequencies above 35 kHz. The results of our island studies indicate that moth hearing persists vestigially in the absence of the selection pressure that originally favored its evolution. The slim cellular investment represented by the two cells in the auditory organ of noctuid moths probably allows this structure to continue. For bat-freed noctuid moths, it appears that two auditory receptor cells are less costly than the behaviors they evoke and, therefore, persist longer.

Is there another reason for the persistence of ears in island moths? One possibility, conspecific social communication, should be especially evident in island moths because bat pressure has been removed. If all of the ears of these moths were now used for social sounds, we would expect to see (and

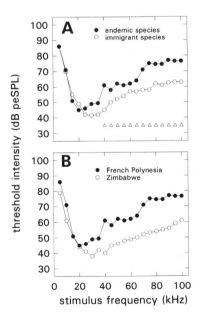

FIGURE 8.22. Endemic noctuid moths from the bat-free islands of French Polynesia are significantly deafer at high frequencies than immigrant species and those sampled in the bat-rich habitat of Zimbabwe. (From Fullard 1994, Birkhäuser Verlag.)

hear) a plethora of singing species. Island moths, however, show no greater tendency for social sounds than do mainland species; in fact, there have been no reports of singing oceanic island moths. Alternatively, the ears of bat-released moths might now serve their owners by allowing them to hear the crackling underbrush of approaching terrestrial predators. The tuning characteristics of these moths, and others for which this function has been proposed, are maladapted for detecting the lower frequency sounds of underbrush (Fullard 1988). In the absence of other uses, it appears that island moths, released from the need to possess bat-detecting ears, have begun a progression towards total deafness. This degeneration appears to begin at the high-frequency end of the spectrum (cf., the aging ears of mammals; Willott 1991) and is more subtle than originally expected. In the next section we see how another group of predator-released moths has experienced a more pronounced version of auditory degeneration that has substantiated these conclusions.

7.2. Temporal Isolation

Because bats are, almost without exception, nocturnal animals, another way that moths could escape them would be by adopting diurnal life histories. Exclusively diurnal moths are rare (probably due to the intense pressure exerted by daytime predators, e.g. birds), but they do exist in scattered locations around the world. One group of temporally isolated moths are the day-flying dioptines (Notodontidae) of Central and South America. These brilliantly colored insects exhibit a number of unmothlike characteristics, in addition to their daytime activities, that have long confounded taxonomists (reviewed in Miller 1991). Certain dioptines are completely earless (Miller, personal communication), while others, examined in western Venezuela, exhibit levels of high-frequency deafness ranging from slight to severe (Fig. 8.23; Fullard et al. 1997).

The bat deafness of certain dioptines (e.g., *Xenorma cytheris*), compared with nocturnal moths, represents a convincing example of complete auditory degeneration for the role of bat defense. Although these moths could not use their ears for an effective escape from an approaching bat, we again observe vestigial sensitivity at the typical antibat best frequencies of 20 to 50 kHz. As with Polynesian noctuids, we assume that Venezuelan dioptines do not use sounds in their social behavior because there is no direct evidence that they do (but see Miller 1989). Also as with island moths, it is possible that the ears of dioptines alert their owners to the sounds of terrestrial predators, but the very high thresholds of these ears make this an even more unlikely purpose.

The reduction of these moths' best-frequency sensitivities, along with the total elimination of their allotonic sensitivities, argues that vestigiality is the most likely explanation for the auditory responsiveness in these and other bat-released moths. It would be interesting to know what has happened to

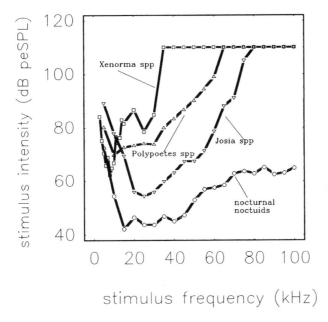

FIGURE 8.23. The median audiograms of day-flying dioptine moths from Venezuela (*Xenorma cytheris, Polypoetes circumfumata* and *Josia radians*) reveal severe high frequency deafness and reduced BF sensitivities compared to sympatric, nocturnal, noctuid moths. (Adapted from Fullard et al. 1997, Springer-Verlag.)

the chordotonal organs of *Dioptis* spp., in which the ears have disappeared, as they have in the Polynesian pyralid, *L. obscura* (Clarke 1971) and in certain flightless geometrids (Sattler 1991). Considering the evolutionary past of moth ears (see Section 4), one prediction is that they have reverted to their original role as wing proprioceptors.

Although dioptines appear to have given up their ears, some moths have retained them, even though their need for bat detection has passed. Male whistling moths, *Hecatesia thyridion* and *H. exultans*, emerge in the afternoon from the Australian coastal healthland to agonistically display to each other (Alcock, Gwynne, and Dadour 1989; Alcock and Bailey 1995) by striking their forewings together to produce trains of 15- to 20-kHz clicks (Bailey 1978) that sound like buzzing to the human ear. The ears of *H. thyridion* are sensitive and are tuned close to the frequencies of the clicks (Surlykke and Fullard 1989). Alcock, Gywnne, and Dadour (1989) showed that males are attracted to these clicks and will enter into aggressive encounters, presumably to take over the site that a resident male is occupying to attract females.

Attraction to ultrasound is a significant reversal of the normal acoustic reaction in moths and suggests that these insects have evolved radically different CNS pathways for processing sounds. However, the agaristines are closely related to noctuids, and it is unlikely that the whistling moths represent an independently evolved line of eared moths. When stimulated with batlike pulses of sound, *H. thyridion* males drop to the ground in a typical antibat fashion (Surlykke and Fullard 1989), so one explanation for the apparent paradox in the ears of these animals is that they use their ears to detect the short sounds of conspecifics produced during courtship but then revert them to a bat-detecting state when the evening arrives. This could explain the mismatch between the best frequency of *H. thyridion*'s ears and the peak frequency of its sounds. If these moths use their ears for two disparate functions, social calling and bat detection, but cannot frequency discriminate, the best compromise might be to adopt a best frequency that encompasses both the frequencies generated by the predatory bat community and the social environment of conspecifics.

Temporal isolation can also exist on a larger scale. Certain species of moths emerge at times in the season that have few or no bats (Yack 1988), and some Nearctic moths have extended this life history strategy to emerge during the winter, when bats have either migrated to warmer habitats or entered hibernation. Surlykke and Treat (1995) described the ears of 11 species of North American wintermoths as possessing best-frequency thresholds comparable with those of summer noctuids and suggested that these moths have retained primitive auditory thresholds from a time when their ears were used against bats. The wintermoth ears also reveal signs of high-frequency deafness, reminiscent of Polynesian endemics (Fullard 1994) so this may be as far as these particular ears have evolutionarily degraded. Another explanation may arise from an examination of the geographical distribution of these moths. If these moths experience gene flow with individuals living in more southern habitats, where bats remain active throughout the winter, the moths analyzed by Surlykke and Treat (1995) may be occasionally refreshed by genetic intermixing with those individuals. As demonstrated by the island studies, achieving a predator-released condition does not necessarily result in rapid sensory changes.

7.3. Behavioral Isolation

Moths reduce their encounter probability with foraging bats by physically concealing themselves (e.g., the reduced flight of earless moths; Morrill and Fullard 1992; Lewis, Fullard, and Morrill 1993). A more extreme form of behavioral isolation, however, is to stop flying altogether. The females of a number of species do not fly and also appear to have lost their ears (Sattler 1991). Female gypsy moths, *Lymantria dispar*, are flightless and are deafer at frequencies above 25 kHz compared with the flighted males (Cardone and Fullard 1988). Females also do not show behavioral responses when exposed to artificial bat sounds (Baker and Cardé 1978), so it appears that these moths, no longer affected by bats, have disconnected their auditory inputs from their motor outputs. Although the females are relatively insensitive at bat frequencies, they do possess best-frequency thresholds that are as sensitive as males. As for island moths, wintermoths, and day-flying moths, the ears of female gypsy moths appear to be vestigial, present only because of the short evolutionary time that this moth has been flightless. One way of testing this hypothesis would be to examine the ears of the Asian *L. dispar*, in which the female has retained her ability to fly (Wallner and McManus 1988).

One explanation for flight loss in females is that this adaptation has increased her capacity for egg production because a part of her body is no longer taken up with flight musculature. Heitmann (1934) proposed that as the female's ovaries evolutionarily increased in size, they invaded the thoracic cavity, resulting in removal of the tympanic air sac and eventual deafness. Another explanation, provided by Sattler (1991), postulates a different sequence of events involving a coevolutionary process. In this scenario, certain moths experienced an adaptive advantage in becoming flightless because this released them from most bats. Once protected from bats, their ears deteriorated and the increased thoracic space was exploited for egg production. The difference between these scenarios lies in the selective circumstances that originally favored earlessness, increased egg production versus antibat behavior. One way of testing these possibilities would be to examine the relative incidence of flightless moths across different habitats. If flightlessness arose from an evolved response to bats, it would be more evident in areas of high aerially hawking bat predation potential (e.g., the tropics).

Moths are not the only insects whose ears show degenerative changes commencing at high frequencies. Yager (1990) describes high-frequency deafness in flightless female preying mantises, whereas Riede, Kämper, and Höfler (1990) demonstrated that two species of nonsinging (i.e., nonauditory) grasshoppers possess audiograms with reduced sensitivities at frequencies above 18 kHz compared with singing species. The similarity of auditory changes across distantly related taxa such as Orthoptera and Lepidoptera suggests that high-frequency insensitivity is a general symptom of early evolutionary stages of apomorphic (i.e., derived) deafness in insects. In turn, the absence of high-frequency deafness might be used to predict that an insect's ear is still functional. Bailey and Römer (1991) describe the sexually dimorphic ears of a zaprochiline tettigoniid in which the males are 2 to 9 dB less sensitive than the females but with no indication that this deafness is more pronounced at high frequencies (cf. female gypsy moths). This suggests that although the ears of the tettigoniid males and females have different overall sensitivities, they are both functional. Similarly, the absence of preferential high-frequency deafness of the noctuid's A2 cell compared with its A1 (see Fig. 8.2) also argues that this is a functional part of the moth's auditory network.

10. Summary

Figure 8.24 describes what we have observed of the evolution of moth auditory abilities, beginning with an earless ancestral state, leading toward maximum sensitivity in tropical moths, and regressing to deafness in species released from bat predation. The data presented here provide evidence that

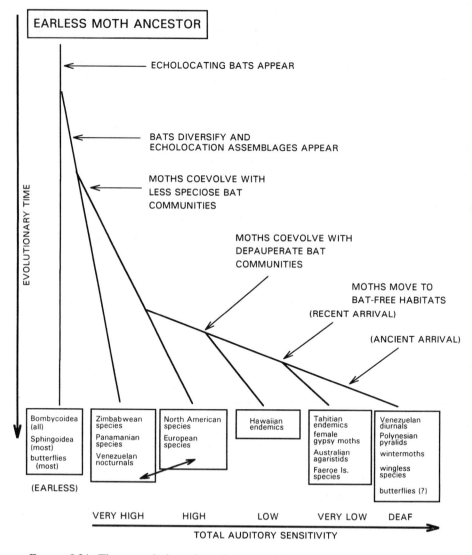

FIGURE 8.24. The coevolution of moth ears and bat echolocation with examples drawn from the research described in this chapter (see text for explanation).

the primary role of moth ears is that of bat detection, but this should not be taken as a denial that other functions exist for these sensory organs, nor should it be taken to mean there is only one direction that these changes can go. For example, Figure 8.24 suggests that reduced sensitivity in temperate moths resulted from a migration away from the bat-dense tropics, perhaps following these moths' reinvasion of temperate zones after the retreat of the Wisconsin ice fields 17,000 to 21,000 years ago. Certain Nearctic moths might have returned to the tropics (the two-headed arrow connecting temperate and tropical boxes in Figure 8.24), but their loss of allotonic sensitivities might have placed limitations on the extent of those distributions.

The evidence provided in this chapter supports the belief that moths evolved ears to detect bats and that some bats have counterevolved echolocation methods to circumvent this defense. Whether this satisfies the precise definition of coevolution may be a question more semantic than real; some bats do enjoy increased foraging success on eared moths because of their countermaneuvers. Regardless of its coincidental versus adaptive basis, there are bats that are now exerting new selective forces on the design of moth ears. It remains to be seen whether this selection has been strong enough anywhere to have resulted in moth countermaneuvers. The antiquity of the arms war between moths and bats, carried out for millions of years in the night skies of almost every habitat in the world, makes it very likely that we will continue to discover novel sensory adaptations in both contestants.

Acknowledgments. Thanks are extended to Drs. M.B. Fenton, D.T. Gwynne, R. Hoy, and A.N. Popper for their helpful comments on this paper. This research was supported by research grants from the Natural Sciences and Engineering Research Council of Canada and the National Geographic Society.

References

Adams WB (1971) Intensity characteristics of the noctuid acoustic receptor. J Gen Physiol 58:562–579.

Adams WB (1972) Mechanical tuning of the acoustic receptor of *Prodenia eridania* (Cramer) (Noctuidae). J Exp Biol 57:297–304.

Adams WB, Belcher EO (1974) Adaptation in the noctuid acoustic receptor. J Acoust Soc Am 56:S40.

Agee HR, Orona E (1988) Studies of the neural basis of evasive flight behavior in response to acoustic stimulation in *Heliothis zea* (Lepidoptera: Noctuidae): organization of the tympanic nerves. Ann Entomol Soc Am 81:977–985.

Alcock J, Bailey WJ (1995) Acoustical communication and mating system of the Australian whistling moth *Hecatesia exultans* (Noctuidae: Agaristinae). J Zool 237:337–352.

Alcock J, Gwynne GT, Dadour IR (1989) Acoustic signalling, territoriality, and mating in whistling moths, *Hecatesia thyridion* (Agaristidae). J Insect Behav 2:27–37.

Avise JC (1994) Molecular Markers, Natural History and Evolution. London: Chapman and Hall.

Bailey WJ (1978) Resonant wing systems in the Australian whistling moth *Hecatesia* (Agarasidae (sic), Lepidoptera). Nature 272:444–446.

Bailey WJ, Römer H (1991) Sexual differences in auditory sensitivity: mismatch of hearing threshold and call frequency in a tettigoniid (Orthoptera, Tettigoniidae: Zaprochilinae). J Comp Physiol A 169:349–353.

Baker TC, Cardé RT (1978) Disruption of gypsy moth male sex pheromone behaviour by high frequency sound. Environ Entomol 7:45–52.

Belwood JJ (1988) Foraging behavior, prey selection, and echolocation in phyllostomine bats (Phyllostominae). In: Nachtigall PE, Moore PWB (eds) Animal Sonar. New York: Plenum Press, pp. 601–605.

Belwood JJ, Fullard JH (1984) Echolocation and foraging behaviour in the Hawaiian hoary bat, *Lasiurus cinereus semotus*. Can J Zool 62:2113–2120.

Black HL (1972) Differential exploitation of moths by the bats *Eptesicus fuscus* and *Lasiurus cinereus*. J Mammal 53:598–601.

Boyan GS (1989) Is there a common "bauplan" for insect auditory pathways? In: Erber J, Menzel R, Pflüger H-J, Todt D (eds) Neural Mechanisms of Behavior. Proceedings of the Second International Congress on Neuroethology. Stuttgart: Thieme, p. 73.

Boyan GS (1993) Another look at insect audition: the tympanic receptors as an evolutionary specialization of the chordotonal system. J Insect Physiol 39:187–200.

Boyan GS, Fullard JH (1986) Interneurones responding to sound in the tobacco budworm moth *Heliothis virescens* (Noctuidae): morphological and physiological characteristics. J Comp Physiol A 156:391–404.

Boyan GS, Fullard JH (1988) Information processing at a central synapse suggests a noise filter in the auditory pathway of the noctuid moth. J Comp Physiol A 164:251–258.

Boyan GS, Miller LA (1991) Parallel processing of afferent input by identified interneurones in the auditory pathway of the noctuid moth *Noctua pronuba* (L.). J Comp Physiol A 168:727–738.

Boyan GS, Williams L, Fullard JH (1990) Organization of the auditory pathway in the thoracic ganglia of noctuid moths. J Comp Neurol 295:248–267.

Brooks DR, McLennan DA (1991) Phylogeny, Ecology, and Behavior. Chicago, IL: University of Chicago Press.

Cardone B, Fullard JH (1988) Auditory characteristics and sexual dimorphism in the gypsy moth. Physiol Entomol 13:9–14.

Clarke JFG (1971) The Lepidoptera of Rapa Island. Smithson Contrib Zool 56:1–282.

Common IFB (1990) Moths of Australia. Leiden: EJ Brill.

Conner WE (1987) Ultrasound: its role in the courtship of the arctiid moth, *Cycnia tenera*. Experientia 43:1029–1031.

Coro F, Alonso N (1989) Cell responses to acoustic stimuli in the pterothoracic ganglion of two noctuoid moths. J Comp Physiol A 165:253–268.

Coro F, Pérez M (1983) Peripheral interaction in the tympanic organ of a moth. Naturwissenschaften 70:99–100.

Coro F, Pérez M (1984) Intensity coding by auditory receptors in *Empyreuma pugione* (Lepidoptera, Ctenuchidae). J Comp Physiol A 154:287–295.

Coro F, Pérez M (1990) Temperature affects auditory receptor response in an arctiid moth. Naturwissenschaften 77:445–447.

Coro F, Pérez M, Machado A (1994) Effects of temperature on a moth auditory receptor. J Comp Physiol A 174:517–525.

Darwin C (1862) The Various Contrivances by Which Orchids Are Fertilised by Insects, 2nd ed. London: Murray.

Dodson S (1988) The ecological role of chemical stimuli for the zooplankton: Predator-avoidance behavior in *Daphnia* Limno. Oceanography 33:1431–1439.

Eggers F (1919) Das thoracale bitympanale Organ einer Gruppe der Lepidoptera Heterocera. Zool Jahrb Anat 41:273–376.

Eggers F (1925) Versuche über das Gehör der Noctuiden. Zeit Vergl Physiol 2:297–314.

Faure PA, Barclay RMR (1994) Substrate-gleaning versus aerial-hawking: plasticity in the foraging and echolocation behaviour of the long-eared bat, *Myotis evotis*. J Comp Physiol A 174:651–660.

Faure PA, Fullard JH, Barclay RMR (1990) The response of tympanate moths to the echolocation calls of a substrate-gleaning bat, *Myotis evotis*. J Comp Physiol A 166:843–849.

Faure PA, Fullard JH, Dawson JW (1993) The gleaning attacks of the northern long-eared bat, *Myotis septentrionalis*, are relatively inaudible to moths. J Exp Biol 178:173–189.

Fenton MB (1990) The foraging ecology of animal-eating bats. Can J Zool 68:411–422.

Fenton MB, Bell GP (1981) Recognition of species of insectivorous bats by their echolocation calls. J Mammal 62:233–243.

Fenton MB, Fullard JH (1979) The influence of moth hearing on bat echolocation strategies. J Comp Physiol 132:77–86.

Fenton MB, Racey P, Rayner JMV (1987) (eds) Recent Advances in the Study of Bats. Cambridge, UK: Cambridge University Press.

Fenton MB, Audet A, Obrist MK, Rydell J (1995) Signal strength, timing, and self-deafening: the evolution of echolocation in bats. Paleobiology 21:229–242.

Forbes WTM (1916) On the tympanum of certain Lepidoptera. Psyche 23:183–192.

Fullard JH (1982) Echolocation assemblages and their effects on moth auditory systems. Can J Zool 60:2572–2576.

Fullard JH (1984a) External auditory structures in two species of neotropical notodontid moths. J Comp Physiol A 155:625–632.

Fullard JH (1984b) Acoustic relationships between tympanate moths and the Hawaiian hoary bat (*Lasiurus cinereus semotus*). J Comp Physiol A 155:795–801.

Fullard JH (1987a) Sensory ecology and neuroethology of moths and bats: interactions in a global perspective. In: Fenton MB, Racey P, Rayner JMV (eds) Recent Advances in the Study of Bats. Cambridge, UK: Cambridge University Press, pp. 244–272.

Fullard JH (1987b) The defensive function of auditory enhancers in the neotropical moth *Antaea lichyi* (Lepidoptera: Notodontidae). Can J Zool 65:2042–2046.

Fullard JH (1988) The tuning of moth ears. Experientia 44:423–428.

Fullard JH (1994) Auditory changes in moths endemic to a bat-free habitat. J Evol Biol 7:435–445.

Fullard JH, Belwood JJ (1988) The echolocation assemblage: acoustic ensembles in a neotropical habitat. In: Nachtigall PE, Moore PWB (eds) Animal Sonar. NATO ASI Series, Series A: Life Sciences, Vol. 156. New York: Plenum Press, pp. 639–643.

Fullard JH, Dawson JD (1997) The echolocation calls of the spotted bat *Euderma maculatum* are relatively inaudible to moths. J Exp Biol 200:129–137.

Fullard JH, Fenton MB (1980) Echolocation signal design as a potential counter-measure against moth audition. In: Busnel RG, Fish JF (eds) Animal Sonar Systems. New York: Plenum Press, pp. 899–900.

Fullard JH, Thomas DW (1981) Detection of certain African, insectivorous bats by sympatric, tympanate moths. J Comp Physiol 143:363–368.

Fullard JH, Yack JE (1993) The evolutionary biology of insect hearing. Trends Ecol Evol 8:248–252.

Fullard JH, Fenton MB, Furlonger CL (1983) Sensory relationships of moths and bats sampled from two Nearctic sites. Can J Zool 61:1752–1757.

Fullard JH, Koehler C, Surlykke A, McKenzie NL (1991) Echolocation ecology and flight morphology of insectivorous bats (Chiroptera) in south-western Australia Aust J Zool 39:427–438.

Fullard JH, Dawson JW, Otero LD, Surlykke A (1997) Bat-deafness in day-flying moths (Lepidoptera, Notodontidae, Dioptinae). J Comp Physiol A 181:477–483.

Ghiradella H (1971) Fine structure of the noctuid moth ear I. The transducer area and connections to the tympanic membrane in *Feltia subgothica* Haworth. J Morphol 134:21–46.

Griffin DR (1971) The importance of atmospheric attenuation for the echolocation of bats (Chiroptera). Anim Behav 19:55–61.

Hasenfuss I (1997) Precursor structures and evolution of tympanal organs in Lepidoptera (Insecta, Pterygota). Zoomorphology 117:155–164.

Haskell PT (1961) Insect Sounds. London: Witherby.

Heitmann, H (1934) Die Tympanalorgane flugunfähiger Lepidopteren und die Korrelation in der Ausbildung der Flugel und der Tympanalorgane. Zool Jahrb Anat Ontogen 59:135–200.

Holloway JD, Bradley JD, Carter DJ (1987) Lepidoptera. Vol 1. In: Betts C.R. (ed.) C.I.E. Guides to Insects of Importance to Man (Wallingford, U.K. CAB International)

Hoy R, Nolen T, Broadfueher P (1989) The neuroethology of acoustic startle and escape in flying insects. J Exp Biol 146:287–306.

Jones G (1990) Prey selection by the greater horseshoe bat (*Rhinolophus ferrumequinum*): optimal foraging by echolocation? J Anim Ecol 59:587–602.

Jones G (1992) Bats vs moths: studies on the diets of rhinolophid and hipposiderid bats support the allotonic frequency hypothesis. In: Horáček I, Vohralík V (eds) Prague Studies in Mammalogy. Prague: Charles University Press, pp. 87–92.

Kennel J von, Eggers F (1933) Die abdominalen Tympanalorgane der Lepidopteren. Zool Jahrb Abt Anat 57:1–104.

Kick SA, Simmons JA (1984) Automatic gain control in the bat's sonar receiver and the neuroethology of echolocation. J Neurosci 4:2725–2737.

Lawrence BD, Simmons JA (1982) Measurements of atmospheric attenuation at ultrasonic frequencies and the significance for echolocation by bats. J Acoust Soc Am 71:585–590.

Lechtenberg R (1971) Acoustic response of the B cell in noctuid moths. J Insect Physiol 17:2395–2408.

Leonard ML, Fenton MB (1984) Echolocation calls of *Euderma maculatum* (Vespertilionidae): use in orientation and communication. J Mammal 65:122–126.

Lewis FP, Fullard JH (1996) The neurometamorphosis of the ear in the gypsy moth, *Lymantria dispar* and its homologue in the earless forest tent caterpillar moth, *Malacosoma disstria*. J Neurobiol 31:245–262.

Lewis FP, Fullard JH, Morrill SB (1993) Auditory influences on the flight behaviour of moths in a Nearctic site II. Flight times, heights and erraticism. Can J Zool 71:1562–1568.

Libersat F, Hoy RR (1991) Ultrasonic startle behavior in bushcrickets (Orthoptera: Tettigoniidae). J Comp Physiol A 169:507–514.

Madsen BM, Miller LA (1987) Auditory input to motor neurones of the dorsal longitudinal flight muscles of the noctuid moth (*Barathra brassicae* L). J Comp Physiol A 160:23–31.

Markl H (1983) Vibrational communication. In: Huber F, Markl H (eds) Neuroethology and Behavioral Physiology. Berlin: Springer-Verlag, pp. 332–353.

Miller JS (1989) *Euchontha* Walker and *Pareuchontha*, new genus (Lepidoptera; Dioptidae) a revision, including description of three new species, and discussion of a male forewing modification. Am Mus Novitates 2938:1–41.

Miller JS (1991) Cladistics and classification of the Notodontidae (Lepidoptera: Noctuoidea) based on larval and adult morphology. Bull Am Mus Natl Hist 204:1–230.

Miller LA, Olesen J (1979) Avoidance behaviour in green lacewings to hunting bats and ultrasound. J Comp Physiol 131:113–120.

Monroe BL, Sibley CG (1993) A World Checklist of Birds. New Haven, CT: Yale University Press.

Morrill SB, Fullard JH (1992) Auditory influences on the flight behaviour of moths in a Nearctic site. I. Flight tendency. Can J Zool 70:1097–1101.

Murphey RK, Bacon JP, Sakaguchi DS, Johnson SE (1983) Transplantation of cricket sensory neurons to ectopic locations: arborizations and synaptic connections. J Neurosci 3:659–672.

Neuweiler G (1984) Foraging, echolocation and audition in bats. Naturwissenschaften 71:446–455.

Neuweiler G (1990) Auditory adaptations for prey capture in echolocating bats. Physiol Rev 70:615–641.

Novick A (1977) Acoustic orientation. In: Wimsatt WA (ed) Biology of Bats, Vol. III. New York: Academic Press, pp. 73–287.

Nüesch H (1957) Die Morphologie des Thorax von *Telea polyphemus* Cr (Lepid). I. Nervensystem. Zool Jahrb 75:615–642.

Obrist MK (1995) Flexible bat echolocation: the influence of individual habitat and conspecifics on sonar signal design. J Comp Physiol A 36:207–219.

Oldfield BP (1985) The tuning of auditory receptors in bushcrickets. Hear Res 17:27–35.

Paul DH (1973) Central projections of the tympanic fibres in noctuid moths. J Insect Physiol 19:1785–1792.

Paul DH (1974) Responses to acoustic stimulation of thoracic interneurons in noctuid moths. J Insect Physiol 20:2205–2218.

Pearson KG, Hedwig B, Wolf H (1989) Are the hindwing chordotonal organs elements of the locust flight pattern generator? J Exp Biol 144:235–255.

Pérez M, Coro F (1984) Physiological characteristics of the tympanal organ in noctuoid moths. I. Responses to brief acoustic pulses. J Comp Physiol A 154:441–447.

Pérez M, Coro F (1986) Effects of picrotoxin on the tympanal organ of a noctuid moth. Naturwissenschaften 73:501–502.

Pérez M, Zhantiev RD (1976) Functional organization of the tympanal organ of the flour moth, *Ephestia kuehniella*. J Insect Physiol 22:1267–1273.

Poché RM (1981) Ecology of the spotted bat (*Euderma maculatum*) in southwest Utah. Utah Dept Natur Res Publ No. 81-1.

Popper AN, Fay RR (1995) (eds) Hearing by Bats. New York: Springer-Verlag.

Pumphrey RJ (1940) Hearing in insects. Biol Rev 15:107–132.

Richards AG (1932) Comparative skeletal morphology of the noctuid tympanum. Entomol Am 13:1–43.

Riede K, Kämper G, Höfler I (1990) Tympana, auditory thresholds, and projection areas of tympanal nerves in singing and silent grasshoppers (Insecta, Acridoidea). Zoomorphology 109:223–230.

Riessen HP, Sprules WG (1990) Demographic costs of antipredator defenses in *Daphnia pulex*. Ecology 71:1536–1546.

Roeder KD (1966) Interneurons of the thoracic nerve cord activated by tympanic nerve fibers in noctuid moths. J Insect Physiol 12:1227–1244.

Roeder KD (1967) Nerve Cells and Insect Behavior. Cambridge, MA: Harvard University Press.

Roeder KD (1970) Episodes in insect brains Am Sci 58:378–389.

Roeder KD (1974) Acoustic sensory responses and possible bat-evasion tactics of certain moths. In Burt MDB (ed) Proceedings of the Canadian Society of Zoologists Annual Meeting, Fredericton, NB, June 2–5, 1974. Fredericton, NB: University of New Brunswick Press, pp. 71–78.

Roeder KD, Treat AE (1957) Ultrasonic reception by the tympanic organ of noctuid moths. J Exp Zool 134:127–158.

Roeder KD, Treat AE (1962) The acoustic detection of bats by moths. Verh XIth Intl Congr Entomol 3:7–11.

Roeder KD, Treat AE, Vandeberg JS (1968) Auditory sense in certain sphingid moths. Science 159:331–333.

Rydell J, Arletazz R (1994) Low frequency echolocation enables the bat *Tadarida teniotis* to feed on tympanate insects. Proc R Soc Lond Biol Sci 257:175–178.

Sanderford MV, Conner WE (1990) Courtship sounds of the polka-dot wasp moth, *Syntomeida epilais*. Naturwissenschaften 77:345–347.

Sattler K (1991) A review of wing reduction in Lepidoptera. Bull Br Mus Nat Hist (Entomol) 60:243–288.

Schiolten P, Larsen ON, Michelsen A (1981) Mechanical time resolution in some insect ears. J Comp Physiol 143:289–295.

Scoble MJ (1992) The Lepidoptera. New York: Oxford University Press.

Simon C (1987) Hawaiian evolutionary biology: an introduction. Trends Ecol Evol 2:175–178.

Spangler HG (1988) Moth hearing, defense, and communication. Ann Rev Entomol 33:39–81.

Surlykke A (1984) Hearing in notodontid moths: a tympanic organ with a single auditory neurone. J Exp Biol 113:323–335.

Surlykke A (1986) Moth hearing on the Faroe Islands, an area without bats. Physiol Entomol 11:221–225.

Surlykke A, Fullard JH (1989) Hearing in the Australian whistling moth, *Hecatesia thyridion*. Naturwissenschaften 76:132–134.

Surlykke A, Gogola M (1986) Stridulation and hearing in the noctuid moth *Thecophora fovea* (Tr.). J Comp Physiol A 159:267–273.

Surlykke A, Miller LA (1982) Central branchings of three sensory axons from a moth ear (*Agrotis segetum*, Noctuidae). J Insect Physiol 28:357–364.

Surlykke A, Treat AE (1995) Hearing in wintermoths. Naturwissenschaften 82:382–384.

Surlykke A, Larsen ON, Michelsen A (1988) Temporal coding in the auditory receptor of the moth ear. J Comp Physiol A 162:367–374.

Swinton AH (1877) On an organ of hearing in insects, with special reference to the Lepidoptera. Entomol Mon Mag 14:121–126.

Thompson JN (1994) The Coevolutionary Process. Chicago: University of Chicago Press.

Tomich PQ (1986) Mammals in Hawai'i. Honolulu, HI: Bishop Museum Press.

Tougaard J (1996) Energy detection and temporal integration in the noctuid A1 auditory receptor. J Comp Physiol A 178:669–677.

Treat AE, Roeder KD (1959) A nervous element of unknown function in the tympanic organs of noctuid moths. J Insect Physiol 3:262–270.

Tuttle MD, Ryan MJ (1981) Bat predation and the evolution of frog vocalizations in the Neotropics. Science 203:16–21.

Wai-Ping V, Fenton MB (1989) Ecology of spotted bat (*Euderma maculatum*) roosting and foraging behavior. J Mammal 70:617–622.

Wallner WE, McManus KA (1988) (eds) Lymantriidae: a comparison of features of New and Old World tussock moths. U.S. Department of Agriculture General Technical Report NE-123.

Waters DA (1996) The peripheral auditory characteristics of noctuid moths: information encoding and endogenous noise. J Exp Biol 199:857–868.

Waters DA, Jones G (1996) The peripheral auditory characteristics of noctuid moths: responses to the search-phase echolocation calls of bats. J Exp Biol 199:847–856.

Whitaker JO, Black HL (1976) Food habits of cave bats from Zambia, Africa. J Mammal 57:199–204.

Whitaker JO, Tomich PQ (1983) Food habits of the hoary bat, *Lasiurus cincereus*, from Hawaii. J Mammal 64:151–152.

White FB (1877) (Untitled). Nature 15:293.

Willott JF (1991) Aging and the auditory system. San Diego, CA: Singular Publishers.

Woodsworth GC, Bell GP, Fenton MB (1981) Observations on the echolocation, feeding behaviour, and habitat use of *Euderma maculatum* in south central British Columbia. Can J Zool 59:1099–1102.

Yack JE (1988) Seasonal partitioning of tympanate moths in relation to bat activity. Can J Zool 66:753–755.

Yack JE (1992) A multiterminal stretch receptor, chordotonal organ and hair plate at the wing hinge of *Manduca sexta*: Unravelling the mystery of the noctuid moth ear B cell. J Comp Neurol 324:500–508.

Yack JE, Fullard JH (1990) The mechanoreceptive origin of insect tympanal organs: a comparative study of similar nerves in tympanate and atympanate moths. J Comp Neurol 300:523–534.

Yack JE, Fullard JH (1993a) What is an insect ear? Ann Entomol Soc 86:677–682.

Yack JE, Fullard JH (1993b) Proprioceptive activity of the wing-hinge stretch receptor in *Manduca sexta* and other atympanate moths: a study of the noctuoid moth ear B cell homologue. J Comp Physiol A 173:301–307.

Yack JE, Roots BI (1992) The metathoracic wing-hinge chordotonal organ of an atympanate moth *Actias luna* (Lepidoptera, Saturniidae): a light- and electron-microscopic study. Cell Tissue Res 267:455–471.

Yager DD (1990) Sexual dimorphism of auditory function and structure in preying mantises (Mantodea; Dictyoptera). J Zool (Lond) 221:517–537.

Yager DD, May ML (1990) Ultrasound-triggered, flight-gated evasive maneuvers in the praying mantis, *Parasphendale agrionina* (Gerst.). II. Tethered flight. J Exp Zool 152:41–58.

Zbinden K, Zingg P (1986) Search and hunting signals of echolocating European free-tailed bats, *Tadarida teniotis*, in southern Switzerland. Mammalia 50:9–25.

Zimmerman EC (1958) Insects of Hawaii, Vol. 7. Macrolepidoptera. Honolulu, HI: University of Hawaii Press.

Index

Note regarding species names. Wherever possible, we have used the scientific names of species since, as described in Chapter 1, common names are often confusing and vary by locale of the investigator (e.g., cricket and bushcricket may just be different common names for the same species). In some cases, there are often no common names and so the literature is accessible only by scientific name.

A1 and A2 cells, central projections, 283–285
  moth, 281ff
Abdominal eardrums, locusts, grasshoppers, and cicadas, 8
Ablation of interneurons, effects on directional hearing, 173
Absorption coefficient, effects of environment, 66
Absorption coefficient, effects on echolocation, 66
Achaeranea sp. (a spider), metatarsal slit organ, 245
Acheta domesticus (a cricket), communication sound
  directional hearing, 175
  temporal pattern, 143
Acoustic communication, evolution, 123
  selection for sounds, 81ff
Acoustic fovea, frequency map in bushcrickets, 160
Acoustic parasites, 198–200
Acoustic particle motion, sound source localization, 19ff
Acoustic signals, evolution, 87–88
  neural processing, 139ff
Acoustics; *see also* Sound
  geometric spreading, 66
  scattering, 67
  and sound source localization, 23ff
Acoustics, stridulation sound, 65

Acridoldea (grasshopper), chordotonal system development, 124
Actias luna (luna moth) 287, 289
Active space, spider, 268–270
Agave sp. (a spider), web, 239
American cockroach; *see Periplaneta americana*
Amplitude modulation, insect song, 73–74
AN1 and AN2 neurons, cricket, 173–174, 186–187
Ancistrura nigrovittata (a bushcricket), temporal pattern in song, 152
Ancistrura nuptialis (a bushcricket), temporal pattern in song recognition, 153
Anidiops sp. (a spider), web type, 235
Antennal receptors, *Drosophila*, 200
Araneus sp. (a spider), 231
Arborization space, competition, 122
Ascending interneurons, AN1 and AN2, 164, 167, 172, 179, 183
Atmosphere, effects on sound, 68–70
Attenuation; *see also* Excess attenuation
  absorption coefficient, 66
  calling song, 73
  environmental causes, 65–66
  excess, 66ff
  frequency dependence, 67–68
  geometric spreading, 66
  ground effects, 87–88

Attenuation (*cont.*)
  spider web, 234
  spider web vibration, 240–241
  turbulence effects, 66–68
  vegetation effects, 66–68
Atympanate insects, chordotonal pathway
    development, 123ff
Auditory afferents, directional hearing,
    214
Auditory behavior, development, 130ff
Auditory deprivation, cricket
    development, 128
Auditory organ, pleural chordotonal
    system, 105
Auditory pathway; *see also* CNS
  bushcricket, 84–85, 87
Auditory receptors, encoding, 155ff
Auditory system, development, 97ff
  bushcricket, 105ff
Axogenesis, auditory system, 97, 103,
    119ff

B cell, 288–290
Backswimmer; *see Notonecta*
Barathra brassicae (Noctuid moth), CNS,
    285
Bat, behavioral isolation from moths,
    316–317
  calls, 306ff
  detection, 139ff
  echolocation, 296–297, 300–301,
    304ff
  evolution of hearing, 9
  foraging behavior, 296ff
  gleaning, 297
  hawking, 297
  moth interactions, 279ff
  predation on moth, 297–299, 305ff
  sonar detection, 19
Behavior, development, 99
Bending waves, spider web, 237–238
Binaural cues; *see also* Directional
    hearing, Sound source localiza-
    tion
  Ormiine flies, 208ff
Bladder cicada; *see also Cytosoma* sp.
Bladder grasshopper; *see also Bullacris*
    sp.
BNC1 and 2 neurons, temporal pattern
    selectivity, 179–180, 181–182
Borneo, sounds in rain forest, 78
Broadcast area, insect song, 87–88
Broadcast height, calling song, 88

Broadcast range, maximizing, 82
Bullacris sp. (a grasshopper), 88–89
  excess attenuation, 69, 70
Bushcricket, *see also Ancistrura
    nigrovittata, Caedicia simplex,
    Conocephalus* sp., *Ephippiger
    ephippiger, Hemisaga* sp., *Isophya
    leonorae, Leptophyes* sp.,
    Metrioptera sp., *Mygalopsis* sp.,
    Poecilimon affinis, Tettigonia sp.
  calling song, 86
  development of sensitivity, 113–114
  development of tonotopy, 105–106
  directional hearing ambiguities, 52–53
  duetting, 152
  ear, 7
  ear location, 102
  frequency coding, 158ff
  hearing trumpet, 31
  peripheral auditory development, 105ff
  phonotaxis, 85
  postembryonic development of ear,
    112ff
  prothoracic ganglion, 158–159
  singing periodicity, 80
  sound discrimination, 78–80
  syllable duration, 152
  temporal pattern recognition, 151ff
  tonotopic organization, 156
  trachea, 32

Caedicia simplex (a bushcricket),
    frequency selectivity in CNS, 166
Calcium currents, role in selective
    attention, 177–178
Calling
  temporal pattern in rain forest, 78
  temporal separation of sound, 77–78
Calling pattern, insect song, 73–74
Calling site, 87–88
Calling song
  attenuation, 73
  bushcricket, 86
  cricket, 42, 140
  degradation, 85
  frequency representation in periphery,
    161
  perception, 83
  phonotaxis in crickets, 142
  pulse rate and intensity in phonotaxis,
    148
  recognition, 146
  redundancy, 83ff

response distance, 88–90
sexual dimorphism, 88–89
spacing, 87–88
syllable onset rate, 151
temporal pattern analysis, 179–180
temporal pattern selectivity, 146ff
Tettigonia sp., 73
Cardioid directional selectivity, 23
Categorical perception, 10
Caterpillar
    chordotonal organ, 292
    CNS, 292
Cell adhesion molecules, fasciclin I,
    109–110
Cell migration, grasshopper development,
    103
Cell surface recognition molecules,
    117–118
Central auditory system, development,
    116ff
Central membrane
    directional hearing, 47–48
    tracheal tubes, 47–48
Central nervous system; *see also* CNS
    development, 117ff
Central processing
    acoustic signals, 163ff
    crickets versus grasshoppers, 187
Cercal synaptic connections, remodeling,
    122, 126
Chordotonal central neurons, activation
    by ground vibration, 130
Chordotonal central pathway, multimodal,
    122
Chordotonal organ, 3
    audition, 292
    caterpillar, 292
    earless moths, 287–288
    larvae, 295
    origin of ear, 287ff
    proprioceptors, 12–13
    and subgenual organ, 110
    wingbeat detection, 292
Chordotonal pathway development,
    atympanate insects, 123ff
Chordotonal projections, thoracic ganglia,
    117
Chordotonal receptor identity, genetic
    regulation, 125–126
Chordotonal sensilla, moth, 287
Chordotonal system
    axogenesis, 119ff
    development in *Drosophila*, 117–118
    development in grasshopper, 121–122

development in tympanate and
    atympanate grasshoppers, 124
postembryonic development, 121ff
regeneration, 126ff
tympanal organs, 98
Chorthippus biguttulus (European
    grasshopper), 36–37
    directional hearing, 169–170
    male–female acoustic interactions,
        150–151
    song, 84–86
    stridulation, 65–66
    temporal gap rejection, 184
Chrysopa carnea (green lacewing), ear, 4
Cicada barbara lusitanica (cicada), 5,
    8–9
    abdominal eardrums, 8–9
    ear, 49
Cicada; *see Cicada barbara lusitanica,*
    Cytosoma sp.
Cicada, vibratory communication, 239,
    240
CNS; *see also* Auditory pathway, Omega-
    neuron
    auditory pathway, 84–85
    developmental comparisons with
        mammals, 121
    directional hearing, 213–215
    embryogenesis, 117ff
    larval moth, 293
    moth, 283–285
    omega neuron, 76–78
    representation of song, 83
    spider orientation, 266
Cockroach; *see also Periplaneta*
    americana
    neuroblast 7–4, 122
    sound sensitivity of subgenual organ,
        161
    vibration sensitivity, 248–249
Cocktail party problem, insects, 177–
    178
Coevolution
    bat and moth, 279–280, 318–319
    moth ear and bat call, 294ff
Communication, 139ff
    cricket, 249
    effects of environment, 51ff
    elevation of caller, 57–58
    long range in insects, 70
    range, 72
    song pattern, 73–74
    spiders, 249ff
    time of day, 70–71, 76

Communication sounds
  chirps, 143, 146
  effects of frequency, 141–142
  pulse period and chirp period, 143ff
  pulse rate tuning, 183
  temporal patterns, 143ff
  trills, 143, 146
Conocephalus sp. (a bushcricket), sound
  orientation, 74
Contralateral inhibition, in mammals,
  170
Courtship, and sound detection, 140
  communication sound frequency
    effects, 141–142
  spiders, 229, 256
  Theridiidae, 252
Courtship signal
  Cupiennius sp., 258ff
  spider, 258ff
Courtship song
  Gryllus bimaculatus, 168
  Teleogryllus oceanicus, 148
Courtship vibration
  Cupiennius sp., 250–252, 256
  spiders, 256
Cricket; *see also Acheta domesticus,*
  Gryllus bimaculatus, Teleogryllus
  commodus, Teleogryllus oceanicus
  AN1 and AN2 neurons, 164, 179, 183,
    186–187
  auditory interneurons and phonotaxis,
    173
  auditory pathway plasticity, 128
  BNC1 cells, 179–180
  calling song, 42, 140
  cercal system development, 122
  communication sounds, 249
  dendritic sprouting, 128
  denervation experiments, 127
  development and auditory deprivation,
    128
  directional hearing, 42ff
  ear location, 102
  ear pressure gain, 32
  escape response development, 130
  evolution of hearing, 187
  foreleg eardrums, 7–8
  foreleg transplantation, 127
  frequency coding, 161ff
  frequency selectivity, 115
  hearing sensitivity, 115
  ON1 interneuron, 161–162, 170
  ON1 neuron in directional hearing,
    171–172

  peripheral auditory development,
    107–108
  phonotaxis, 140
  positive and negative phonotaxis,
    167–168
  pressure gradient receiver, 42ff
  relationship to parasitic flies, 198–199
  remodeling of cercal synaptic
    connections, 122
  sound localization strategy, 23
  sound in rain forest, 78
  spiracle and trachea, 42
  stridulation, 7, 65
  temporal pattern analysis, 179–180
  tonotopic organization, 156, 161
  tympanal membrane and spiracle, 32
Cricket versus grasshoppers, essential
  differences in processing, 187
Crista acustica, bushcricket, 106–107,
  158ff
  field cricket inner ear, 11
  postembryonic development, 112ff
Critical phase, development of auditory
  CNS, 127–128
Cupiennius sp. (wandering spider)
  courtship signal, 258ff
  courtship vibration, 250–252, 256
  metatarsal slit organ, 245
  orientation, 264ff
  plant vibration, 269
  vibration discrimination, 253–255
  web, 239
Cut gene, *Drosophila* nervous system,
  125
Cuticular hairs, 244
Cuticular structures of bushcricket ear,
  postembryonic development, 112ff
Cuticular structures, postembryonic
  development, 110–111
Cydnid bugs, vibratory communication,
  239, 240
Cystosoma saundersii (a cicada)
  calling site, 87–88
  communication sound frequency
    effects, 142
  song choice, 75

Deafened animals, unilateral, 153–154
Dendritic sprouting, auditory system
  development, 128
Denervation
  cricket, 127
  grasshopper, 127

Detection range, increasing, 88
Development
  auditory behavior, 130ff
  auditory CNS, critical phase, 127–128
  bushcricket auditory system, 105ff
  central projections, 116ff
  central tonotopy, 116
  cercal systems, 122
  chordotonal pathway, 123ff
  CNS, 117ff
  cricket peripheral auditory system,
    107–108
  G neuron of grasshopper, 117–118
  insect auditory system, 97ff
  insect central nervous system, 101
  insects, 99ff
  neuroblast, 118–119
  peripheral auditory system, 102ff
  postembryonic, 97–98
  role of fasciclin, 120
  role of pioneer neurons, 119
  sensitivity, 113–114
  timing of ear components, 102
  tonotopy, 105–106
  tympanal projections to CNS, 108ff
Dichotic stimulation, temporal pattern
    analysis and directional hearing,
    185–186
Dichotic stimuli, locust, 175–176
Diffraction
  sound source localization, 18ff
  sphere, 26
Diptera; see also Ormiine flies, parasitic
    flies, 197ff
  ears, 200ff
  evolution of hearing, 218ff
  far-field detection, 221–222
  forms of hearing, 200ff
  near-field hearing, 200
  nontympanal hearing, 200
Directional cues
  degradation by environment, 51ff
  versus directional hearing acuity, 41
Directional hearing; see also Sound
    source localization, 153ff
  ablation of interneurons, 173
  ambiguities, 52ff
  AN1 and AN2 neurons, 172
  binaural cues, 206ff
  CNS, 213–215
  cricket auditory interneurons, 173
  different mechanisms, 222–223
  far-field, 49–50
  near-field, 49–50

ON1 neuron, 171–172
Ormia sp., 222–223
  ormiine fly, 49, 199, 206–207, 208ff
  performance by grasshopper, 55ff
  sensitivity, 211–212
  size, 222–223
  size effects, 206–207
  small animals, 198ff, 222–223
  tympanal mechanisms, 208ff
  tympanum, 211–213
  vertebrate, 222
Directional hearing cues, 169ff, 206–207
  amplitude vs phase, 54–55
  cricket, 43
  degradations for grasshopper, 55ff
  frequency effects, 57–58
  grasshopper, 52, 54–55
Directional sensitivity of ear, calculations
    for cricket, 46–47
  calculations for locust, 38ff
Directionality of ears, and frequency
    analysis, 51
Discrimination
  bushcricket, 78–80
  sound, 76–77
Distance calling, bushcrickets, 73
Distance determination, spider, 268
Distance perception, cues, 160
Dolomedes, distance determination,
    268
Dolomedes sp. (fishing spider), vibration
    discrimination, 255
  prey capture, 269–270
  surface wave detection, 244
  water surface orientation, 267
Drosophila
  antennae, 200
  chordotonal receptor development, 125,
    200ff
  chordotonal system development,
    117–118
  neuroblast 7–4, 122
Drosophilidae; see also Diptera

Ear
  connections between ears, 217
  convergence in evolution, 220–221
  Diptera, 200ff
  evolution, 3
  evolution of bat-detection, 292–294
  Lymantria dispar, 317
  moth after release of selection pressure,
    311ff

Müller's organ, 110
Ormiine flies, 200ff
postembryonic development, 110ff
proprioceptor origin, 288, 291
proprioceptors, 12–13
tympanal, 205–206
tympanate vs. nontympanate, 201ff
Ear evolution, chordotonal organ, 287ff
Ear location
 diversity, 101–102
 insects, 6
 various species, 102
Earless moths, 287–288
 chordotonal, 287–288
Echolocation
 bat, 300–301t, 304ff
 effects of absorption coefficient, 66
 feeding behavior, 307
 high-frequency, 304ff
 low-frequency, 307ff
 of moth, 296–297
 spectra, 301
 vibratory, 230
Ecology, insect, 63ff
Elephant, infrasound, 70
Emblemasoma sp. (a fly), tympanal ear,
  205–206
Embryogenesis
 central nervous system, 117ff
 peripheral auditory system, 102ff
Engrailed gene, *Drosophila* nervous
  system, 125
Environment
 absorption coefficient, 66
 and communication, 51ff
 degradation of directional cues, 51ff
 effect on attenuation, 68ff
 effects on sound source localization,
  75–76
 sound attenuation, 65–66
Ephippiger ephippiger (a bushcricket),
  postembryonic ear development,
  112ff
Eptesicus fuscus (bat), moth foraging,
  296ff
Escape behavior, earless moth, 310–311
Escape responses, developing crickets,
  130
Eumasta coldea, chordotonal system
  development, 124
European grasshopper; *see Chorthippus
  biguttulus*
 male and female songs, 150–151

silent gaps in syllables, 151
 temporal pattern and temperature,
  150–151
Evolution; *see also* Selection
 acoustic communication, 123
 acoustic signals, 87–88
 bat-detecting moth ear, 292–294
 coevolution between bat and moth,
  318–319
 frequency tuning, 299ff
 hearing, 5, 130, 197–198
 hearing in bats, 9
 hearing in insects, 187
 hearing in Ormiine flies, 218ff
 hearing organs in Sarcophagidae,
  221–222
 moth ear, 285ff
 tympanal ears, 12–13, 197ff, 287
Excess attenuation; *see also* Attenuation
 evolution of insect sounds, 87–88
 frequency filtering, 72–73
 ground effects, 70–72
 sound, 67–68
 stratified environments, 68ff
 time of day effects, 70–71

Far-field detection, Diptera, 221–222
 detection, tympanal organ, 221–222
 directional hearing, 49–50
 hearing, 4
 sound, 65–66
 spiders, 229
Fasciclin I
 auditory system development, 109–110
 role in development, 120
Feeding behavior, bat, 307
Feltia heralis (Noctuid moth), 289
Field cricket; *see also* Cricket and
  bushcricket, *Gryllus* sp.
 ear, 7
 parasitism on, 220–221
 spherical treadmill, 21
Filtering, frequency, 72–73
Fishing spider; *see Dolomedes*
Foraging, bat, 296ff
Frequency analysis
 AN1 and AN2 neurons, 164, 167
 bushcrickets, 158ff
 CNS, 163ff
 CNS role for inhibition, 165ff
 grasshoppers, 157–158
 peripheral, 155–156

Frequency coding
    crickets, 161ff
    grasshoppers, 156ff
Frequency discrimination, 155
Frequency, effects on directional hearing
    cues, 57–58
Frequency map, anisotropy or acoustic
    fovea, 160
Frequency range of hearing, 11, 140ff
    postembryonic development of crista
        acustica, 113ff
Frequency representation in ear, calling
    song frequency, 161
Frequency selectivity
    grasshoppers, 157–158
    in hearing, 140ff
Frequency tuning
    development in cricket, 115
    grasshoppers, 157–158
    moths, 299ff
    sharpening, in bushcrickets, 166
    sharpening in CNS, 166–167
Frog, vibration sensitivity, 249
Fruitflies; see also Drosophila sp.,
    Diptera near-field hearing, 50
Fura-2 dye, intracellular calcium,
    177–178

G neuron
    development of branching, 120
    of grasshopper, development, 117–
        118
Genetic regulation, chordotonal receptor
    identity, 125–126
Grasshopper; see also Chorthippus
        bigutullus, Bullacris, 7, 36ff
    abdominal eardrums, 8
    audiogram, 317
    auditory development compared to
        mouse, 103
    auditory pathway plasticity, 128
    cercal system development, 122
    cercal system regeneration, 126
    chordotonal system development,
        121–122
    denervation experiments, 127
    directional hearing ambiguities, 53–
        54
    directional hearing cues, 52
    directional sensitivity, 38ff
    ear location, 102
    evolution of hearing, 187

frequency coding, 156ff
frequency tuning and analysis, 156ff
G neuron development, 117–118
intensity coding, 168–169
interaural amplitude and phase cues,
    28
interaural intensity threshold, 52
neuroblast, 118–119
notch gene, 130
parallel processing, 185–186
peripheral auditory development,
    103ff
postembryonic development compared
    with mammals and birds, 112
postembryonic development of ear,
    110ff
scolopidium, 110
sound localization strategy, 20
sound source localization, 19
temporal gap rejection, 184
temporal pattern recognition, 150–151,
    183ff
time difference cues, 19
tympanal organ receptor cells, 110
without ears, 123–124
Green lacewing; see Chrysopa carnea
Gromphadorina portenosa (giant hiss-
    ing cockroach of Madagascar),
    ear, 4
Ground attenuation, insect song, 87–88
Ground cricket, 7
Gryllus bimaculatus (a cricket),
    communication sound
    calling range, 72
    communication sound frequency,
        141–142
    courtship song, 168
    ear, 42–43
    parasitism on, 220–221
    peripheral frequency analysis, 161
    temporal pattern, 143–144, 148–149
Gryllus campestris (a cricket), calling
    song phonotaxis, 142ff
Gypsy moth; see Lymantria dispar

Habitat, constraints on sound source
    localization, 20
Habituation, 10
Halysidota maculatum, response to bat
    sound, 309
Hawaiian moths; see Lasiturus sp.
Hawkmoths, 310

Hearing
  biomechanics, 208ff
  chordotonal organ, 292
  coevolution between moth and bat,
    318–319
  development, 130f
  Diptera, 200ff
  directional, 206–207
  distance, 88–90
  diurnal vs nocturnal moths, 314ff
  evolution, 197–198, 218ff
  evolution in Ormiine flies, 218ff
  forms in Diptera, 200ff
  grasshoppers, 317
  Hecatesia sp., 315, 316
  Lymantria dispar, 316–317
  mechanoreception, 2
  moth, 299ff
  moth audiogram, 305
  moth loss of high frequency, 315
  moths after release of selection
    pressure, 311ff
  nontympanal in Diptera, 200
  organ's location on body, 222
  praying mantis, 317
  pressure-difference receivers, 222–
    223
  relation to stridulation, 187
  sensitivity convergence with vibration
    sensitivity, 131
  spiders, 3, 229
  tympanal, 2–3
  tympanum, 197ff, 211–213
Hecatesia sp. (whistling moths), high-
  frequency hearing, 315, 316
Heide amiculi (morabine grasshopper),
  chordotonal system development,
  124
Heliothis virescent (Noctuid moth), CNS,
  285
Hemimetabolous metamorphosis,
  grasshopper, 99–100
Hemisaga sp. (a bushcricket), sound
  discrimination, 78–80
Hissing giant cockroach of Madagascar; see Gromphadorina portenosa
Hoary bats; see Lasiurus sp.
Holometabolous metamorphosis,
  Drosophila, 99–100
House cricket, 7
Human, vibration sensitivity, 249
Hygrolycosa sp., vibration production,
  252

Impedance, induced by ground, 70–72
Information transfer, insect song, 73–74,
  83ff
Infrasound, elephant, 70
Inhibition
  frequency analysis in CNS, 165
  frequency-specific, 167
  intensity coding, 168–169
Intensity coding, inhibition, 168–169
Interaural cues, CNS, 213–215
Interaural hearing cues; see Binaural cues
Interaural intensity differences, 153
  contralateral inhibition, 170
  detection and amplification, 169–170
  directional hearing cues, 169–170
  and latency, 175ff
Interaural intensity threshold,
  grasshopper, 52
Interaural response latency, directional
  hearing in locust, 175–176
Interaural sound transmission, locust and
  grasshopper, 36
Interaural spectral differences, 18ff
Interaural time difference, 18ff, 169
Interaural time processing, 29
Interference, sensory, 76ff
Intermale spacing, role of calling song,
  146
Interneuron 501, moth, 285–286
Interneurons
  moth CNS, 283–286
  vibration sensitivity, 257
Intersegmental nerve, development of
  projections to, 108
Intraspecific interference, singing, 80–81
Isophya leonorae (a bushcricket),
  temporal pattern in song recognition,
  153

Johnston's organ, 200
  mosquito, 50
Jumping spiders, 229

Katydid, 7
  foreleg eardrums, 8
  tonotopy, 12–13

Lacinipolia illaudabilis, response to bat
  sound, 309
Larvae, chordotonal organ, 295
Laser vibrometry, 34

Lasiurus sp. (Hoary bats), 306, 312
Latency, and interaural intensity
    differences, 175ff
Latrodectus sp. (a spider), web type, 235
LBN-ei neuron, directional processing in
    brain, 173
Leptodactylus sp. (White-lipped frog),
    vibration sensitivity, 249
Leptophyes punctatissima (Phaneropterine
    bushcricket), calling song, 86
    sexual dimorphism in song, 89
    temporal pattern in song, 152–153
Localization of host, parasite, 9–10
Localization, sound sources, 10–11
Locust; *see also Schistocerca gregaria*, 5,
    36ff
    directional sensitivity calculations, 38ff
    interaural amplitude cues, 28
    interaural response latency, 175–176
Locusta migratoria (a grasshopper),
    chordotonal system development, 124
    CNS frequency analysis, 164
    directional hearing, 170–171
    frequency analysis, 156ff
    intensity coding, 168–169
Locusts
    abdominal eardrums, 8
    cicadas, 8–9
Longitudinal waves, spider web, 236
Loudness, selection for, 81–83
Luna moth; *see Actias luna*
Lycosa sp. (a spider), vibration
    production, 252
Lymantria dispar (Gypsy moth)
    ear, 317
    hearing, 316–317
    ontogeny of auditory system, 292

Malacosoma disstria (tent caterpillar
    moth), ontogeny of auditory
    system, 292
Manduca, neuroblast 7–4, 122
Manduca sexta (Tobacco hornworm
    moth), 287
Mantids
    ear location, 102
    frequency analysis, 156
    neuroblast 7–4, 122
Masking, 76ff
    effects on insect sounds, 76ff
    interspecific interference, 76ff
Mate attraction, using sound, 19ff
Mating call, grasshopper, 70

Mating systems, 64
    insect, 64
Meta sp. (a spider), 231
Metatarsal lyriform organ, 244–245
    sensitivity, 245ff, 269–270
Metatarsal vibration receptor, signal
    amplitude, 262
Metathorax, CNS, 294
    eared vs earless moth, 289
Metrioptera sp. (a bushcricket), sound
    production, 67
Microphone probe tube, 27
Microphones, principles of positioning,
    27
Miopharus sp., ear structure, 201f
Mole cricket; *see also Scapteriscus* sp., 8
    burrows, 82–83
Morabine grasshopper; *see Heide amiculi*
Mosquito, Johnston's organ, 50, 200
Moth
    A1 and A2 receptors, 281ff
    attraction to ultrasound, 315–316
    audiogram, 282, 305
    Barathra brassicae, 285
    bat interactions, 279ff
    behavioral isolation from bats, 316–317
    caterpillar, 292
    chordotonal organ, 292
    chordotonal sensilla, 287
    CNS, 283–285
    coevolution of hearing with bat,
        318–319
    detecting bat biosonar signal, 30–31
    ear, 281ff
Moth ear
    central projections, 283–285
    CNS, 283–285
    ear location, 102, 287–288
    ear ontogeny, 291ff
    ear and peripheral nervous system,
        281–283
    ear polyphyletic origin, 285–287
    ear tuning, 299ff
    eared, 287
    earless, 287, 310–311
    evolution, 285ff
    evolution of bat-detecting ear, 292–294
    hearing, 299ff
    Heliothis virescens, 285
    interneurons, 283–286
    island, 311ff
    larval CNS, 293
    metathorax, 289
    metathorax CNS, 294

Moth ear (*cont.*)
   nocturnal vs diurnal, 314ff
   predation by bat, 297–299, 305ff
   proprioceptor, 287, 292
   response to bat sound, 308, 309
   ring tract, 284
   selective pressures for hearing, 311ff
   sensory neurons, 281–283
   singing, 280
   temporal isolation from bats, 314ff
   vibration detection, 287
Movement receiver, 23
Musa sp. (a spider), web, 240
Mygalopsis marki (a bushcricket), central
      tonotopic organization, 159–160
   CNS frequency analysis, 164
   intensity coding, 169
   sound discrimination, 78–80
Myotis sp. (long-eared bats)
   gleaning, 306
   moth foraging, 296ff
Müller's organ, ear, 110

Near-field, Diptera, 200
   directional hearing, 49–50
   hearing, 4–5
Near-field hearing, fruit flies, 50
   role of antennae, 200
Near-field sound, 65–66
Nephila sp. (a spider)
   metatarsal slit organ, 245
   orb web, 263
   web type, 234
   web vibration, 268
Net-casting spiders, 229
Neural processing, acoustic signals, 139ff
Neuroblast, CNS development, 118–119
   grasshopper, 118–119
Noctuid moth; *see* Moth
Nonchordotonal receptor, moth, 282, 284
Notonecta (backswimmer)
   orientation, 265
   water surface orientation, 267
   web type, 234

Odotopic and tonotopic central
      organization, 160
Oecanthus sp. (South African tree
      cricket), song loudness, 82–83
Oecobius sp., web, 235
Omega-neuron, calling sounds, 78–80
Omidia sp., phonotaxis, 199

ON1 interneuron
   directional hearing, 170, 171–172
   frequency analysis, 161–162
   selective attention, 177, 180
Ontogeny, moth ear, 291ff
Opisthomal signals, spider, 250–252
Orb weaver, orientation, 263–264
Orb web, 231ff, 234
Orientation
   cues, spider, 264ff
   orb weaver, 263–264
   to sound, bushcricket, 74
   spider, 263ff
   vibration, 263ff
   water surface, 267
Ormia sp. (a tachinid fly); *also see*
      Ormiine flies
   attraction to host, 198
   and cricket, 9–10
   directional hearing, 222–223
   ear comparison with nontympanate
      species, 201ff
   evolution of parasitoid behavior,
      220–221
   far-field detection, 221–222
   reproduction, 220
   reproductive behavior, 199
Ormiine flies, Ormiine flies; *see also*
      Diptera, parasitic flies, *Ormia* sp.,
      197ff
   CNS, 213–215
   connections between ears, 215–217
   convergence of ear, 220–221
   directional hearing, 199, 206–207,
      211–212
   ear, 200ff
   evolution of hearing, 218ff
   hearing, 208
   hearing biomechanics, 208ff
   intertympanal coupling, 215–217
Orthopteran, vibration sensitivity,
      248–249

PABN2, central auditory interneuron, 167
Parasite, localization of host, 9–10
Parasites, acoustic, 198–200
Parasitic flies; *see also* Diptera, Ormiine
      flies
   directional hearing, 206–207
   life style, 206
   size limitations, 206
Parasitism, *Ormia ochracea*, 9–10
   sound detection, 139–140

Particle displacement, vs pressure, 65
Paruractonus, orientation, 265
Perception, and hearing, 10
Periodicity, insect singing, 80
Peripheral auditory system, development, 102ff
Peripheral nervous system, see PNS
Periplaneta americana (American cockroach), sense of hearing, 4
Phase cues, directional hearing in locust and grasshopper, 27ff
Phase difference
  spiracles, 45–46
  caused by central membrane, 48
Phonotaxis, 19ff
  and AN2, 173–174
  calling song, 89
  crickets, 140
  cricket auditory interneurons, 173
  insect, 85
  and ON1, 174–175
  Ormia sp., 199
  parasitic flies, 199
  positive and negative, 140–141, 167–168, 174–175
  pulse rate and intensity, 148
Pioneer cells, grasshopper chordotonal system, 126–127
Pioneer neurons, role in development, 119
Pirata sp. (wolf spider), water surface, 244
Plant
  propagation velocity, 239–240
  Rayleigh waves, 241
  spider webs, 235ff
  vibration and courtship, 269
  vibration orientation, 264ff
Plasticity
  cricket auditory pathway, 128
  grasshopper auditory pathway, 128
Pleural chordotonal system, auditory organ, 105
Poecilimon affinis (bushcricket), peripheral auditory development, 106–107
Polichne sp. (a bushcricket), frequency analysis, 160
Postembryonic development
  bushcricket, 112ff
  central chordotonal system, 121ff
  ear, 110ff
  grasshopper comparisons with mammals and birds, 112
  insect auditory system, 97–98
  sound sensitivity, 111

Pox neuro gene, *Drosophila* nervous system, 125
Praying mantis, hearing, 317
Precedence effect, 10
Precopulatory behavior, spiders, 229
Predation, bat on moth, 297–299, 305ff
Predator detection, 9
Pressure difference receiver, 3, 33ff
Pressure gain, cricket ear, 32
Pressure gradient receiver, 33ff
  cricket, 42ff
  insects, 19ff
  phase differences, 37–38, 40
  principles of operation, 23ff
  transmission gain, 34, 37
Pressure receiver, 23, 30ff, 222–223
  tympanal auditory organs, 222–223
Pressure, vs particle displacement, 65
Pressure-difference receivers, 222–223
  tympanal auditory organs, 222–223
Prey capture, spider, 229, 269–270
Prey detection, adaptations, 198ff
Propagation velocity, spider web, 239–240
Proprioception, and hearing, 12–13
Proprioceptor
  chordotonal organs, 12–13
  ear origin, 288, 291
  moth, 287, 292
Proximal sensilla, crickets, 161
Psychoacoustics, 10
Pulse interval analysis, delay lines, 183–184

Rain forest, cricket sounds, 78
Range of communication, insect, 66ff
Rayleigh waves, plant vibration, 241
Receptor development, timing, 102
Receptor physiology, general features, 163
Receptors, vibration, 244ff
Redundancy, insect sound, 83ff
Regeneration
  chordotonal system, 126ff
  insects compared with vertebrates, 129–130
  tympanal organ, 127
Reproductive behavior, *Omidia* sp., 199
Respiratory pathways, hearing, 35
Reverberation, caused by forest, 73
Rhinolophus sp. (bats), 306, 307

Ring tract
  central tonotopy, 117
  moth CNS, 294
  tympanal afferents, 123
  Roeder KD, 280, 294

Sarcophagid flies, tympanal ear, 205–206
Sarcophagidae, evolution of hearing
    organs, 221–222
Scapteriscus sp. (Mole cricket), selection
    for sound level, 81–82
Scattering, sound, 67
Schistocerca gregaria (a grasshopper),
    36–37
  chordotonal system development, 124
  frequency analysis, 156ff
  transmission gain, 37,40
Scolopale cell, 3
Scolophorous subgenual organ,
    Gromphadorina sp., 4
Scolopidial organ, 3
Scolopidium, grasshopper, 110
Scorpion; see Paruractonus
  vibration sensitivity, 248
Segestria sp., web type, 235
Seismic communication, frog, 249
Selection; see also Evolution
  bats on moths, 297–299, 305ff
  insect sounds, 81ff
  pressure on moths, 311ff
  relaxation of, 311ff
Selective attention, ON1 neuron, 177, 180
Sense of hearing, evolution, 49
Sensilla, chordotonal, 287
Sensitivity development
  bushcricket, 113–114
  in cricket, 115
  metatarsal lyriform organ, 269–270
  postembryonic development of crista
    acustica, 113ff
Sensory hairs, directional hearing, 49
Sensory interference, insect calls, 76ff
Seothyra sp., web, 235
Sexual dimorphism, calling song, 88–89
Sexual selection, 64
  insect, 64
Short-horned grasshoppers, 7–8
Signal degradation, bushcricket, 85
Signal encoding, auditory periphery,
    155ff
Signal production, spiders, 249ff
Silkworm moths, 310

Singing
  intraspecific interference, 80–81
  periodicity, 80
Slit sensilla, 244
Sonar signal detection, 19
Song, calling pattern, 73–74
Song discrimination, frequency-specific
    inhibition, 167
Song recognition
  neural circuit, 148–149
  versus sound source localization, 22
Song; see also Insect song
  temporal overlap, 80
  time domain, 73
Sound, and vibration sensitivity, 160–161
  bat, 306ff
  discrimination in insect, 76–77
  effects of environment on sound level,
    68–70
  effects of vegetation, 66–68
  excess attenuation, 67–68
  far-field, 65–66
  frequency discrimination, 155
  frequency selectivity in grasshoppers,
    157–158
  ground impedance, 70–72
  near-field, 65–66
  propagation, 66ff
  propagation velocity in tracheal tube,
    44
  reverberation, 73
  temporal separation of calling sounds,
    77–78
  turbulence effects, 66–68
  types used in communication, 11
Sound level, calling distance, 88–90
  selection for, 81–83
Sound localization; see Sound source
    localization, Directional hearing
Sound pressure level, enhancements by
    environment, 68–70
Sound shadow, grasshopper, 39
Sound source distance perception, 160
Sound source, insects, 67
Sound source localization; see also Di-
    rectional hearing, 10–11, 18ff,
    153ff
  and acoustics, 23ff
  environmental effects, 75–76
  and habitat characteristics, 20
  insect, 75–76
  unilateral deafening, 153–154
  versus song recognition, 22

Sound temporal pattern, recognition, 142ff, 146
Sound transmission, insect, 66ff
    time of day effects, 70–71, 76
Spacing, calling song, 87–88
Spatial isolation, moths, 311ff
Spider; *see also Achaeranea* sp., *Agave* sp., *Anidiops* sp., *Araneus* sp., *Cupiennius* sp., *Dolomedes* sp., *Latrodectus* sp., *Lycosa* sp., *Meta* sp., *Musa* sp., *Nephila* sp., *Pirata* sp., *Uroceta* sp., *Zygiella* sp., 228ff
    active space, 268–270
    brain, 257
    CNS, 266
    communication, 249ff
    courtship on plants, 235–236
    courtship signal, 258ff
    distance determination, 268
    hearing, 3, 229
    jumping, 229
    mating on plants, 235–236
    net-casting, 229
    on plants, 235ff
    orientation, 263ff
    orientation cues, 264ff
    orientation in web, 263–264
    plant surface orientation, 264ff
    prey capture, 229, 269–270
    signal production, 249ff
    subesophageal ganglionic mass, 257
    substrate type, 228
    vibration discrimination, 253ff
    vibration orientation, 263–264
    vibration production, 250ff
    vibration sensitivity, 248
    water surface orientation, 267
Spider web, 230ff
    individual radii, 233
    propagation velocity, 239–240
    signal fiber, 231–232
    signal thread, 235
    types, 231ff
    vibration, 268–270
    vibration attenuation, 240–241
    vibration transmission, 234
    vibration types, 236ff
Spiracles, cricket ear, 32, 42, 44ff
Stridulation, 7–8,11
    acoustic components, 65
    relation to hearing, 187
    substrate vibration, 65

Subgenual organ; and chordotonal organ, 110
    bushcrickets, 158
    sound sensitivity, 161
    tympanal organs, 107
Substrate, effect on vibration, 230ff
Substrate vibration, stridulation, 2–3, 65

Tachinid fly; *see Ormia ochracea*
Teleogryllus commodus (a cricket), calling song
    chirps and trills, 146–147
    courtship song, 148–149
    ON1 interneuron, 161–162
    peripheral frequency analysis, 161–162
    phonotaxis, 142, 147
    positive and negative phonotaxis, 140–141
    role of calcium in selective attention, 177–178
    selective attention of CNS neurons, 180
    two-part song, 146–147
Temporal gap rejection, AN4 neuron in grasshoppers, 184
Temporal isolation, bats and moths, 314ff
Temporal pattern
    analysis, 179ff
    cricket brain and behavior, 181–182
    cricket sounds, 78
Temporal pattern recognition, and sound source
    bushcricket, 151ff
    grasshoppers, 183ff
    localization, 185ff
Temporal pattern selectivity, BNC1 and 2 neurons, 181–182
    vertebrate brain cells, 181–182
Tent caterpillar moth; *see Malacosoma disstria*
Territoriality, and sound detection, 140
Tettigonia viridissima (a bushcricket), 52
    calling song, 73
    CNS frequency analysis, 164
    environmental effects on directional hearing, 52
    ON neuron and selective attention, 180
Tettigoniid, vibration sensitivity, 248–249
Theridiidae, stridulatory courtship behavior, 252
Thoracic ganglia, chordotonal projections, 117

Time difference cues, arthropods, 18
Time of day, communication, 70–71,
    76
Tobacco hornworm moth; see *Manduca
    sexta*
Tonotopic and odotopic central
    organization, 160
Tonotopic organization, central, 159
    CNS, 164–165
    cricket ear, 161
    ears of crickets and bushcrickets, 156
Tonotopy, 12–13
    central nervous system in grasshopper
        and bushcricket, 116–117
    development in bushcricket, 105–106
    development in central nervous system,
        116
    grasshopper neuropil, 158–159
    ring tract, 117
    ventral intermediate tract, 117
Trachea
    bushcricket, 32
    cricket ear, 42
    sound propagation velocity, 44
Tracheal chamber, 3
Tracheal tube, hearing trumpet in bush
    cricket, 31
Transduction, 4
Transmission gain, pressure gradient
    receivers, 34, 37
    Schistocerca gregaria, 37, 40
Transplantation, cricket foreleg, 127
Transverse waves, spider web, 236
Treadmill, spherical, 21
Tree cricket; see also *Oecanthus*, 7
Tuning
    moth ear, 299ff
    pulse rate, 183
Turbulence, effects on sound attenuation,
    66–68
Tympanal auditory organs, pressure-
    difference receivers, 222–223
Tympanal development, timing, 102
Tympanal ear, 197ff
    Emblemasoma sp., 205–206
    morphology, 197ff
    Sarcophagid flies, 205–206
Tympanal hearing, 2–3
Tympanal hearing organ, characteristics,
    3
Tympanal membrane, cricket, 32
Tympanal nerve, intersegmental nerve,
    108–109

Tympanal organ, and subgenual organs,
    107
    chordotonal system, 98
    development, 98–99
    far-field detection, 221–222
    receptor cells frequency response,
        110ff
    receptor cells, grasshopper, 110
    receptor cells, types, 110
    regeneration, 127
    segmental specialization, 117
Tympanal projections to CNS,
    development, 108ff
Tympanal receptors, development in
    cricket, 107–108
Tympanic membrane, moth, 281
Tympanum
    directional hearing, 208ff
    dynamics, 211–213
    evolutionary origin, 287
    intertympanal coupling, 215–217

Ultrasonic hearing, bats, 9
Ultrasound, attraction of moths, 315–
    316
Uroceta sp. (a spider), web type, 235

Vegetation, effects on sound attenuation,
    66–68
Velocity propagation, plants, 239–240
Ventral intermediate tract
    central tonotopy, 117
    chordotonal system development, 19
    tympanal afferents, 123
Vibration
    active space sensing, 268–270
    attenuation, 234, 269
    attenuation in spider web, 240–241
    background noise, 253ff
    behavior, 229–230
    cicada, 239, 240
    cydnid bugs, 239, 240
    discrimination, 253ff
    distance determination, 268
    echolocation, 230
    measurement, 230
    moth detection, 287
    orientation, 263ff
    plants, 235ff
    propagation velocity, 239–240
    Rayleigh waves, 241

and sound sensitivity, 160–161
spider web, 230ff, 234
substrate, 230ff
transmutation, 268
types, 253
water surface, 242–244
water surface orientation, 267
wave types, 236ff
Vibration production
arthropod, 252
Hygrolycosa sp., 252
Lycosa sp., 252
opisthosoma movement, 250–252
spiders, 250ff
Vibration reception, 228ff, 244ff
Vibration receptor
metatarsal lyriform organ, 244–245
sensitivity, 245ff
types, 244–245
Vibration sensitivity, convergence with
hearing sensitivity, 131
human, 249

orthoptera, 248–249
scorpion, 248
spiders, 248
Vibration transmission, spider webs, 234

Wandering spider, see *Cupiennius*
Water surface
orientation, 267
spider webs, 235ff
surface vibration, 24–244
Web; *see* Spider web
Whistling moths; *see Hecatesia* sp.
White–lipped frog; *see Leptodactylus*
Wind-sensitive receivers, 49
Wolf spider; *see Pirata*

Zotheca tranquilla, response to bat sound,
308
Zygiella sp. (a spider), 231
orb web, 263